The World Grain Trade

The World Grain Trade

Grain Marketing, Institutions, and Policies

EDITED BY

Michael J. McGarry
and Andrew Schmitz

Westview Press

BOULDER • SAN FRANCISCO

Pinter Publishers

LONDON

Copyright © 1992 by Westview Press, Inc.

Published in 1992 in the United States of America by Westview Press, Inc., 5500 Central Avenue, Boulder, Colorado 80301-2847

Published in 1992 in Great Britain by Pinter Publishers Limited, 25 Floral Street, London WC2E 9DS

Library of Congress Cataloging-in-Publication Data
The World grain trade : grain marketing, institutions, and policies /
  edited by Michael J. McGarry and Andrew Schmitz.
    p.   cm.
  Includes bibliographical references.
  ISBN 0-8133-8336-6 (U.S.)
  1. Grain trade—Government policy—Case studies. 2. Grain trade—
Case studies. I. Schmitz, Andrew.
HD9030.6.W67   1992
380.1′4131—dc20   91-41737
                                                                          CIP

A CIP catalogue record for this book is available from the British Library.
ISBN 1-85567-069-0 (U.K.)

Printed and bound in the United States of America

The paper used in this publication meets the requirements
of the American National Standard for Permanence of Paper
for Printed Library Materials Z39.48-1984.

10   9   8   7   6   5   4   3   2   1

Contents

Preface

Grain markets around the world have always been of interest to agricultural economists because grain represents the single most important component of world food consumption, accounting for about 60 percent of the calories consumed. Total global production of wheat, coarse grains, and rice is approximately 1.89 billion metric tons per annum, with some 220 million metric tons (mmt) traded (about 12 percent). Oilseed production is around 215 mmt and approximately 18 percent of this is traded. Because grain is one of the world's key staple products, it is a highly political commodity. It has been used in the past as an economic weapon (e.g., the 1980 United States embargo on grain sales to the Soviet Union) and it receives special economic and political status in both developed and developing countries. Grain is a special commodity because the possibility of a grain shortage is real, although the probability of one is very low; world carryover stocks average about two to three months supply, more or less the same as thirty years ago.

This book aims to explain the operations, the economics, and the politics of the grain markets in five key producing countries: Argentina, Australia, Brazil, Canada, and the United States. The world is highly dependent on grain and oilseed output from these five countries. As a group, they grow about 25 percent of the world's wheat, coarse grains, and rice, and they export over 60 percent of all the grain traded internationally. These five countries supply about one-half of the world's oilseeds and account for over three-fourths of the trade.

Currently, there are several key issues and questions associated with the grain markets, and this book sheds light on many of these. One of the major issues at present is the market impact of the ongoing move toward deregulation of the grain industry in many different countries around the globe. Pressure to deregulate is coming from both multilateral and bilateral trade negotiations, along with other sources. Australia is deregulating its grain industry, and this is due mainly to internal pressure. However, the United States-Canada free trade agreement has resulted in the removal of some important Canadian grain regulations. Multilateral negotiations may have a larger impact than bilateral agreements. For example, if the

GATT negotiations are successful and the United States reduces subsidies to grain farmers, will total production rise or fall? As we learn from Part Five on the United States, the answer is not straightforward. If Brazil and Argentina were to restructure their marketing systems and policies to remove many of the inefficiencies, how much additional production would result? How important is macroeconomic policy in terms of its impact on the grain markets? Macroeconomic policy seems to be of utmost importance in Argentina and Brazil. Which countries are the low-cost producers, and how will production patterns change under deregulation? Where have the largest productivity gains been made in the past, and where are they likely to be realized in future years? A thorough understanding of marketing systems and policies is necessary before we can begin to analyze the above questions and issues.

This book describes the grain marketing systems and policies in three developed countries—Australia, Canada, and the United States—and in two developing countries—Argentina and Brazil. As a result there are some interesting similarities and contrasts. In developing countries overall economic performance is closely tied to growth in agricultural production. In developed countries this is not the case. Argentina and Brazil both have great potential to augment the world's supply of grains and oilseeds, but this potential is held back by economy-wide problems.

The overall purpose of this book is to uncover aspects of the grain marketing systems in the developed countries that might be helpful to Argentina and Brazil. The thesis is that many of the "problems"with the Latin American marketing systems lie outside of agriculture—they are economy-wide (i.e., macroeconomic problems). Each part of the book is laid out in a very similar fashion and adheres to a well-structured framework. First, there is a short discussion of the overall economy along with summary statistics. An overview of the agricultural sector is then presented. This is followed by a section on grain production and marketing. Each country Part also provides an analysis of public policy.

The United States is by far the largest grain producing and exporting country—it alone produces over 15 percent of the world's grain supply and exports about 40 percent of total grains traded. It is, therefore, not surprising that the United States is the center of the world grain market. However, the United States grain market is somewhat schizophrenic. On the one hand, the United States is the "last bastion" of free enterprise—the government marketing boards are absent, and the United States is home of the large futures markets. On the other hand, in many years farmers receive a government target price that is well above market-clearing levels so they basically produce for the government. Government programs not only affect production but they also drive exports through generous export subsidies.

In Part Five on the United States, Lowell Hill describes in detail how grain is marketed in the United States and how the grain-related policies operate. This provides a good reference point for the others because, as Professor Hill points out, the technical aspects of the United States marketing and transportation system are highly efficient. The United States is a large and reliable grain exporter, has an efficient grain handling infrastructure, and is very responsive to market signals. However, this does not mean the United States system is without weaknesses. For example, the laissez-faire grading system is one of the principal drawbacks of the United States system and reduces somewhat the ability of the United States to compete in quality conscious markets. Almost all of the increased production in the United States over the past several years has come from yield increases. There is strong evidence that indicates that United States grain farmers would be worse off under multilateral freer trade, and this presents an interesting dichotomy for United States policy makers.

Compared to the United States, Canada has some important geographical and technical liabilities. The cost of handling and transporting grain is higher in Canada than in the United States; consequently, the Canadian handling and marketing system is not as responsive as in the United States. However, as explained in Part Four on Canada by Schmitz and Furtan, one of Canada's biggest assets is its stringent grading system. Canada is well known for grain exports that are consistent from cargo to cargo and of the highest quality. This is a distinct marketing advantage. Canadian grain farmers have fewer production alternatives than their counterparts in either the United States or Australia. As a result, the Canadian acreage data show less variability across crops and across years. Total production of Canadian grain has not grown as rapidly as in many other exporting countries. Canadian grain marketing is handled by cooperatives, private firms, and one large public agency, the Canadian Wheat Board. The Canadian Wheat Board markets more grain than any other public or private firm in the world. One of the big issues in Canada is whether the government should continue to try to match subsidies provided to grain farmers in the European Community and the United States. In addition, the United States–Canada free trade agreement is placing strong deregulatory pressure on certain segments of the Canadian industry.

The Australian grain industry is the third cousin of the United States-Canada–Australia triopoly. The Australian government has been less generous to its grain farmers, compared to either Canada or the United States. In fact, Part Three on Australia suggests that protection in the manufacturing sector has acted as a significant tax on agriculture. Whether the grain sector is subsidized or taxed on a net basis is still debatable. As is the case for Canada, Australia's grain yields have not shown much growth over the past thirty years. Yields in the United States have grown much faster.

The technical marketing and handling system in Australia has many of the same inefficiencies found in Canada. The Canadian Wheat Board was patterned after Australia's wheat marketing authority, so there are many similarities between the Australian Wheat Board and the Canadian counterpart. As in Canada and the United States, Australia has taken steps towards deregulating its grain industry. However, it has gone further in this direction than either of the other two countries, and where this will place Australia in the market in future years is an interesting question. Piggott, Fisher, Alston, and Schmitz raise this issue, and they point out that it is one of the most important questions currently facing the Australian grain industry.

Relative to the situation in the United States, Canada, and Australia, grains play a much larger role in both the Brazilian and Argentine economies. Brazil is not an exporter of cereal grains but is a large exporter of soybean meal and soybeans. Part Two on Brazil by McGarry et al. emphasizes the rapid growth in grain and oilseed production that has taken place in Brazil over the past twenty-five years. Most of this production growth is due to area expansion. They point out that government programs have played an important role in this expansion of grain. McGarry et al. point out that Brazilian grain producers operate in a domestic market environment that is characterized by much greater uncertainty compared to the Australian, Canadian, or American markets. Macroeconomic policies are highly volatile in Brazil. In addition, the physical grain marketing infrastructure is inferior and highly inefficient. McGarry et al. argue that if the grain markets were returned to the private sector, Brazil would be better off.

Argentina is the other Latin American success story in terms of growth in grain and oilseed output—the biggest surge has come in soybeans. Combined soybean output in Argentina and Brazil, as a percent of United States production, went from 7 percent in 1970 to 60 percent in 1990. Lacroix et al. point out that Argentina's yields of wheat and soybeans are second only to the United States among the five countries considered in this book. It not only has the second highest yields, but it is also the lowest-cost producer. However, of the five countries considered, Argentina is the only one in which it is clear that grain farmers are taxed rather than subsidized. This means the potential for further grain output growth may be the greatest in Argentina if government policies are rationalized. Argentina also shares the same Latin American problem as Brazil, namely, a very unstable macroeconomic environment. Its grain marketing infrastructure is also in disrepair. Lacroix et al. reason that more stable macroeconomic policies and a more efficient marketing system could enable Argentina to double its grain output in a relatively short period of time.

It is unavoidable that in a book of this sort some important countries are left out. However, the strength of this volume is that it does contain an in-depth analysis of developments that have taken place in some of the major grain producing countries around the world. This book reminds us that as determinants of future grain availability, marketing and policy constraints are at least as important as biological or ecological constraints.

Using its own staff and consultants, the World Bank managed and completed a study of the grain sub-sectors of Argentina, Brazil, Australia, Canada, and the United States upon which this book is based. Nevertheless, the views expressed are solely those of the authors and do not necessarily represent those of the World Bank or the institutions with which the different authors are affiliated.

Colin A. Carter
University of California, Davis
College of Agricultural &
Environmental Sciences
Department of Agricultural Economics

About the Contributors

JULIAN M. ALSTON is an associate professor in the Department of Agricultural Economics at the University of California, Davis. Previously he was the chief economist in the Department of Agriculture in Victoria, Australia. He obtained his masters degree in agricultural science (economics) from LaTrobe University and his Ph.D. in economics from North Carolina State University. Dr. Alston's research interests include domestic and international markets and policies for agricultural commodities and agricultural inputs, food demand analysis, and the economics of agricultural R&D. He has published articles on these topics in economics and agricultural economics journals.

COLIN CARTER obtained a B.A. in economics and an M.Sc. in agricultural economics from the University of Alberta. His studies continued at the University of California at Berkeley, where he obtained an M.A. in economics and, in 1980, a Ph.D. in agricultural economics. He is professor of agricultural economics at the University of California, Davis. His published work has covered such topics as import tariffs in the world wheat market, grain export cartels, futures market efficiency, United States wheat policy, China's trade in grains, and Japanese trade policy. He recently held a three-year fellowship from the Kellogg International Fellowship Program in food systems.

BRIAN FISHER completed his undergraduate training in agricultural science and his Ph.D. in agricultural economics at the University of Sydney, Australia. He became professor of agricultural economics and dean of the faculty of agriculture at that institution. Currently he serves as executive director of the Australian Bureau of Agricultural and Resource Economics. He has been coeditor of the *Australian Journal of Agricultural Economics* and president of the Australian Agricultural Economics Society. Fisher's research interests are in the areas of agricultural marketing, agricultural policy, and resource economics.

W. H. FURTAN is professor of agricultural economics at the University of Saskatchewan. He received his B.S.A. and M.Sc. at the University of Saskatchewan and his Ph.D. at Purdue University. His research is in the area of agricultural policy and technological change. He has published numerous articles in such journals as the *Journal of Political Economy* and the *Quarterly Journal of Economics*.

THOMAS C. GLAESSNER is currently a financial economist and financial sector specialist in the Trade and Finance Division of the Latin America and Caribbean Technical Department of the World Bank. Mr. Glaessner received his B.A. in economics and mathematics from Kenyon College and is a member of Phi Beta Kappa. He completed his Ph.D. in economics at the University of Virginia and also undertook graduate work in economics and finance at the University of Wisconsin, Madison. Prior to working for the World Bank, Mr. Glaessner worked for five years with the International Finance Division of the Federal Reserve Board. At present, his work is focused on financial sector reform, design of technical assistance programs, and on links between reforms to agricultural marketing systems and agricultural credit. Mr. Glaessner has also worked in the Bank's Financial Risk Management Unit with responsibility for evaluating, measuring, and monitoring different risks and for dealing with other issues relating to World Bank financial policies.

LOWELL D. HILL is L. J. Norton Professor of Agricultural Marketing at the University of Illinois at Urbana-Champaign. He is the editor of several books, including *Corn Quality in World Markets, World Soybean Research*, and *Role of Government in a Market Economy*. He is the author of the book *Grain Grades and Standards: Historical Issues Shaping the Future*, and has authored chapters in fifteen books plus numerous articles in journals. He is recognized as a leading authority in the area of grain marketing, particularly as it relates to quality and grading standards. His work in the area of transportation and market structure has also been recognized for its contribution to the improved efficiency of marketing of agricultural products.

RICHARD L. J. LACROIX is an independent consultant, living in Boston and working worldwide on various aspects of agro-industrial and economic development. He has a graduate engineering degree from the Delft Institute of Technology in the Netherlands and a management degree from Massachusetts Institute of Technology (MIT). He lived and worked in Peru, Bolivia, Brazil, the United Kingdom, and Egypt. In the early to mid-1970s he was a member of the staff of Arthur D. Little, Inc., of Cambridge, Massachusetts, working in that company's agri-business and international operations groups. His work on the grain trade in Argentina and Brazil was done, originally, in the context of World Bank related activities. Among Mr. Lacroix's earlier work on Latin America is a study of the concept and practice of "Integrated Rural Development" on that continent, published as a World Bank staff working paper. Recently, he has contributed to the "Country Economic Memorandum"on India, a forthcoming public document of the World Bank.

MICHAEL J. MCGARRY is division chief of the Agriculture Division, Technical Department, Latin America and the Caribbean Region of the World Bank. Mr. McGarry received his undergraduate degree in agricultural science with honors from the National University of Ireland, Dublin, and his M.S. and Ph.D. in economics from the University of Wisconsin, Madison. Prior to joining the World Bank through its Young Professional Program, Mr. McGarry worked with the Agricultural Research Institute in Ireland. At the World Bank he has worked in various capacities and in many regions of the world, including East Asia, Latin America, and the Caribbean. His work has focused principally on the policy issues affecting the development of agriculture in different country settings.

MATTHEW A. MCMAHON is a senior agriculturalist with the World Bank working in the Latin America and Caribbean Technical Department. Dr. McMahon received his undergraduate degree in agricultural science from the National University of Ireland, and his Ph.D. in agronomy from the University of Kentucky in 1973. From 1973 to 1978 he served as an agronomist in Mexico with the Centro Internacional de Mejoramiento de Maiz y Trigo (CIMMYT), working on all aspects of wheat agronomy. From 1978 to 1986, Dr. McMahon served as regional wheat agronomist for CIMMYT in the Southern Cone Region of Latin America. While in that position he developed, in collaboration with national scientists, research programs in wheat agronomy in Argentina, Chile, Paraguay, and Uruguay. From 1986 to 1989 he was head of CIMMYT's Wheat Agronomy Program with responsibility for supervision of research activities worldwide. At present his interests are focused on the organization and effectiveness of national agricultural research systems.

ROLEY PIGGOTT completed undergraduate and masters degrees in agricultural economics at the University of New England, Australia, and his Ph.D. in agricultural economics at Cornell University. Currently he is senior lecturer in agricultural economics at the University of New England. He was a Kellogg International Fellow in food systems and a National Agricultural Biotechnology Council Joyce Fellow at the University of California, Davis. Piggott was a coeditor of the *Australian Journal of Agricultural Economics* and is the current president of the Australian Agricultural Economics Society. His research interests are in the areas of agricultural price analysis, marketing, and policy.

JOSEPH C. REID is currently an assistant professor of law at the University of Baltimore School of Law. Prior to joining the faculty of the University of Baltimore, Professor Reid served as an adjunct professor in commercial law at Pace University School of Law, as an instructor at

Baruch College of the City University of New York, and as an associate at the New York law firm of Brown & Wood, specializing in representing commercial and investment banks in banking, corporate, commercial, and securities matters. Before joining Brown & Wood, Professor Reid served as an attorney in the legal department of the Federal Reserve Bank of New York. As a member of the New York Bar, Professor Reid is a graduate of Harvard College and the University of Pennsylvania School of Law. From time to time, Professor Reid consults with the World Bank and various commercial and investment banks.

ANDREW SCHMITZ is a graduate of the University of Saskatchewan, Department of Agricultural Economics, and of the University of Wisconsin, Department of Economics. His area of emphasis is international economics and agricultural policy. He is chairman of the Department of Agricultural and Resource Economics, University of California at Berkeley, where he holds the George W. and Elsie M. Robinson Endowed Chair. He is also an adjunct professor, Department of Agricultural Economics at the University of Saskatchewan in Saskatoon. Schmitz is a fellow of the American Agricultural Economics Association. He has published numerous papers and books and has received several awards for his research, including the Harold Groves Doctoral Dissertation Award for the best thesis in the Department of Economics at the University of Wisconsin, five awards from the American Agricultural Economics Association on research discovery and quality, and an award for a book on grain export cartels. In 1986, at the University of Saskatchewan, he was the first recipient of the Department of Agricultural Economics Hadley Van Vliet Chair and also received the Outstanding Graduate Award for the College of Agriculture's 75th Anniversary. Schmitz has consulted for numerous organizations, including many law firms that focus on antitrust and price-fixing considerations, as well as for the World Bank, Canada Department of Agriculture, U.S. Department of Agriculture, U.S. Central Intelligence Agency, National Grain and Feed Association, International Trade Commission, Potash Corporation of Saskatchewan, and U.S. Office of Technology Assessment.

PART ONE

Argentina

Grain Marketing, Institutions, and Policies

Richard Lacroix, Michael J. McGarry,
Matthew McMahon, and Lowell D. Hill

1

Background

The Argentine Economy

The most striking feature of Argentina's economy during the twentieth century is its decline in importance on the world stage. At the end of the 1920s, Argentina was one of the ten wealthiest countries in the world. By 1988, its GNP of US$83.0 billion placed it 24th among the world's economies, behind Norway (US$84.2 billion), but ahead of South Africa (US$77.8 billion) and Indonesia (US$76.0 billion). However, Argentina has the third-largest GNP of the Latin American countries, following Brazil (US$328.9 billion) and Mexico (US$151.9 billion).

In 1988, Argentina's per capita GNP of US$2,640 placed it 28th among the countries of the world (excluding those with a population of less than 3 million). Among the Latin American countries it has the second-largest per capita GNP, after Venezuela (US$3,170), but ahead of Brazil (US$2,280) and Mexico (US$1,820).

In the period between 1980 and 1988, Argentina's real per capita income declined by 1.6 percent per annum, but between 1986 and 1988, the rate of decline accelerated to 2.0 percent per annum. Inflation rose sharply in the early 1980s, from 105 percent in 1981 to 688 percent in 1984, and reached an annualized rate of 1,129 percent by June 1985. Despite a series of stabilization efforts in the 1984–1989 period, by the time of President Carlos Menem's inauguration in July 1989, inflation had surpassed 100 percent per month, output was stagnating, unemployment was increasing, and the Government was unable to obtain credit from either domestic or foreign sources.

Argentina's population grew at an annual rate of 1.4 percent in the 1980s, reaching a level of 31.5 million by 1988. Life expectancy in 1988 was 71 years, compared with 75 in the United States. There has been a strong trend towards urbanization since the 1940s. By 1986, the urban population was 84 percent of the total population, compared with 47 percent in 1947. An astounding 46 percent of the total urban population is concentrated in Buenos Aires. For the year 2000, the Argentine census bureau projects an overall population of some 36 million, and an urban population of 32 million, 89 percent of the total.

The Agricultural Sector

Since the 1940s, successive Argentine governments, both military and civilian, have protected industry at the expense of agriculture, and provided subsidies to the urban population at the expense of the rural population. Nevertheless, the agricultural sector has continued to grow and provide support to other areas of the economy in the form of cheap food and substantial export earnings. For example, during the 1980s, while overall GDP fell by an average annual rate of about 1.6 percent, agricultural output increased by about 2.3 percent per annum. This positive performance of agriculture was achieved at a time when real world prices for the country's main agricultural exports were declining. A much higher rate of agricultural growth could have been achieved if producers had been operating within a favorable business climate, free from the urban, industrial bias of government policies.

In spite of the protection afforded to industry, Argentina has not substantially broadened its manufacturing base. Also, unlike the other major grain-exporting countries, it has been unable to diversify its agriculture away from a traditional reliance upon grains and livestock. The economy, and government's fiscal revenues, are unusually dependent upon, and subject to, the vagaries of the world market for a few commodities—beef, wool and grains. Grains accounted for almost half of all export earnings between 1980 and 1987, while the agricultural sector as a whole provided about 70 percent.

In contrast with the pattern observed in many other countries, where the relative contribution of agriculture has declined over time, in Argentina, agriculture's share in GDP has marginally increased in the last 15 years, while industrial development has stagnated. In 1988, agriculture contributed 13 percent of GDP, equivalent to about US$10.0 billion. The agricultural sector also had a 13 percent share of total employment.

TABLE 1.1   The Agricultural Sector in Five Grain-Exporting Economies

	% of GDP	% of Total Export Earnings	% of Employment
Argentina	13	70	13
Brazil	12	44	25
Australia	4	29	n.a.
Canada	3	18	4
USA	2	10	3

Sources: The World Bank Atlas, 1989.

Table 1.1 compares the importance of the agricultural sector to the overall economy in five of the world's leading grain exporters—Argentina, Brazil, Australia, Canada, and the United States. One of the most striking features of the table is that agriculture's share of GDP, export earnings and employment is much higher in Argentina and Brazil than in the other countries considered.

Geography and Land Use

Distinct climatic zones, from sub-tropical to arctic, allow for the production of a wide range of agricultural commodities. However, the bulk of Argentina's arable land is in a temperate zone. Argentina is self-sufficient in most foodstuffs as well as in raw materials for a diversified agro-industrial sector.

Sixty-five percent of Argentina's 279 million hectares of land is suitable for either crop agriculture or livestock. Another 16 percent is forested and the remainder is unsuitable for either cultivation or forestry. Of the 183 million hectares of arable land, about 12 percent (23 million hectares) is under annual or perennial crops, 8 percent (15 million hectares) is under improved pasture, and 80 percent (145 million hectares) is in natural pasture.

The major geographic zones are: the Pampas, the Chaco, the Andean zone, and Patagonia (see map: Grain Areas).

The Pampas. The Pampas, the largest ecological zone, is the dominant grain-producing area. This broad, herbaceous plain of some 45 million hectares radiates out from the city of Buenos Aires and covers most of the provinces of Buenos Aires, Cordoba, La Pampa, Santa Fé and Entre Rios. The area has a temperate climate, between 750 and 1,000 mm of rainfall annually, and contains about 50 percent of Argentina's arable land. Two thirds of the Pampas area has deep fertile soils similar to the great plains of the United States.

The Chaco. The Chaco comprises the northern lowlands along the River Plate. This area, about one third the size of the Pampas, has heavier rainfall and poorer soils, especially in the northeast, and contains important forest resources. The principal agricultural products are cotton, nuts, tea, rice, tobacco, citrus and cross-bred cattle.

The Andean Zone. This large area west and northwest of the Pampas is hilly to mountainous, poorly watered and, except in some valleys, can only be cultivated under irrigation. The principal agricultural products are sugar, tobacco, citrus, grapes for wine, and miscellaneous cash crops, including beans.

Patagonia. This area, about the same size as the Pampas, receives only minimal rainfall, is sparsely populated, and is devoted principally to sheep ranching and to the production of some high quality temperate-zone fruits.

Land Occupancy

The productive agricultural areas are occupied by about 553,000 holdings of various sizes. Nearly 50 percent of these are in the Pampas. Table 1.2 shows, in the Pampas and nationwide, a distribution of these holdings by size.

Nationwide, 75 percent of the land area is in holdings of 1,000 hectares or more, held by less than 6 percent of all farmers. In the Pampas, 50 percent of the land area is in similar-sized holdings, and is held by less than 6 percent of landowners. Nationwide, 50 percent of all farms are 50 hectares or less, while in the Pampas only 38 percent are.

TABLE 1.2   Size Distribution of Land Holdings (in percentages)

Farm Size	No. of Farms		Total Area	
(hectares)	Pampas	Nationwide	Pampas	Nationwide
0 - 50	38	50	3	2
51 - 100	18	13	5	3
101 - 200	18	11	10	4
201 - 500	15	16	16	9
501 -1000	6	5	16	7
1001 - 2500	3	3	18	15
2501 - 5000	1	1	13	12
5001 - 10000	<1	<1	11	14
> 10,000	<1	<1	8	34

Source: Confederation of Agricultural Cooperatives, 1988

In the Pampas, about 75 percent of the productive agricultural land is privately owned. The other 25 percent is held by corporations or the State; about half of it is rented out to farm operators, and the other half is operated under other types of arrangements, including contract farming and corporate management.

Production

In Argentina, production of cereal grains and oilseeds is the leading agricultural activity: in the period 1970 to 1979 it accounted for about 26 percent of agriculture's total value added, and in the period 1980 to 1987, this increased to 37 percent, mostly due to the growth in oilseeds (Table 7.1). Cereal grains and oilseeds account for about 80 percent of the planted area, the majority of which is in the Pampas.

About 70 percent of the beef herd of some 51 million head is found in the Pampas. Argentina is South America's largest milk producer with an annual production of between 5.5 and 6.0 million liters. Most dairying is pasture-based, employing modern technology. There are also many small family-operated dairy farms that rely partly on rented pasture. On these farms, milking is done once daily by hand and the operators generally do not use supplemental feed.

Production of sheep for wool and meat is a major agricultural activity. The current sheep herd is estimated at 24 million head, 50 percent of which are pastured in the Patagonia region where the primary end product is wool and the typical flock consists of several thousand animals.

Most swine production is not carried out on a large scale, but as an adjunct activity on small family farms. The animals forage for their feed, and maize is used as a supplement for fattening.

Poultry production technology in Argentina has paralleled the evolution of poultry production in North America during the last two decades. Broiler production, the largest segment of the industry, is dominated by three vertically integrated firms that together claim 90 percent of the domestic market. Another 35 to 40 small firms account for the other 10 percent. Poultry is the substitute of choice for beef; therefore, the profitability of poultry production depends to a large extent on the price of beef. While Argentina is a large exporter of feed ingredients, including soybean meal, its poultry exports are small.

The horticulture sub-sector of Argentina's agriculture serves primarily the domestic market. By and large, Argentina produces grains and beef for export and all other products for domestic consumption. The country's vast potential for exports of non-traditional products, such as fruits, vegetables or poultry meat, remains largely untapped.

2

The Grain Sub-Sector

Sub-Sector Performance

The grain sub-sector is the most important sub-sector of Argentina's agriculture. In the eighties, it contributed about 37 percent of the gross agricultural product, and 65 percent of the value of all crops. Within the larger context of the Argentine economy, grains, edible oils and oilseed by-products accounted for approximately 47 percent of average annual foreign exchange earnings during the period 1980 to 1987 (Table 7.7). Argentina is an important supplier to the world market of wheat, maize, sorghum and soybeans, and of soybean, sunflower and linseed derivatives.

Between 1960 and 1985, the average annual growth rate of grain production was 3.8 percent, a rate of growth well above that of the national economy. A level of 5.8 percent was reached between 1980 and 1985. Exports of grain also increased during this five-year period, attaining an average growth rate of 5.5 percent. In 1985, a record harvest of 44 million metric tons of grains, including oilseeds, was obtained, and a record 27 million metric tons of grains and crushing by-products were exported.

The main changes accompanying the growth in Argentina's grain production in the last 25 years have been:

- increased role of private firms (particularly seed companies);
- rapid adoption and spread of improved varieties;
- substantial yield increases for all grains;
- expansion of the area cropped at the expense of pasture;
- focus on wheat, maize, soybeans, sunflower and sorghum;
- increased importance of crops of a higher unit value, such as the oilseeds;

- increased preference for continuous cropping over crop rotation;
- double-cropping of soybeans with wheat, or use of a wheat-soybeans-maize rotation;
- increased share of industrial inputs and capital goods in total production cost; however, fertilizer use is still low;
- increased sophistication in production, and increased marginalization of small and medium-scale producers (this group, about 75 percent of all farmers, accounts for about one third of total production).

Table 2.1 gives the key indicators (production, yield, area) for the principal crops.

In the mid-1980s, Argentina was among the world's lowest-cost grain producers, as demonstrated by comparison of total production costs for Argentina, the United States, Canada and Australia (Table 2.2). However, not too much significance should be attached to this since average country production costs, at least in the medium- to long-term, are partly a function of the prices received by grain producers, and Argentine producers have traditionally received the lowest farmgate prices for their grains.

That Argentina has a strong comparative advantage in grain production is borne out by examining the cost of domestic resources required to generate a unit of foreign exchange. More specifically, Domestic Resource Costs (DRCs) were calculated according to the formula: DRC = sum of non-tradeable costs (the primary factors of production, valued at domestic prices, converted into US$ using the official exchange rate) divided by net

TABLE 2.1   Argentina—Key Indicators for the Principal Crops, Selected Years

	Wheat	Maize	Soybeans	Sunflower	Sorghum
Production ('000 mt) (1985–89)	10,220	9,425	7,160	3,040	3,540
% of world production (1985–89)	2.0	2.1	7.2	14.9	5.8
Prod. growth % p.a. (1970–79)	3.5	-1.1	39.5	5.1	7.0
Prod. growth % p.a. (1980–89)	0.2	-0.3	8.8	7.4	-9.9
Exports ('000 mt) (1985–89)	4,740	4,000	1,780	0.0	1,200
Exports as % of national production (1985–89)	46.3	42.4	24.9	0.0	33.9
Exports as % of world trade (1985–89)	5.0	6.6	6.6	0.0	14.3
Area ('000 ha) (1985–89)	5,750	2,894	3,670	2,265	1,180
Growth in area % p.a. (1970–79)	1.4	-5.0	32.3	2.7	1.4
Growth in area % p.a. (1980–89)	-1.5	-1.3	10.2	4.0	-9.1
Yield (kg/ha) (1985–89)	1,900	3,400	1,960	1,340	2,980
Growth in yield % p.a. (1970–79)	2.0	4.0	7.2	2.2	5.7
Growth in yield % p.a. (1980–89)	1.7	1.0	-1.5	3.4	0.5

Source: USDA, *Agricultural Statistics*, Washington, D.C., various issues.

TABLE 2.2   Indices of Total Production Costs for Major Grains and Major
Producers, Selected Years

	Wheat (1985)	Wheat (1986)	Maize (1985)	Soybeans (1985)
Argentina	100	100	100	100
Australia	170	144	n.a.	n.a.
Canada	229	196	n.a.	n.a.
USA	243	193	143	184

Sources: Gerald F. Ortman, *Analysis of Production and Marketing Costs of Corn, Wheat and Soybeans among Exporting Countries*, Ohio State University symposium, 1986. C. Lewis, *Unpublished monograph on Argentinian Grain Trade*, undated (probably 1987).

traded benefits (FOB (Free On Board) Buenos Aires prices minus the cost of tradeable production inputs). For the period 1981/82 to 1984/85, the domestic resource cost ratios for Argentina, based on official exchange rates, averaged 0.41 for wheat, 0.34 for maize, 0.43 for sorghum, 0.24 for-soybeans and 0.34 for sunflower (Table 2.3).

The Principal Cereals and Oilseeds

As stated above, during the last 25 years the main focus of growth in the grain sub-sector has been on wheat, maize, soybeans, sunflower and sorghum. The pattern of development for each of these crops has been slightly different. Each crop will be examined separately here. Wherever possible, changes in production and export patterns are compared to

TABLE 2.3.   Argentina—Domestic Resource Cost Coefficients for Grain
Production, 1981/82 to 1984/85

Grains	1981–82	1982–83	1983–84	1984–85
Wheat	0.38	0.34	0.43	0.48
Maize	0.34	0.31	0.35	0.36
Sorghum	0.43	0.41	0.50	0.39
Soybeans	0.31	0.14	0.32	0.18
Sunflower	0.33	0.41	0.31	0.31

Source: F. Cirio and M. Regunaga, *La Producción de Granos Argentina Frente a Las Condicionamientos del Mercado Mundial: Situación Actual y Perspectivas*. Presented at an IICA Seminar, Buenos Aires, November 1986.

changes occurring among a group of Argentina's principal competitors: Brazil, Australia, Canada and the United States.

Wheat

During the years 1985-1989, Argentine wheat production has averaged 10.22 million metric tons, 2.0 percent of total world production (Table 2.1). Almost half of the wheat produced (46.3 percent) is exported. Exports have averaged 4.74 million metric tons annually. In the 1985–1989 period, Argentina accounted for 5.0 percent of total world trade in wheat, compared with 8 percent thirty years earlier in the period 1956–1960; the volume of exports more than doubled in that time. In the periods 1924–1933 and 1937–1938, Argentina's share of world wheat exports were 18 percent and 23 percent, respectively. Due to competition from the United States and the European Economic Community, Argentina's export markets have recently become fragmented, and the country has had to look for ever more distant buyers. Approximately equal shipments of wheat were sold to eleven major destinations in 1980/81, but to 22 destinations in 1985/86. The principal buyers for the 1985/86 harvest were Brazil, Peru and Iran.

Wheat, the first crop grown in the Pampas area of Argentina, was and still is Argentina's most important crop. It is still grown predominantly in the Pampas. Over 5.75 million hectares have been planted to wheat over the past five years, 38 percent of the total area sown to grain and oilseed crops.

Wheat is sown in the fall and harvested in the late spring and early summer. All of Argentina's wheat, except for that grown in small irrigated areas in the Northwest, is rainfed. The rainfall in the wheat region ranges from 800 mm to 600 mm, decreasing from east to west. Soils are deep and fertile, but fertility levels are declining as a result of continuous cropping; responses to nitrogen are almost universal, and the phosphorous-deficient area is increasing.

The typical production unit for wheat is a family-owned farm of between 75 and 100 hectares, engaged in mixed farming.

Yield increases over the past 20 years have been modest, averaging less than 2 percent per year. Much of this increase is due to the use of improved varieties—the use of chemical inputs for wheat production is still quite limited. It is estimated that in 1985 there were only 13 kg nutrients/ha applied to wheat.[1] Further data on how Argentina's production, yields and exports compare with those of its main competitors are given in Table 2.4.

The most significant event in wheat production in the past 20 years was the introduction of semi-dwarf wheats in the mid-1970s. An estimated 96 percent of the area used for wheat is now sown to these high-yielding va-

TABLE 2.4   Key Indicators for Wheat, Major Grain Producers, Selected Years

	Argentina	Brazil [a]	Australia	Canada	U.S.A.	World
Production ('000 mt) (1985–89)	10,220	4,698	14,600	24,420	57,000	513,800
% of world production (1985–89)	2.0	0.9	2.8	4.8	11.1	100
Yield kg/ha (1985–89)	1,900	1,510	1,470	1,780	2,370	2,290
Growth in yield % p.a. (1980–89)	1.7	9.9	2.8	-1.7 [b]	0.6	2.8
Exports ('000 mt) (1985–89)	4,740	0.0	13,000	18,240	34,000	95,300
Exports as % of production (1985–89)	46.3	0.0	89.0	74.7	59.7	18.6
Exports as % of world trade (1985–89)	5.0	0.0	13.6	19.1	35.7	100

[a] 1985–1989
[b] 1980–1988

Sources: USDA, *Agricultural Statistics*, Washington, D.C., various issues; Canadian Wheat Board.

rieties. The greatest advantage derived from the introduction of the semi-dwarf varieties is an earlier maturity of the crop. Use of early-maturing varieties has enabled farmers to double-crop wheat with soybeans. As a result, the practice of double-cropping spread rapidly and initially led to more than 90 percent of the wheat stubble being sown to soybeans in the soybean-producing areas. This rotation, however, was found to be too intensive in zones with limited water; in these zones, other rotations involving three crops every two years (wheat-soybeans-maize) have been introduced.

About half the wheat varieties released since 1972 have been developed by private firms. Argentina is the only developing country in which the private sector is the major source of semi-dwarf wheat seed.

Maize

In the period 1985–1989, Argentine maize production averaged 9.4 million metric tons, 2.1 percent of total world production. Almost half of the maize produced (42.4 percent) is exported. Exports have averaged 4.0 million metric tons and account for 6.6 percent of total world trade in maize (Table 2.5). Argentine maize has had the reputation of high quality, especially with respect to the percentage of high endosperm starch. The traditional premium paid in Europe for Argentine maize virtually disappeared during the time when Argentina shifted its destination to the U.S.S.R. (during the United States embargo on U.S.S.R. sales) and southern Europe found they could supplement yellow dent corn with sufficient carotene to

TABLE 2.5   Key Indicators for Maize, Major Producers, Selected Years

	Argentina	Brazil [a]	U.S.A.	World
Production ('000 mt) (1985–89)	9,425	24,580	186,360	452,120
% of world production (1985–89)	2.1	5.4	41.2	100
Yield kg/ha (1985–89)	3,400	1,830 [b]	6,880	3,640
Growth in yield % p.a. (1980–89)	1.0	0.8 [c]	0.4	n.a
Exports ('000 mt) (1985–89)	4,000	0.0	44,940	60,729
Exports as % of production (1985–89)	42.4	0.0	24.1	13.4
Exports as % of world trade (1985–89)	6.6	0.0	74.0	100

[a] 1985–1989
[b] 1984–1988
[c] 1980–1988

Sources: USDA, *Agricultural Statistics*, Washington, D.C., various issues; Canadian Wheat Board.

produce the yellow color desired in broilers and egg yolks. About the only place where Argentine hard endosperm corn is still preferred is in the dry milling industries producing flaking grits.

Maize production has decreased over the past 20 years, but with extreme year-to-year variability as a result of weather conditions. It reached a high of 12.9 million metric tons in 1981. Although Argentina's share of total world production has remained relatively stable, its share of world exports has diminished (Table 2.6).

In the early 20th century, Argentina played a much more important role in world maize export markets than it does now. In the period 1911–1912

TABLE 2.6   Annual Maize Exports and Market Shares: The United States, Argentina and the World (1911–1989)

Year	Export Volume ('000 mt)			Market Share (%)	
	World	Argentina	U.S.	Argentina	U.S.
1911–13	8,780	4,820	1,210	54.9	13.8
1956–60	9,107	1,778	4,391	19.5	48.2
1966–70	28,700	4,622	13,877	16.1	48.2
1976–80	71,980	6,433	53,611	8.9	74.5
1985–89	60,720	4,000	44,940	6.6	74.0

Source: Adapted from Hill and Paulsen, "Maize Production and Marketing in Argentina", Bulletin 785, College of Agriculture, University of Illinois at Urbana-Champaign, 1987; and USDA, *Agricultural Statistics*, Washington, D.C., various issues.

to 1913–1914, average annual maize exports were 4.82 million metric tons, accounting for 54.9 percent of world trade (Table 2.6). U.S. exports at that time were 1.21 million metric tons, only 13.8 percent of world trade. While production and exports increased in both countries over the next 75 years, they increased at a much faster rate in the United States than in Argentina. By the 1950s, the relative positions of the two countries in world markets had been reversed. Since then, the United States has continued to gain market share at the expense of Argentina. In the most recent period for which statistics are available (1985–1989), their respective shares were 74.0 percent and 6.6 percent.

The area of harvested maize has been declining over the past 20 years, but with wide fluctuations from year to year. During the period 1985–1989, it averaged 2.89 million hectares. The decline in harvested area has been primarily due to the increase in wheat-soybean cropping. Maize production is very concentrated geographically—over 80 percent of the maize grown comes from northern Buenos Aires, southern Santa Fe and southeastern Cordoba. In this zone, soils are deep and fertile, although nitrogen is universally deficient as a result of many years of cropping (see Map: Grain Areas).

Though the area harvested to maize has been declining, yields have been increasing, as a result of several factors, including the introduction of hybrids and use of improved agronomic practices such as planting at higher densities and increased use of fertilizer, especially nitrogen. Improved maize hybrids were introduced in the 1950s and the process of adoption of the new hybrids continued through the 1960s. Since the 1970s, all Argentine maize production has been based on hybrids, all of which are produced by private-sector seed companies, mostly multinational firms.

The growth in maize yields experienced in Argentina since the 1950s following the introduction of hybrid maize is comparable to that experienced in the United States during the 1930–1950 period when hybrids were introduced. Over the past 20 years, Argentine maize yields have increased steadily, reaching a rate of growth of 4.0 percent per annum in the period 1970–1979. Yields of maize have averaged 3.4 metric tons per hectare over the past 5 years (see Table 2.1). The main limiting factors to yields are drought and high temperatures at pollination. These factors account for the year-to-year variability in the harvest.

For many years, Argentina has been the primary source of flint corn (also called Duro Colorado or Plate Maize). Plant breeders have been introducing higher-yielding dent corns at a time when the special market for Plate Maize, notably Italy, has been reduced in size.

Soybeans

Argentina is one of the world's four main producers of soybeans, together with the United States, Brazil and China. During the past five years (1985–1989), exports of beans have averaged 1.78 million metric tons, accounting for 24.9 percent of national production and 6.6 percent of total world trade (Table 2.7).

Soybeans are a relatively new crop in Argentina –they did not become economically important until the mid-1970s. Soybean production technology was introduced from the United States by Argentine farmers, with commercial firms providing seed; the Argentine farmers generally did their own adaptive research. All the new varieties used (mainly Lee and Bragg) were imported from the United States. These varieties belong to maturity classes VI and VII, which were developed in the southeastern United States.

Initially, the area sown to soybeans was concentrated in the most fertile part of the "Pampa Humeda," which comprises northern Buenos Aires, southern Santa Fe and southeastern Cordoba. Although this area still has the highest level of production and yields, soybeans have spread both west and south. Initially, 90 percent of the soybeans were double-cropped with wheat. Double cropping is still the norm where annual rainfall is above 800 mm. However, single-crop soybean culture prevails in the drier area where most of the expansion has occurred in recent years.

Over the last five years (1985–1989), Argentina has harvested an average of 3.7 million hectares of beans. Since 1980, the average rate of expan-

TABLE 2.7   Key Indicators for Soybeans, Major Producers, Selected Years

	Argentina	Brazil [a]	U.S.A.	World
Production ('000 mt) (1985–89)	7,160	16,464	51,470	100,200
% world production (1985–89)	7.2	16.4	51.3	100
Yield (kg/ha) (1985–89)	1,960	1,700	2,160	1,800
Growth in yield % p.a. (1980–89)	-1.5	0.0 [b]	1.4	n.a.
Exports ('000 mt) (1985–89)	1,780	3,286	18,608	26,800
Export as % of production (1985–89)	24.9	17.4	36.2	26.6
Export as % of world trade (1985–89)	6.6	12.1	69.4	100

[a]   1984–1988
[b]   1980–1988

Source: USDA, *Agricultural Statistics*, Washington, D.C., various issues.

TABLE 2.8   Average Soybean Yields in Selected Countries, 1985–1989

Country	Yield (mt/ha)
Argentina	2.0
Brazil	1.7
China	1.4
Paraguay	1.6
U.S.A.	2.2

Source: USDA, *Agricultural Statistics*, Washington, D.C., various issues.

sion of the harvested area has been 10.2 percent; this expansion in area accounts for the increase in production. Expansion is being achieved at the cost of reducing areas sown to maize, sorghum and permanent pasture.

Even though most of the technology used for soybean production in Argentina has been imported and adapted, yields compare very favorably with those attained by the country's main competitors (Table 2.8). The United States leads, but it is only slightly higher than second-place Argentina.

Most of the increase in yields took place in the period 1970–1979, when an average yield increase of 7.2 percent per year was achieved. This was a period when soybean technology was being introduced and adapted to Argentinian conditions; the increases are typical of a period when improved technology is being adopted by farmers. Production was stimulated by worldwide demand for vegetable oils and oilseed meals, ready availability of agricultural technology and increasing sophistication of marketing techniques and infrastructure.

Since 1980, yields have been declining at an average rate of 1.5 percent per year. The decline in yields is due to a buildup of pests and diseases in the more traditional areas and to the expansion of soybean cropping into areas less favorable to soybean production. Despite the reduction in yields, total production increased at a rate of 8.8 percent per year during the period 1980–1989 (Table 2.1). This increase was totally due to the expansion of the area sown to beans.

Argentina exports both oil and meal. Over the last five years, it has exported an average of 824,000 metric tons of oil, accounting for 86.7 percent of national production and 22.2 percent of world trade. During the same period, exports of soybean meal have averaged 4.18 million metric tons, accounting for 93.9 percent of national production and 16.3 percent of world trade. Most of Argentina's soybean crop is locally crushed. Whole beans and meal are mainly exported to Europe and Japan, while soybean oil goes primarily to developing countries (Table 2.9).

TABLE 2.9   Production and Exports of Soybeans and Soybean Products, World and Argentina, 1985–1989 (million metric tons)

	Soybeans		Oil		Meal	
	World	Argentina	World	Argentina	World	Argentina
Production	100.2	7.2	14.9	1.0	65.8	4.5
% world production	100.0	7.1	100.0	6.4	100.0	6.8
Exports	26.8	1.8	3.7	0.8	25.7	4.2
% world production	26.8	1.8	24.9	5.5	39.0	6.4
% national production	n.a.	24.9	n.a.	86.7	n.a.	93.9
% world exports	100.0	6.6	100.0	22.2	100.0	16.3

Source: USDA, *Agricultural Statistics*, Washington, D.C., various issues.

Sunflower

Sunflower is a traditional crop in Argentina. Production was stimulated in the 1970s by the same factors that stimulated soybean production. Since 1980, production of sunflower has increased rapidly—by 7.4 percent per year. A record high of 4.1 million metric tons was reached in 1986. In the period 1983–1988, Argentina produced an average of 3.04 million metric tons of sunflower, 14.9 percent of the world total. In recent years, all of Argentina's domestic production has been crushed—Argentina has become a major exporter of sunflower oil. Lately, it has also made some direct exports of sunflower seed.

Sunflower production began in southern Buenos Aires Province. In the 1980s, the area sown to sunflower expanded northwards. In the period 1980–1989, the rate of expansion was 4.0 percent per year. The increase in area sown to sunflower was in response to improved technology, especially the introduction of superior hybrids.

Over the period 1985–1989, yields of sunflower have averaged 1,340 kg/ha. Over the 10-year period 1980–1989, the average annual rate of increase was 3.4 percent. The increase in yields has been a major technological occurrence in Argentine agriculture over the past decade.

Sorghum

Sorghum production in Argentina has averaged 3.54 million metric tons over the past five years (1985–1989), roughly 5.8 percent of world production. But it is declining very rapidly; production for 1990 is estimated at only 1.4 million metric tons compared to a high of 8.1 million metric tons in 1983. This decline in production is attributable to a decline in the area sown. Farmers switched to other grains when world sorghum prices fell.

Exports over the 1985–1989 period have averaged 1.2 million metric tons, 33.9 percent of national production and 14.3 percent of total world trade. This percentage is high considering that the FOB discount of Argentine sorghum is large, because of Argentina's transportation disadvantage to the distant markets of Japan and the U.S.S.R., and Argentina's sorghum is of lower quality than that produced in the United States.

Sorghum is normally grown in the drier areas of the Pampas towards its western limits. The area sown to sorghum has declined from a high of 2.51 million hectares in 1983 to only 700,000 hectares currently. Sorghum is being replaced by soybeans and sunflower as the adaptation of these latter crops is broadened.

The average yield of sorghum over the past five years has been 2.98 metric tons/ha. Sorghum yields have followed a similar pattern of evolution to maize. Hybrids introduced in the sixties and seventies resulted in yield increases in the 1970s of 5.7 percent per year. Throughout the eighties, yields have remained virtually stagnant.

Notes

1. *Wheat Facts and Trends*, Mexico, CIMMYT, 1988.

3

Policies Affecting
the Grain Sub-Sector

Policy Objectives

Politics in Argentina may be viewed, since the 1930s at least, as having been dominated by a battle over the allocation of economic rents from the extraordinarily rich land resources with which nature endowed the country. The two protagonists were the rural landed classes and the urban workers. The static and inward-looking view of politics and agriculture implicit in this formulation of national politics had dire consequences for the grain sub-sector—and indeed for Argentina. More specifically, governments, particularly the Peronist-oriented governments which ruled the country from 1943 to 1955 and again from 1973 to 1976, pursued policies toward the grain sub-sector which attempted: (1) to channel a major portion of the profits from grain production (often seen as rent on a national resource, land) to urban people; (2) to use taxes on the grain sub-sector, both direct and indirect, as a funding source for industrialization and expansion of the public sector role in society; (3) to keep the price of bread low; and (4) to maintain the technical base of the grain sub-sector.

Three summary measures of the effects of government policies on the grain sub-sector are: (1) effects of direct price interventions on the relative prices of the principal grain crops; (2) effects of direct and indirect price interventions on the relative prices of the principal grain crops; and (3) producer subsidy equivalents. The measure of direct price interventions on relative prices shows the prices of each grain crop—appropriately adjusted for domestic transportation and distribution costs to ports—relative to the prices that would have prevailed if they had been determined by international prices converted to domestic prices at actual current exchange rates. The data show that every grain crop has been significantly taxed,

TABLE 3.1   Argentina—Effects of Direct Price Interventions on Relative Prices of
Principal Grain Crops, 1960–1985 (in percentages)

	Wheat	Maize	Sorghum	Soybean	Sunflower
1960	-37.3	-31.3	-30.4		
1961	-23.5	-16.3	-36.2		
1962	-21.8	-2.3	6.7		
1963	-13.3	4.7	-7.3		
1964	-1.8	1.0	-19.2		
1965	-18.7	-8.7	-7.0		
1966	-6.9	10.6	-22.2		
1967	8.3	-34.5	-28.8		
1968	-27.0	-20.4	-19.4		
1969	-17.8	-1.0	-4.2		
1970	-18.3	-15.1	-22.0		
1971	-12.2	-17.2	-26.0		
1972	-49.2	-28.7	-37.9		
1973	-43.1	-24.9	-28.6		
1974	-64.3	-27.7	-19.9		
1975	-60.2	-45.6	-35.2		
1976	-40.6	-58.3	-53.4		
1977	-13.1	-20.2	-24.7	-17.6	-23.2
1978	-19.4	-10.0	-12.3	-15.4	-33.3
1979	-12.2	-12.1	27.6	-12.1	-21.8
1980	-9.7	6.2	-9.2	-4.8	-24.3
1981	-3.1	-16.0	-19.4	-11.6	-9.1
1982	-14.5	-15.9	-17.4	-16.4	-27.8
1983	-28.7	-29.1	-28.4	-30.2	-34.8
1984	-24.9	-23.9	-30.7	-26.2	-24.4
1985	-29.8	-22.7	-31.2	-27.0	-26.8

Source: Adolfo C. Sturzenegger, *Trade, Exchange Rate, and Agricultural Pricing Policies in Argentina*, World Bank Comparative Studies, World Bank, Washington, D.C.,
1990.

and at very high rates in some years. The effects of direct price interventions on the principal grain crops are given in Table 3.1.

Discrimination against the grain sub-sector, however, has gone well beyond that captured in Table 3.1. In addition, the grain sub-sector was indirectly penalized by over-valuation of the currency, which depressed the
price of tradables relative to non-tradables. The ratios of the actual relative
prices (grain prices deflated by the non-agricultural price index) to the estimated relative prices that would have prevailed in the absence of these
interventions (i.e. international prices converted at the sustainable equilibrium free-trade exchange rate) show the impact of all direct and indirect
price interventions on the relative prices of the principal grain crops. Table

3.2 shows the result of such direct and indirect effects for each grain crop over the period 1960–1985.

The effects of direct and indirect price interventions shown in Table 3.2 for the grain sub-sector are generally greater (in absolute value) than the effects of direct price interventions shown in Table 3.1, indicating that the effects of direct and indirect price interventions have been mutually reinforcing. The relative prices of grains appear to have been discriminated against by nearly 50 percent during the period 1960 to 1985, meaning that without such discrimination relative prices of grains would have been nearly 100 percent higher than those that prevailed. However, there is less variation from year to year in Table 3.2 than in Table 3.1, indicating that the

TABLE 3.2 Effects of Direct and Indirect Price Interventions on Relative Prices of Principal Grain Crops, 1960–1985 (in percentages)

	Wheat	Maize	Sorghum	Soybean	Sunflower
1960	-48.4	-43.3	-42.7		
1961	-45.5	-41.2	-55.7		
1962	-39.2	-24.6	-18.0		
1963	-26.0	-10.7	-21.3		
1964	-19.8	-18.2	-35.1		
1965	-44.4	-37.9	-37.1		
1966	-39.4	-28.3	-49.8		
1967	-14.9	-48.2	-43.9		
1968	-45.5	-41.1	-40.8		
1969	-41.9	-30.7	-33.9		
1970	-36.6	-34.4	-40.3		
1971	-30.0	-34.1	-41.5		
1972	-48.1	-27.2	-36.4		
1973	-45.8	-28.7	-31.9		
1974	-66.3	-31.6	-23.9		
1975	-68.2	-56.8	-48.7		
1976	-37.9	-56.7	-51.3		
1977	-17.6	-24.0	-28.0	-22.1	-27.2
1978	-39.3	-32.1	-34.7	-35.2	-49.1
1979	-44.1	-45.4	-22.8	-42.6	-49.0
1980	-58.1	-53.2	-61.1	-56.4	-65.2
1981	-55.1	-61.9	-64.3	-57.4	-55.9
1982	-37.9	-39.6	-41.3	-39.0	-47.2
1983	-44.9	-45.3	-44.9	-45.7	-49.3
1984	-49.1	-48.8	-53.9	-49.3	-47.9
1985	-51.5	-46.4	-52.6	-48.5	-48.2

Source: Adolfo C. Sturzenegger, *Trade, Exchange Rate, and Agricultural Pricing Policies in Argentina,* World Bank Comparative Studies, World Bank, Washington, D.C., 1990.

TABLE 3.3   Argentina—Producer Subsidy Equivalents, Major Crops, 1982–1986
(percentage of producer price)

	1982	1983	1984	1985	1986	1982–86 (average)
Wheat	-46	-63	- 87	- 69	-35	- 54
Maize	-61	-55	- 71	- 87	-46	- 65
Sorghum	-71	-82	-123	-130	-92	-111
Soybeans	-52	-49	- 91	- 86	-69	- 76

Source: USDA-ERS-Agriculture and Trade Analysis Division, "Estimates of Producer and Consumer Subsidy Equivalents," *Government Intervention in Agriculture, 1982–86*, Washington, D.C., April 1988.

direct measures (mainly export taxes) have been adjusted to some extent to offset changes in the indirect measures (mainly over-valuation of the currency).

Still another measure, producer subsidy equivalents (PSEs) are given in Table 3.3, and confirm again the discrimination by government against the grain sub-sector.

Policy Instruments

The policy instruments used by the various governments to pursue their objectives vis-a-vis the grain sub-sector were principally: (1) use of multiple exchange rates; (2) export taxes; (3) export quotas; (4) import tariffs; (5) creation of transport monopolies; (6) use of untargeted food subsidies; and (7) use of regulations and direct government participation in grain trading through the creation of a National Grain Board. Each of these policy instruments is briefly discussed below.

Exchange Rates

Two-tier and multi-tier exchange rates, often different for different grains, have been used off and on as an instrument to extract rent from the grain sub-sector for over half a century. For example, the dual exchange rate system was eliminated in 1976, brought back in 1978, eliminated again in 1983, and reintroduced in August of 1988. The use of this instrument creates enormous uncertainties for investors in the grain sub-sector.

Table 3.4 illustrates how application of such an exchange rate system over a seven month period in late 1988, early 1989, opened up a huge gap between the free market exchange rate and that obtained for grain exports.

TABLE 3.4 Argentina—Monthly Average Exchange Rates, October 1988 to March 1989 (in Australes/U.S. dollars)

	Sept.88	Oct.88	Nov.88	Dec.88	Jan.89	Feb.89	Mar.89
Official Rates							
Commercial [a]	11.9	12.1	12.6	13.1	13.6	14.4	15.2
Free Market [b]	14.3	14.9	15.4	15.8	16.8	25.0	40.4
Special						18.4	19.0
Exchange Rates for Grain Exports [c]							
Wheat and Maize	11.8	11.9	12.3	12.8	13.3	14.6	19.8
Sunflower Seed	10.5	10.7	11.1	11.5	11.9	13.3	17.8
Soybeans	10.4	10.6	10.9	11.3	11.8	13.2	17.6
Linseed	10.2	10.4	10.8	11.2	11.7	13.4	15.5
Others	11.7	11.9	12.3	12.7	13.3	15.3	20.2

[a] Benchmark rate applicable to official exports and imports.
[b] Applicable to all tourist-related exchanges and freely available in commercial banks and exchange houses.
[c] Includes applicable export taxes: 1.5 percent contribution to INTA, 1.2 percent commission on foreign exchange transfers, and 3 percent contribution to the statistics fund.
Source: National Grain Board.

Export Taxes

Another policy used by the Peron regime and most succeeding administrations to tax the grain sub-sector is imposition of export taxes on grain. Export taxes have been high and have varied considerably over the years, creating a large measure of uncertainty for grain producers.

The government relies heavily on export taxes as an easy way to collect revenue. The agricultural sector, as the principal exporter, bears the heaviest burden. Though the tax burden to the grain sub-sector varies according to the tax regime imposed by the government of the day, the grain sub-sector's contribution overall has been disproportionate to that of the other sectors of the economy. In 1983 and 1984, for example, export taxes accounted for about 11 percent of fiscal revenues, or about 1.2 percent of GDP. Soybeans, maize and wheat contributed over 60 percent of these revenues. Since 1981, maximum/minimum export taxes for grains as a percentage of FOB have ranged from 28.5 percent to 0.0 percent for wheat, from 31.0 percent to 0.0 percent for maize and sorghum, from 32.5 percent to 0.0 percent for soybeans and sunflower seeds, and from 38.0 percent to 0.0 percent for linseed (Table 7.9).

Since export taxes on agricultural commodities reduce domestic prices, they are used as a price control device that increases real wages in the urban sector at the expense of agricultural producers. Taxes on grains also provide an advantage to processing industries that use grains as raw materials. From the producer's point of view, the taxes reduce price incentives for use of inputs and on-farm investments.

Export Quotas

The government uses the export registry operated by the National Grain Board as a primary instrument to regulate supply to the domestic market. Periodically, the registry is closed, effectively inhibiting exports. At other times, the export registry price is set, on purpose, above the market price, thus obviating export sales.

Import Tariffs

Import tariffs are imposed on production inputs to protect domestic suppliers. In May 1988, for example, import tariffs were set at up to 38 percent of Cost, Insurance, Freight (CIF) for the imported components of small and heavy tractors and other locally-produced agricultural machinery, at up to 31 percent for the imported raw materials needed for locally-produced fertilizers, and at up to 38 percent for the imported ingredients used in some locally-produced insecticides (Table 7.10).

State Trucking Cartels

The central government has little direct influence over the trucking of grain. The states have effectively established state-centered trucking cartels, which determine cost and availability of trucking for grain transportation. Usually, out-of-state truckers are not allowed to transport grain. Transportation tariffs are set by truckers' interest associations with obvious adverse consequences for the grain sub-sector.

Railroads

All foreign-owned railroads and port facilities were nationalized during the first Peron regime (1943–1955). The railroads have remained nationalized. The State railroad management has become entrenched—government exerts little authority and does not hold management responsible for profitable operations. Priority is given to passenger transport. Cargo transport, including grains, carries the full burden of inefficiency and lack of maintenance. Customarily, not more than about 10 percent of the number of operational locomotives is dedicated to grain transport at the peak of the harvest.

Untargeted Consumer Subsidies

Estimated consumer subsidy equivalents, resulting from many government policies, are given in Table 3.5.

The overvalued exchange rate, export taxes and export prohibitions discussed above are the main sources of the untargeted food subsidies received by all consumers. Other policies contributing to these consumer subsidies are: the earmarking of a certain quantity of wheat for bread and subsidized credit for storing wheat.

The government, through the National Grain Board, annually earmarks about 1.7 to 2 million metric tons of wheat for the domestic market in order to subsidize the price of bread, which contributes about 7 percent to the Consumer Price Index. These earmarked wheat stocks, which are sold by the grain board to the wheat millers at cost, supply about 75 percent of the nation's wheat consumption. Because this food subsidy is untargeted, it is an inefficient and costly way to help the poor.

Furthermore, when calculating its costs, the grain board charges a subsidized interest rate (12 percent) on the loans used to finance the storage of wheat stocks, a rate well below the rate farmers must pay (typically 15 percent to 20 percent and more). Since interest charges account for about 55 percent of carrying costs, this is a significant subsidy. The government then places price controls on bread to ensure that the benefits from subsidized interest rates on the wheat stocks are passed on to the consumer.

TABLE 3.5   Argentina—Estimated Consumer Subsidy Equivalents, 1982–1986 (as a percentage of consumer price)

	1982	1983	1984	1985	1986	1982–86 (average)
Wheat	41	55	76	60	33	45
Maize	47	42	58	68	39	51
Sorghum	45	50	80	82	61	70
Soybeans	47	45	82	77	62	68

Source: USDA-ERS-Agriculture and Trade Analysis Division, "Estimates of Producer and Consumer Subsidy Equivalents," *Government Intervention in Agriculture, 1982–1986*, Washington, D.C., April 1988; and Adolfo C. Sturzenegger, *Trade, Exchange Rate, and Agricultural Pricing Policies in Argentina*, World Bank Comparative Studies, World Bank, Washington, D.C., 1990.

In the early 1980s, the government used the wheat sales made by the grain board as an instrument to subsidize the bread price directly, through sales to the millers below total carrying costs. The losses incurred at that time, over US$ 100 million, are still on the books of the grain board as an obligation.

Grain Regulation and Trading

The government has collected rents from the grain sub-sector through government monopolies of trade, government furnishing of services (through terminal or port elevators) and through export controls and quotas. The government's principal agent for intervening in the grain sub-sector is the National Grain Board.

Since its beginnings in the 1950s, the board has been either a direct arm of the government (a state monopoly) or a weak autonomous institution subject to political manipulation. It has been used both as a trader and trade controller, and has been expected to both promote producer and consumer interests. These conflicting functions have made it impossible for the grain board to carry out any policies consistently, other than the unstated policy of diverting revenues from the grain sub-sector to the industrial and urban economy.

The board is currently used to: (1) regulate the grain trade (through use of an export registry); (2) control the movement of grain (through reporting requirements) and set grain grading standards; (3) ensure domestic supplies of wheat and prevent monopolization of the wheat market (by export quotas on wheat); (4) ensure a supply of grain to fulfill government international commitments; (5) stabilize prices (by fixing support prices for grains, principally wheat); (6) provide subsidies to farmers in marginal areas of the Northwest and Northeast (by purchasing, at its posted minimum price, all grains offered in these areas; and (7) promote the use of fertilizer (by serving as the Ministry of Agriculture's executing agency for its fertilizer program).

4

Institutions Affecting
the Grain Sub-Sector

A number of public and private institutions affect the operation and development of the grain markets in Argentina. Those institutions that have had the largest effect on grain markets are: the National Grain Board, the private seed companies, the multinational exporters of grain, and the grain exchanges in Rosario, Buenos Aires and Bahia Blanca.

Public Institutions

The National Grain Board

The National Grain Board is an autonomous agency of the Secretariat of Agriculture, Livestock and Fisheries (SAG). There have been a succession of boards since the 1940s. The present board is based on Decree Law Number 6698–63, which establishes it as an autonomous institution under the umbrella of the Ministry of Agriculture.[1] As stated in the previous chapter, the board is the principal agent of the executive for intervening in the grain sub-sector.

History. The board has its origins in the Grain Regulatory Board created in 1933, ostensibly to control grain quality, but indirectly to provide subsidies to producers. Under the Peron regime (1943–1955), this regulatory board became the Argentine Trade Promotion Institute (IAPI), a statutory monopsony. IAPI's unofficial purpose was to channel export proceeds into industry and keep food prices low by depressing producer prices. Physical handling of grain stocks was left to private sector groups, which operated on a fixed commission of sales proceeds.

The military government that succeeded Peron disbanded the IAPI and replaced it with the National Grain Board, which was responsible for trade regulation, administration of port facilities, and trading. State monopoly over trade was abolished; instead, the board offered the farmers support prices. These were based on international market prices rather than on local production costs. In effect, the board became a buyer of last resort.

After Peron's return to power in 1973, the National Grain Board became a public monopoly and the private sector again acted as agent for the government on a commission basis.

The board lost its monopoly during a period of economic liberalization that followed the end of Peronist rule in 1976. It returned to offering support prices to farmers.

The Alfonsin government (1983–1989), continuing the market orientation of its predecessors, further privatized grain trading. However, an uninterrupted economic crisis absorbed most of the government's attention. The Menem government replaced it in July 1989. Although the new government is Peronist in inspiration, its thinking and policies appear to steer away from traditional Peronism. The role the grain board will play under the Menem regime remains to be decided.

The Board's Functions and Operations. The National Grain Board has four main functions:

- to act as a regulatory and control agency;
- to collect and disseminate information on grain production and trade;
- to act as trader; and
- to operate physical, trade-related infrastructure.

All participants in the grain trade are required to register with the board. The board regulates all aspects of the trade, including types and formats of purchases and sales contracts, sale and delivery procedures. Essentially all grain shipments, whether between farmer and country elevator or between country elevator and miller or exporter, are accompanied by documentation standardized by the board. All cooperatives, country elevators and industries must submit monthly grain movement reports to the board; exporters must submit weekly purchase reports. However, newly emerging integrated trading channels are increasingly dispensing with standards, regulations and reporting requirements. These new structures are self-contained. Inasmuch as they do not need the board, they see no reason to fulfill its requirements.

The board supervises all export sales. All signed export contracts have to be registered within 24 hours; the FOB (Free On Board) price of the contract, applicable taxes and other conditions of sale are determined and fixed at the time of registration. Grain shipments may by law be registered several months prior to actual shipment. Failure to effect a shipment, once it has been registered, carries a penalty of 15 percent of the total value of the shipment as registered. For all grains, the board determines daily minimum export prices, below which shipments cannot be registered. The main motives for requiring registration are to avoid capital flight and establish the tax base.

The board operates a wheat export quota system. Currently, 31 entities hold quotas under this system, including the board itself. Periodically, the board opens the wheat export registry, for those who obtained a quota. The board determines the quotas on the basis of past export performance. The two cooperatives (ACA and FACA—see discussion of cooperatives on page 35) are an exception. They are assured of an automatic 10 percent of available quota.[2]

The board establishes the grading standards (Normas de Clasificacion) which provide the grades and grade factors for maize, soybeans and wheat. These standards are used throughout the industry and by law must be used with an official inspection, even at the point of first purchase from farmers. The board also inspects all export shipments and issues Certificates of Quality. Board regulations prohibit the movement of grain in commercial channels unless it meets condition "Camara", which essentially means it must meet all the factor limits for one of the numerical grades specified in the official standards. The Arbitration Chamber (Camara Arbitral) located within the grain exchanges provides grading services complete with laboratories. However, the exchanges must use the grades developed by the board and, in general, their laboratory technicians are trained in the school conducted by the board's inspection department.

The Board's Organization and Resources. The board's law of establishment prescribes a directorate of nine members, five from the public and four from the private sector. For the public sector, the Ministry of Agriculture appoints the President and Vice-President and the Ministries of Finance and of Trade and Transport appoint two other members. The four private sector members include one each from the farming sector, the cooperative sector, grain millers, and the trading sector. However, it is a government prerogative to name all board members, and the Menem government has used this prerogative to appoint a board made up fully by members from the private sector.

In practice, the private sector has had more representation on the board than the law calls for. The board's two immediate past presidents were from the cooperative movement and the current chief executive is from a private consulting company—all of these appointments were made by the Minister of Agriculture.

The board currently has eight departments, five divisions and one general secretariat, all of the same hierarchical level. Personnel number about

4,000, considerably fewer than the 7,000 employed in the mid-1970s. The board has 34 district offices in all grain-producing areas in Argentina.

The board was financially self-sufficient until about 1987. Since then, it has been increasingly short of funds.

Until 1981, the two main sources of income for the board's recurrent, operational budget were: (1) fixed levies on the grain trade, and (2) income from services provided. The fixed levies, consisting of up to 1.5 percent of the FOB, or delivered industry value, of all grains marketed, were abolished during the period of military rule prior to 1983 as part of a move to eliminate all special funds and fundings. Service income was, and still is, derived from grain testing, elevation, fines, donations, rents, etc.

Until recently, the board was able to generate over 98 percent of its resources from its own income, and did not have to rely on the government's annual budget. Income from the elevator function has accounted for an average of over 80 percent of total annual revenues since 1980.[3] The board also derives income from its charges for services rendered in laboratory and facilities inspections. Between 1974 and 1985, total income fell, in constant terms, more than 40 percent, due mainly to a drop in income from the elevator function as private elevators began to cut into the board's market share.

Since 1980, from 44 percent to 62 percent of the board's budget has gone to personnel remuneration: wages and salaries, overtime, and the cost of meals. The nominally non-salary part of remuneration (meals and overtime) have, in combination, formed from a low of 30 percent of total remuneration in 1980, to a high of 45 percent in 1983.

For many years, only a fraction of the board's budget has been used for capital investments, including periodic equipment renewal, or for maintenance. The largest capital investments –for country elevators –were funded out of World Bank credits.

Weaknesses of the National Grain Board. The law that established the board is vague about its ultimate role. Nowhere is a clear definition to be found of its objectives, of policies and priorities. The law as such does not guide the development of a long-term strategy, and the board does not have one. Over the years, the board appears to have responded in an ad-hoc manner (though often very effectively) to perceived or imposed problems. The only programs that have been developed and that have a certain measure of continuity and coherence are a price support program, a fertilizer distribution program, and support for marginal areas. Even with respect to these programs, it is difficult to see an underlying leitmotif guiding their development in a unifying structure.

Unlike most other public sector institutions, the board deals with often rapidly changing conditions. It is confronted with issues of commercial importance that have a high political visibility, with interest associations and groups that are vocal and have political influence, and it has to deal with issues that are at the heart of the country's macroeconomic policies. Thus, the institutional context in which it has to operate is complex. In spite of the importance and the multifaceted nature of the board's actions, its formal contacts with other public sector institutions such as the Central Bank, the Ministry of Foreign Affairs or the Ministry for Industry and External Commerce remain limited and inconsequential. The board's important role of organizing the logistics of grain transport, both on land and at sea, is stymied by its limited interaction with either the transportation ministry or the port authorities, in spite of the representation of the former on the board's board. The new port law proposed by the Ministry of Public Works has been written without input from the board, the major user of Argentina's port facilities.

The changing responsibilities implicitly or explicitly assigned to the board over time are reflected in the multiplicity of technology and expertise found within the organization. Its somewhat turbulent history is also reflected in its multi-departmental, overlapping organizational structure with redundancies and duplications of efforts. The political nature of the board has given rise to frequent turnover of its directorate with resultant instability of management, and lack of long-term direction and developmental strategy. Recent fiscal policies, the privatization drives of successive governments, and the long erosion of efficiency in the elevator complex, have emaciated the board and have led to a financial and organizational crisis.

As a result of its political nature, of the high visibility of its actions, and the aggressiveness of some of its interlocutors, the board has seen a remarkably high turnover of top management. In the last 50 years, the board and its predecessors have had 54 different presidents, caretakers or national directors. It has had chief executive officers whose average stay in office was less than one year. Apart from this frequent turnover at the top, the rest of the directorate has not always been sufficiently competent professionally, or personally interested, to make substantive contributions to the work or the future of the board.

Though the number of persons employed by the board has decreased considerably since 1974, departmental reductions are not always explainable on the basis of departmental restructuring or demonstrated need. There is an imbalance between reductions in field offices and those in the central office. Salaries have suffered over past years, with substantial reductions in take-home pay. The effect of the lower salaries has been partly offset by promotions, which have led to a proliferation of posts at the top, by increased and institutionalized overtime, and by remunerations in kind. Thus, personnel reductions, in large part occasioned by dwindling income of the board, have led to organizational changes that are not related to operational requirements and to distortions in the hierarchical structure of the institution that are not conducive to harmonious development and good labor relations.

In the past, overtime was related to the need for work to be done. It has become an institutionalized part of remuneration with quotas (cupos) assigned to each department. For instance, the administrative personnel in headquarters has an assigned overtime quota of about 15 days per month. Overtime itself is paid at 1.5 times normal hourly rates. The available quota of overtime varies up or down with the availability of funds. In other words, the personnel of the board receives a fixed basic salary and a variable component depending on the availability of funds.

The Secretariat of Agriculture, Livestock and Fisheries (SAG)

While in principle the Secretariat of Agriculture, Livestock and Fisheries (SAG) is responsible for establishing and implementing agricultural policy, and the Secretary of Agriculture has cabinet rank, in practice, SAG is under the purview of the Ministry of Economy.

The National Institute of Agricultural Technology (INTA)

The National Institute of Agricultural Technology (INTA) is Argentina's main agricultural research agency. INTA was established in 1956 to: (1) improve farm management techniques; (2) continue ongoing research projects and expand their scope; and (3) develop effective extension services. INTA's primary innovations include fostering improved agro-economic techniques, encouraging the domestic manufacture of farm machinery, and improving seeds and agrochemical products.

INTA was reorganized in 1984, but proposed changes were not fully implemented until 1988. INTA reports to SAG, but does not depend on it for funding. Since 1984, it has been funded through a surcharge of 1.5 percent on primary agricultural exports. This levy is theoretically supplemented from government budgetary sources. The Interamerican Development Bank (IDB) has also provided financial support. Available operating funds have fluctuated extremely from year to year, causing disruptions in technology transfer, and sometimes morale problems.

INTA's principal administrative offices are located in Buenos Aires. There are also 15 regional offices. Each region has at least one major station as well as secondary sites where experiments are carried out. There are 37 experimental stations in all. The national center provides more basic and supporting studies. Personnel include 1,818 professionals and 3,579 supporting staff.

INTA performs research on a number of farm crops and publishes technical bulletins containing its research findings and recommendations. Its research activities are an important resource for private and public extension technicians who service producers and producer organizations. INTA also provides training, and exchanges information with other research centers.

INTA developed the first Argentine wheat varieties using Mexican germplasm. Two private enterprises later joined the effort. Applied research into soybean cultivation resulted from a joint effort between INTA and the University of Buenos Aires. Hybrid seed development and other technical innovations are increasingly occurring outside the public sector—the result of deliberate public policies to: (1) transfer genetic and human resources to the private sector; (2) release hybrid seed research results generated by the public sector into the private domain; and (3) enable private researchers to profit from innovations. INTA, however, continues to be an important developer of new wheat varieties. It is of minor importance in the distribution of seed for other grains and oilseeds.

Private Institutions

Private sector groups include: (1) commercial seed companies; (2) rural cooperatives; (3) other farmer associations; (4) the grain exchanges; and (5) the multinational trading companies.

Commercial Seed Companies

About 30 private seed companies operate in Argentina, 8 of them multinationals. Six local firms at least can be considered major enterprises likely to have research units capable of creating new hybrids and varieties. Commercial enterprises dominate the production and distribution of hybrid seeds (maize, sorghum and sunflower). They also contribute half or more of the major self-pollinated species (including wheat and soybeans).

The commercial seed industry became established within the last 15 years, although some firms began earlier. A period of depressed commodity prices has led to consolidation for some firms and closure of the weaker ones. However, commodity prices have since risen and government controls on retail seed prices have been lifted, causing some renewed optimism in the industry.

Cooperatives

The cooperatives are the most visible, if not the most important, of the private institutions that affect the grain sub-sector in Argentina. There are approximately 1,278 individual cooperatives organized under 15 national associations. It is estimated that some 459,000 producers, about 85 percent of the nation's 538,000 farms, are members of cooperatives. There are two second-tier cooperative organizations, the Association of Argentine Cooperatives (ACA) and the Argentine Federation of Agricultural Cooperatives (FACA), and one "super-association," the Intercooperative Agricultural Confederation (CONINAGRO).

ACA was formed in 1921 as an umbrella organization for a number of primary cooperatives established primarily by European settlers. These early cooperatives supplied inputs and organized marketing. ACA contributed to improving grain elevator capacity. FACA, established in 1947, is an important supplier of inputs; it also markets grains internationally. CONINAGRO, formed during the first Peron administration, embraces 13 secondary cooperative organizations.

ACA and FACA jointly own a large deepwater elevator complex in Quéquèn, from which they export grain. Between 1980 and 1985, they were responsible for about 8 percent each of yearly grain exports. Their combined percentage share of grain exports grew from an average of 13 percent between 1980 and 1982 to 18 percent between 1983 and 1985. Based on the labor hours per ton of grain exported, the cooperative export facility is dramatically more efficient than the grain board elevator nearby.

The cooperatives benefited from government support after the Alfonsin government came to power in 1983. For example, ACA and FACA were able to acquire 4 port elevators from the grain board, a member of the cooperative was elected as the board's president, and the cooperatives were given an automatic 10 percent of the wheat quota. However, their position has since deteriorated, as they have suffered grain trading losses. Their financial difficulties have prevented them from expanding their activities.

The cooperatives are involved mainly in the purchase, storage and transport of commodities; giving technical advice; disseminating marketing information; bulk purchasing and retail sales of farm inputs; and promoting the social and political interests of their members. Cooperatives handle one third of the 75 percent of grain that moves from the farm to country elevators in the first stage of marketing. They purchase grain for cash, and store grain for a fee. The cooperative system has about 6.7 million metric tons of storage capacity.

The cooperatives generally have a good relationship with the National Grain Board. One cooperative member is automatically included in the grain board's directorate—the grain board's two most recent presidents have been from the cooperative movement. The grain board has certified a portion of the cooperatives' storage capacity under its "1825" system (see Chapter 5). By the end of 1987, 370 cooperatives had 1825-certified storage.

Cooperatives are part of the grain reporting system administered by the grain board, and must send monthly reports of grain movements to the board.

Trade between the cooperatives and exporters or millers takes place on the grain exchanges, which are located in Buenos Aires, Rosario and Bahia Blanca. Like other groups, the cooperatives generally do not trade on the futures markets.

Other Farmer Associations

There are four farmer organizations, the Argentine Rural Society (SRA), the Argentine Rural Confederation (CRA), the Argentine Agrarian Federation (FAA), and the Regional Agricultural Experimentation Groups (CREA) that, like the cooperatives, are sometimes able to influence the grain sub-sector.

The SRA, established in 1866, is an association of large farmers and landowners. It now has about 10,000 members, including important grain producers. The CRA evolved from a network of rural associations, developed between 1880 and 1940, which advocated a larger role for government in the agricultural sector. It has about 100,000 members, mainly medium- to large-scale farmers. The FAA, an association of small- to medium-sized farmers, has about 70,000 members, and a network of 400 branches, 130 youth centers, 185 affiliates and 30 associated entities. CREA is an advisory service that functions like a cooperative. It has 176 groups, divided into 15 regions, and 2,012 members. It provides the equivalent of extension services through a cooperative association of farmers, and thereby influences the efficiency of production and marketing for its farmer members. It does very little lobbying.

In spite of their large combined membership, producer associations in Argentina, particularly the cooperatives, have never achieved the overriding economic importance that similar associations have elsewhere, particularly in Europe. Because they lack cohesion, they have been unable to establish a unified rural political force that could counterbalance the urban industrial interests.

The Grain Exchanges

Argentina has grain exchanges in Rosario, Buenos Aires and Bahia Blanca. These exchanges provide price determination by public outcry just as trading is conducted on the Chicago Board of Trade. Both cash and futures transactions are conducted in all three exchanges, although the relative importance of the two markets differs among the three exchanges. Rosario is recognized as the largest cash market while Buenos Aires is recognized as being relatively more important in the futures market. With the lack of speculators in the market, none of the three exchanges conducts a large volume of business in the futures markets. The cash price of grain is reported following the close of the exchange. The Arbitration Chamber (Camara Arbitral) located within the exchanges has a committee that meets in the morning following the close of each day's cash market, decides upon the effective price established during the previous day's trading, and announces the price through the media such as radio and wire services. The announced price becomes the basis for price offers to producers at the country elevators. All segments of the trade, past the first handler, are represented on the grain exchanges. Quality, quantity, and price disputes, or disputes relating to trading rules are handled by the arbitration group within the Arbitration Chamber.

Grain exchanges in many countries are an important element in providing market transparency. In Argentina, however, much of the newly-developing trade bypasses the grain exchanges. Sales from country elevators to exporters are sometimes handled by integrated firms including multinationals, the grain board, and FACA. Country elevators also sell directly to exporters and processors without going through the cash or futures market represented in the grain exchanges. In most cases, however, they do rely upon information from the grain exchange as a basis for price quotations. Those firms that are integrated from country elevator through export shipments have no need of the grain exchange for transactions but would still rely on the exchange either for hedging futures transactions, establishing forward sales prices, or for keeping abreast of market activity. Large integrated firms would rely heavily on the Chicago Board of Trade supplemented by information in the Argentine grain exchanges. A commission of five members is elected on a weekly basis to represent each of the sectors and to maintain an equilibrium between buyers and sellers, and thus avoid any bias in price formation.

The average price of grain at the previous day's close is announced as the base price for the following day. The posted price determined by the Arbitration Chamber on the grain exchanges becomes the base price or benchmark for price setting by the country elevators on their cash transactions with farmers. Farmers, however, have other alternatives. A common one is the use of prices to be determined in the future (precios a fijar). This is similar to the delayed price option in the United States although the Argentine country elevators seldom use the hedging techniques for protecting against risk of price changes. This is not a futures market transaction on the farmers' part, but it gives them the opportunity of speculating on price increases in the future without actually holding physical title to the grain. At the time the farmer chooses to price the grain, the posted price at the grain exchange on that day is taken as the benchmark for settlement. As the exchanges lose business, and more and more sales are made outside the grain exchange, these posted prices are becoming less reliable in terms of representing true market transactions. All three grain exchanges are taking action to increase their volume and all are confident that cash volume through their markets will increase in the future. Market transparency requires that a significant volume of the transactions move through the market where public prices are announced.

Multinational Trading Companies

All major multinational grain trading companies are represented in Argentina. The five largest, Bunge & Born, Cargill, Continental Grain, Dreyfus and NIDERA, were responsible for about 33 percent of all grain exports from Argentina between 1980 and 1985 (Table 7.8). The major multinational trading companies have their headquarters in Buenos Aires but maintain facilities at most of the major ports and in the country. Most of them are also engaged in processing, either directly or indirectly, with joint ventures or part ownership in some of the Argentine companies. These multinational exporters have several advantages over Argentine domestic firms, cooperatives, or government exporters. One of these is their flexibility to originate grain from whichever country provides them the greatest economic advantage. If the freight rates, price, and delivery terms are better in New Orleans than in Buenos Aires, the multinational can originate grain from New Orleans. The opposite is also true. Their size and locations throughout the world also provide them access to additional information and advantages in transportation.

In the case of operations in Brazil and Argentina, one of the multinationals' biggest advantages is their ability to operate in international financial markets, using changes in exchange rates to their advantage. Interviews with multinationals in these countries indicate that many sales are made with no profit margin on the grain because of the opportunities for profit in the offsetting financial markets. The multinationals' ability to maintain an adequate staff to understand government policies and financial markets throughout the world gives them another advantage.

Because of their size and efficiency, the multinationals provide strong competition for domestic and government firms in Argentina. The net effect is probably a more efficient marketing system and a reduction in the overall average marketing margin. In order to maintain market share, domestic and government exporters in Argentina must keep their buying and selling prices in line with those offered by the multinationals. In some cases, this may result in a loss for inefficient firms with outdated facilities such as those operated by the grain board.

Notes

1. For the legal history and framework of the board see: Walter A. Villegas, *Regimen Juridico del Control del Comercio de Granos*, Buenos Aires, 1982 (particularly Chapter III as of page 21).

2. In June, 1989 the Menem government allowed advance registration of 125,000 metric tons of wheat to be exported during the following season, i.e., in December 1989/January-February 1990, by each registered exporter at the prevailing futures price of wheat of about US$150/metric ton. As a quid pro quo, exporters had to deposit the value of their registrations in a 90 day fixed term account with the government-owned Banco de la Nación Argentina (BNA). In this way, the newly-elected government raised an estimated US$450 million in short-term funds.

3. It is said that the board's need for income, rather than actual operating costs, determines the elevator tariff. These above-cost tariffs were also the benchmark for tariff setting by newly built private elevators in the early eighties, making the latter particularly profitable. A severe fall in exports since 1985 has increased the competition for elevator business and has led to sharp drops in service fees charged by the new private facilities.

5

Grain Marketing

Marketing Channels

Marketing of grains in Argentina typically passes through four stages. In the first stage, the grain moves from the farm to the country elevators, which are owned by farmers through cooperatives (about 1,200), by private independent firms such as Continental Grain, and by the National Grain Board (64 in 1981, and about 30 today). All three of these ownership patterns result in a country elevator system where one elevator operates very much the same as every other one. In the second stage, the grain moves to millers or exporters. In the third stage, the grains acquired by the millers are either processed into food or are sold to exporters; some grain sold to exporters in stage two may be resold to other exporters in stage three. Finally, it passes through the export stage. This division into stages is, of course, somewhat of a simplification, since there are many shortcuts and other parties, such as brokers, may play a role in marketing the grain.

About 75 percent of the grain produced moves from the farm to country elevators. The country elevators owned by private firms and the grain board take about two-thirds of this amount, and those owned by the cooperatives take the other third. The remaining 25 percent is either consumed on the farm or sold directly to millers and exporters.

The country elevators, whether private or cooperative, are essentially rural traders. They receive grain from the farmer and perform the traditional marketing services of assembly, conditioning, storage, buying, selling, and transporting the grain. They provide a number of services to farmers including supplying inputs, financing, and occasionally selling consumer goods. The country elevator generally acts as the farmer's agent. The farmer delivers grain that may or may not be clean and dry. The elevator's responsibility is to put that grain into "condition Camara" (required by law prior to entering the market channel) and to deliver the grain to the next stage in the market channel. Since the grain the farmer delivers is often not in "condition Camara", the elevator manager does not take title to it until it has been cleaned, dried, and made ready to be stored or delivered. The price generally shown on the settlement sheet is the price quoted at the nearest grain exchange. The settlement sheet then shows subtractions (or charges) for transporting from farm to elevator, transporting from elevator to port destination, cleaning (including shrinkage), drying (including shrinkage), and storing. In addition, there may be charges for such services as fumigation, discounts for weed seeds that must be removed, and, in most cases, a commission charged by the elevator.

In the case of drying, the grain board provides a shrink table which appears to be universally used throughout the industry for adjusting inbound grain weights as received from farmers down to the base moisture content of 14.5 percent (the table slightly exceeds the actual water loss calculated mathematically and, therefore, is similar to the "shrink factors" used by U.S. grain traders to allow some slack to cover other costs). Screening is also allocated a fixed percentage of shrinkage regardless of the quantity or weight removed. Thus, most of the charges to the farmers for services performed by the country elevator are relatively constant among elevators. The greatest variation appears to be in the commission charge for the handling services, the transportation charge (based on destination, distance, and mode of transport), and a charge by some of the cooperatives for capital replacement.

The second step of grain marketing, between the country elevators and cooperatives and the exporters or millers, takes place mostly on the grain exchanges in Buenos Aires, Rosario or Bahia Blanca. In 1984/85, about one third of the grain traded on the exchanges was acquired by millers, and two thirds by exporters. Argentina has an estimated 99 wheat mills, with an annual throughput of over 4 million metric tons, representing 60 percent of milling capacity. In early 1989, there were also 27 sunflower and 26 soybean mills.

The third stage of grain marketing is called the FOB (Free On Board) market. This market is made up of all the multinational grain trading companies and approximately another 100 local traders, some of them quite small. The trade is essentially conducted offshore, in foreign currency, between industries and exporters of grains and milling products such as soybean meal.

The final stage of marketing, which handles a major portion of Argentina's grain crop, is the export market. About 15 to 16 traders usually account for 85 to 90 percent of all export shipments (Table 7.8), but the largest five handle about 40 percent. Most of the private companies use offshore intermediary companies to negotiate sales with the ultimate buyer. The National Grain Board was a substantial exporter until 1985 –between 1983 and 1985 it handled about 10 percent of all shipments. The board's exports originate in government-to-government agreements. Such agreements are principally made for wheat. Although the board is the nominal exporter, agreements entered into between government agencies have, in effect, been executed by one or another of the major international grain trading companies.

Before being loaded onto ships, the grain passes through terminal or port elevators. Until 1980, when a 1979 law allowed transfer of government-owned port elevators to cooperatives or other producer organizations and construction of new elevators by private parties, the grain board was the sole owner and operator of these port elevators. Since then, ten private elevators have been built on the Paraná river, and the ACA and FACA, Argentina's main cooperative groups, have acquired an elevator complex at the deepwater ocean port of Quequén. The private sector export complexes have very little storage capacity. They essentially operate as throughput houses.

Since the new elevators came into service, third party use of private elevators has increased substantially, and the share of board elevators in total export shipments has dropped—from 76 percent in 1982 to less than 56 percent in 1985. The board system in several river ports is either idle or used as static storage. An exception is the elevator in Bahia Blanca, Argentina's sole deepwater grain port, which is still operating, thanks to its monopoly position.

The loss in business at the public sector facilities is most severe at the ports along the Paraná river. Capacity utilizations of public elevators in 1987 ranged from 8.5 percent in Santa Fé to 34 percent in Bahia Blanca to 44 percent in San Nicolas.

The private sector has taken business away from the public sector for four reasons: (1) the public sector is inefficient; (2) the public sector's handling and port-related charges are not competitive; (3) the share of oilseeds and oilseed by-products in grain exports has risen (from 26 percent in 1981/82 to 46 percent in 1986/87), and the public sector is not equipped to handle crushing by-products, which make up the bulk of the shipments; (4) the private complexes, owned by major exporters, are integrating their grain handling via bulk transportation contracts all the way back to country elevators and large producers; and (5) the private sector export facilities, through superior scheduling, minimize transit time, both for land transport and for grain-loading ships.

Grain Handling and Transportation

Argentine institutional and physical infrastructure has been able to cope remarkably well with the high growth in grain production and exports since 1970. Laws, regulations and standards have been established to support the expanding commerce. The grain board has taken a leading role in establishing the regulatory framework for the trade. The private sector has been responsible for rapidly expanding storage, truck transportation capacity, and port elevators. The downturn in production and exports that occurred in 1986 has put great strains on both the producers and the marketing infrastructure.

Drying and Storage

Most grain drying is done at the country elevators or processing plants. A few of the medium-sized farms have on-farm drying and storage facilities, but the large farmers and ranchers tend to prefer to move the marketing responsibilities off-farm. The smaller farmers cannot afford a drying/storage setup for their small volume. Most elevators have a high temperature dryer for maize and soybeans delivered at moisture levels above the maximum allowed in the official standards. Several elevators and processing plants are considering low-temperature or multiple-pass types of drying systems in order to achieve a higher quality. Drying capacity appears to be adequate in most years for the quantity of grain and the moisture content delivered. The supply of and demand for drying services is balanced by the fact that the grain harvest can be slowed if the drying capacity at the local elevator is not adequate to handle the volume. When the truck lines at the elevator become too long, farmers slow down the harvesting process, allowing the maize or soybeans to dry further in the field, thus, increasing the effective capacity of existing dryers. In 1984/85 it has been reported that 30 percent of the wheat, 71.2 percent of the maize, 59.2 percent of the soybeans, and 60.7 percent of the sorghum was dried artificially. The total of all grains dried was 21.3 million metric tons. Drying capacity reported in 1984 for Argentina was 44,870 metric tons per hour.[1]

Primary storage is sufficient for current conditions and for the medium-term future. Argentina has an almost 1:1 relationship between yearly grain harvests and storage capacity. For example, the 1986/87 grain harvest was about 32 million metric tons; total static storage capacity, according to a 1985 survey done by the grain board, is about 31 million metric tons. Since a large part of production is exported shortly after harvest, there appears to be no foreseeable lack of storage capacity. Much on-farm storage remains idle, while storage capacity can be purchased for 50 percent or less of replacement costs.

Over the past decade, private development of storage infrastructure has kept pace with increasing production. Storage capacity has been estimated as follows: country elevators, 18 million metric tons; private storage at ports, 4 million metric tons; grain board, 1 million metric tons; on-farm, 6 million metric tons.

Under its "1825" rule (see page 47), the grain board has certified about 450,000 metric tons of storage capacity (a very small proportion of the total) to serve as bonded warehouses where grain can be stored before it is sold to exporters.

Truck and Rail Transport

The transport situation is less favorable. Argentina has an inadequate and increasingly deficient highway system, a trucking fleet of which more than three quarters is over 10 years old, and several obsolete, incompatible and costly national rail systems. The precarious situation of public sector finances in Argentina over the last several decades has led to limited, if any, new investments in transportation infrastructure and to a virtual abandonment of preventive maintenance. The railroads are government-owned.

About 80 percent of all of grain transported is now moved by truck, about 18 percent by rail, and less than 2 percent by barge. Trucks are preferred because: (1) the average transportation distance is low (the average transportation distance between farmgate and millers or export port is 200 km for trucks, and about 350 km for rail transport); (2) the railroads are unreliable; and (3) many inland elevators are not directly serviced by rail, while rail unloading at many ports is slow and wasteful.

Between 1974 and 1985, an average of about 6 million metric tons of grain was shipped annually by rail. Although the total amounts shipped remained fairly constant, the share of rail transport in shipments destined for exports decreased from over 55 percent to less than 29 percent. During the same period, grain production increased from 24 million to 41 million metric tons, while exports increased from 9 million to 25 million metric tons.

Port-Related Costs

In Argentine ports, grain handling and shipping costs are high, because of high administrative and service charges and physical restrictions on shipping.

High administrative and service charges are largely attributable to heavy public sector involvement with ports and port-related matters—6 different ministries and 18 public-sector enterprises are involved. Much of the equipment used is obsolete, and loading is slow and inefficient.

According to a 1986 study, port entry costs in neighboring countries are a fraction of those charged in Buenos Aires.[2] For Montevideo (Uruguay), port access charges were 18 percent of the costs of entering Buenos Aires, for Santos (Brazil) they were 12 percent, and for Rio Grande and Paranaguá (also in Brazil) they were 11 percent. Total port-related charges in Argentina were also much higher than those charged elsewhere –total costs in Santos were 17 percent, in Rotterdam 33 percent and in Yokohama 36 percent, of the costs in Buenos Aires. These costs include entry to, handling within, and exit from the port. They do not include the costs of topping off.

Physical restrictions are a result of the characteristics of the Paraná river and its access channels, particularly shallow draft and sinuous access channels (see Map: Argentina—Port and Rail Transport Network). The Paraná has been the principal outlet for shipments of grain from the grain-producing areas in Santa Fé and Cordoba provinces since the early years of this century. Draft restrictions have led to diversion of traffic to other ports. Large vessels top off at floating transfer stations in the mouth of the river Plate and in deepwater ports in Argentina, Uruguay and Brazil.

Up to 1976, access to Argentina's river ports was via the Martin Garcia channel, which forms the border between Uruguay and Argentina. Its draft limit is 7.3 meters. Dredging for greater draft would be feasible, but would require rock removal. Uruguay and Argentina have longstanding disputes about the borderline and hence about jurisdiction over dredging and rock removal. In 1976 the Mitre channel was opened, linking the access channel of the port of Buenos Aires with the ports up-river through a sinuous route following the southern branch of the Paraná. The Mitre channel is usually at 9.75 meters draft. Manoeuvering problems in the tight turns of the Paraná river limit a ship's length to 225 meters. The Mitre channel is tedious and costly to travel because the Paraná carries a large load of silt.

In 1982/83, unusually heavy rainfall and widespread flooding led to silting of the Mitre channel, reducing its draft to 5.15 meters, and the Martin Garcia channel had to be used. Shippers whose vessel sizes conformed to the Mitre channel's draft of 9.75 meters had to resort to emergency topping up in Bahia Blanca, with demurrage times of 45 to 60 days. Demurrage charges at that time were about US$5,000 per day. The total extra cost of that year's grain shipments, due to draft restrictions in the access channels of the river ports, was calculated at US$160 million. Yearly dredging costs for all access channels to Argentine ports are over US$100 million.

Ocean Shipping Rates and Port Tariffs

Only a fraction of the rate differentials for grain shipments from Gulf as compared with Argentine ports to Europe is due to the larger transportation distances from Argentina. In the mid-1980s, freight rates from Gulf ports to Rotterdam were about US$6 per ton, while those from Argentina were US$15 per ton (Table 7.21). The difference in transportation distance is about 1,300 miles or 3 days' voyage. At an average daily cost of US$3,000 for a 35,000 ton grain carrier, the cost differential due to the difference in transportation distance is only US$0.26 per ton, 3 percent of the rate differential. The higher costs of ocean freight from Argentine ports as compared with United States ports are primarily due to limitations on vessel size. A 50,000 ton Panamax vessel from New Orleans will have transportation rates per ton that are about 50 percent of the rate in the 16,000 to 25,000-ton vessels going out of United States lake ports without the expensive top-off procedure in Argentina. The Rosario and Buenos Aires ports are limited by draft to about 30,000-ton vessels. Plans for deepening that draft are being discussed, with the objective of loading Panamax vessels in Rosario. If these plans are implemented, the cost of transportation will be dramatically reduced.

The self-contained private ports along the Paraná river circumvent many of the administrative and service costs charged at the less-modern public sector ports and thus operate well below their tariffs. In 1988, for example, the National Grain Board's elevator in Rosario, a major river port, was charging US$4 per ton of throughput, while private elevator complexes nearby were charging US$2 per ton.

Methods of Sale and Payment

There are six main ways of trading in grain in Argentina. These are: (1) storage of grain under an "1825" guarantee from the grain board, with immediate payment and a board guarantee of delivery within 30 to 45 days; (2) immediate delivery of the grain at a price to be determined in the future at the seller's option (precio a fijar), with payment made at the time the price is set; (3) cash payment for grain upon delivery; (4) immediate price setting for future delivery of grain, under a futures contract, with payment made upon delivery; (5) the barter system under which the grain board receives grain at harvest in exchange for fertilizers it supplies at time of planting; and (6) payment for grain upon delivery under the grain board's price support system. Table 5.1 shows how the six ways of selling grain are organized in the primary and secondary channels, from farmer to country elevator and from there to either industry or exporter.

The "1825" System. In 1977, the grain board introduced the "1825" regulation, under which a seller of grain (e.g., grain for export) can receive payment for it even though it is stored in his own facilities or in a country elevator until delivery occurs. The owner of the warehouse space issues to the buyer a non-negotiable 1825 certificate of deposit, accredited by the grain board, which guarantees delivery of the grain within a specified period of time, usually 30 to 45 days. If either the owner of the grain or the owner of the silo where it is stored defaults on delivery, the board becomes responsible for delivery of the grain to the buyer.

In return for its guarantee under the 1825 program, the grain board takes a lien on the assets of the warehouse operator or accredited producer. A mortgage on the grain is not considered sufficient in Argentina, where enforcement of claims, or obtaining proof of fraudulent practices, may be difficult. The board uses registration of grain under its 1825 system as the basis for granting daily delivery allowances from the various silos to the exporter's ship.

The 1825 system was introduced because delivery of grain for export against immediate payment had led to unmanageable traffic congestion in ports, where the entire pressure of grain sales was felt at harvest time. The system was immediately popular (by 1980, about 75 percent of all grain exports were being handled under the 1825 regulation) but has become less

TABLE 5.1   The Six Main Ways of Selling Grain in Argentina

Delivery	Price Setting	Time of Payment	Instrument and Names Used
In future (up to 45 days)	Immediate (within 48 hrs)	Immediate (98%)	Grain Board 1825 Certificate
Immediate	In future, at seller's option	When price set	Price to be fixed (à fijar)
Immediate	Immediate	Upon delivery (7 to 15 days)	Cash Sale
In future at fixed date	Immediate, according to futures market	Upon delivery	Futures Contract
At harvest	Barter terms fixed at planting with Grain Board	At planting	Barter
At harvest (mainly wheat)	Immediate	Upon delivery	Price support system

Sources: Adapted from Patricio Lamarca, *Diagnostico de Situación del Sistema Comercial Argentino de Granos y Subproductos*, Buenos Aires—Undated, probably 1987.

so (presently about 40 percent of exports are handled this way). There are two main reasons for the system's decline in popularity. First, the board's requirements with respect to underlying guarantees have become more stringent. Second, similar operations effected without the grain board guarantee, outside the strict administrative boundaries of the regulation but similar in spirit, have increased. The main reason for going outside the grain board's guarantee is to avoid the costs, including the administrative costs, of the 1825 system and to prevent all assets from being mortgaged. Mortgage of assets freezes a potential basis for trade or production credit.

By the end of 1987, 370 cooperatives, 1,075 country elevators, 190 farmers and 45 mills had storage certified as usable under the 1825 system. But the storage capacity certified under the system is only a small proportion of the total amount of storage.

The Precio à Fijar System. The precio à fijar system (price to be determined in the future) is a procedure by which the farmer can deliver his grain and continue to speculate on the price. It is similar to the delayed pricing system in the United States. The ownership of the grain transfers at the time the farmer delivers it. The price is set any time in the future at the farmer's option. A prepayment can sometimes be made to the farmer, which gives him an advantage of some cash at harvest time, with the balance to be paid at the time the price is selected at some point in the future. That future price is whatever price the elevator is paying that day. Therefore, the Arbitration Chamber establishes that daily price as it does for a cash sale. The strategy gives the farmer greater flexibility in pricing and speculation on price while giving the country elevator control over the grain to meet current needs. When little grain is moving in the market, the ability to establish "true" prices may be more difficult and the country elevator has greater flexibility in adjusting its commission rate.

Cash Sale. The grain is paid for in cash upon delivery. The farmer retains ownership of the grain until he sells it to the elevator, from his own warehouse or from the elevator's warehouse. Payment will generally be upon delivery or within a few days, depending upon the elevator's arrangements and contracts. The farmer has retained physical and financial control of the grain, and has absorbed all price risks.

Futures Trading. The futures markets in Argentina are used much as were the futures markets in the United States in the early part of this century—to contract for actual future delivery rather than to hedge price changes. The amounts traded are only a fraction of the total crop. In the United States, by contrast, only about 3 percent of futures contracts culminate in delivery—the total amount traded far exceeds the total crop.

Although a futures contract for grains was introduced in Buenos Aires as early as 1907, its use has remained limited. Reasons given for this limited use are high inflation, uncertainties in relation to exchange rates, the high cost of financing margins, and unpredictable government policies.

An example of the latter is the cancellation by government fiat of all out-standing futures contracts at the onset of the Austral plan in 1985. However, a stronger reason is that most agents actively engaged in the trade –farmers, country elevator managers, and cooperative manage-ment—do not understand the operation of futures markets. Outside Buenos Aires, where the main grain exchange is located, the futures mar-ket is not easily accessible to the trader or to the farmer, and there is a gen-eral distrust of the market as a place where fortunes have been lost.

Advance Purchases and Barter. An increasing number of grain sales are made prior to harvest, and even prior to planting. Early contracts for this mode of sales, entered into in 1985, were sales for cash of part of the future crop. The seller of the grain had to give a bank guarantee or other guaran-tee instrument to the buyer. When grain prices firmed up during the peri-od of this outstanding obligation (up to early 1986), farmers found that they had sold their crop too cheaply. This led to defaults, to contract ad-justments, and to numerous disagreements.

The barter system was introduced by the grain board in 1984 as a means to promote the use of fertilizer. Under the board's barter agreements, which have become widespread, the farmer obtains fertilizer, gas, oil or seed in exchange for a specific amount of his future crop. The real yearly interest rate implied in these barter agreements, based on product values in mid-1988, ranges from 28 to 35 percent (Table 5.2).

Most of the funds used for advance purchases originated with exporters who were securing future export positions with funds borrowed abroad against these positions. Until the end of 1988, when political uncertainty associated with the pending election dried up the flow of funds, both bar-ter and advance purchase operations were becoming increasingly com-mon.

The Support Price System. The support price (precio de sosten), the government's most important price stabilization measure, is a farmgate price, guaranteed by the National Grain Board, based on the prevailing FOB price minus the calculated average costs incurred between farmgate and export port. It supposedly protects the farmer against the trade during the period of his highest vulnerability, i.e., during the early part of the har-vest of wheat, the first grain crop of the year. Lately, the grain board has been purchasing about 30 percent of the annual wheat crop and small amounts of the other grains at the support price.

The grain board has also operated income support programs for grain farmers in distant provinces, mainly in the Northeast and the Northwest, where it subsidizes the costs of transporting grain for smallholders and for rice farmers.

TABLE 5.2   Argentina—Barter Arrangements and Implied Interest Rates, 1988–1989

	Typical Private Arrangement in Wheat Growing Area of Bahia Blanca:	
	Fertilizer	*Wheat*
Barter relation	1 mt [a]	4.5 mt
Time of delivery	June 88	Jan. 89
Prices at time of delivery	US$ 321/mt [b]	US$ 90/mt [c]
Values at time of delivery	US$ 321	US$ 405
Implied interest rate		
per month	2.95%	
per year	35.38%	
	======	
	Barters Offered by the Grain Board in Early May of 1988	
	Fertilizer	*Wheat* [d]
UREA		
Barter relation	1 mt	2.80 mt
Time of delivery	June 88	Jan. 89
Prices at time of delivery	US$ 207/mt [e]	US$ 90/mt [c]
Values at time of delivery	US$ 207	US$ 252
Implied interest rate		
per month	2.49%	
per year	29.87%	
	======	
SUPER PHOSPHATE		
Barter relation	1 mt	3.40 mt
Time of deliveryJune 88	Jan. 89	
Prices at time of delivery	US$ 254/mt [f]	US$ 90/mt [c]
Values at time of delivery	US$ 254	US$ 306
Implied interest rate		
per month	2.36%	
per year	28.27%	
	======	

[a] Super Phosphate.
[b] Port of Nicochea, June of 1988.
[c] Best estimate future price January 1989 (May 13/88 futures market quotation Buenos Aires was US$ 87.50/mt for January delivery).
[d] Delivered up to 200 km from a port, i.e., typically at a country elevator, or cooperative.
[e] The price of urea to the grain board, in March 1988, was US$ 239.09/mt. A subsidy of US$ 31.73/mt applied, resulting in an actual cost of US$ 207.36/mt.
[f] The price of super phosphate to the grain board, in March 1988, was US$ 294.34/mt. A subsidy of US$ 40.06/mt applied, resulting in an actual cost of US$ 254.28/mt.
Source: AMEC, Inc.: *Argentina—Its Grain Trade* (Consultant's Report to World Bank), Boston, June 1988.

Price Determination in the Argentine Markets

Price formation starts with the FOB price, denominated in foreign currency, usually US dollars. The difference between the FOB (Free On Board) price and the FAS (Free Alongside Ship) cost, denominated in Australes, has three components: taxes, handling charges and miscellaneous marketing costs (Table 5.3). Under conditions applying in mid-1988, for a typical wheat shipment, handling costs made up 67 percent of the total FOB/FAS difference, taxes 32 percent, and miscellaneous marketing costs 1 percent. If the FAS cost, calculated backwards from the FOB price, is above the prevailing price for the product in question at the grain exchange, there is a profit between FAS and FOB price. If the FAS cost is below the prevailing quotation on the grain exchange, a loss (contramargen) obtains.

An important factor in the conversion from FOB to FAS is the profitability of capital market operations using the cash flow of the grain trade. Profits thus made are a major determinant of what might be called the actual exchange rate, the rate at which the FOB price is translated into Australes.[3] Although the applicable exchange rate, at any point in time, is known and is the same for all exporters, its foreign exchange position may be different from one company to another. Companies can achieve different foreign exchange positions by importing foreign exchange (for up to 540 days in early 1989) before actual exports are made, ostensibly as a means to finance production through advance purchase of future crops. In Argentina, this is called the prefinanción issue.

The profitability of the grain trade, expressed in the actual conversion rate applicable for each exporter between FOB and FAS, depends upon amounts and timing of the company's imports of foreign exchange, the extent to which the foreign exchange is used for advance crop purchases, and the extent to which some of it is applied in the local capital market at effective interest rates well in excess of those the exporter has to pay on the same funds to the financing agency from which he borrowed.

Temporarily imported foreign funds are customarily invested in short-term obligations of public enterprises and thus carry government guarantees. The system has been likened to a Ponzi scheme used to bankroll the deficit public enterprise sector. The yearly amount involved, up to early 1989, was reportedly about US$ 2 billion. Profits made in this way also determine the opportunity cost of the same capital when it is used to finance barter deals (see Table 5.3).

The ability to import foreign exchange under the prefinancing schemes greatly complicates the calculation of average exchange rates applying to each exporting company and makes the FOB market opaque.[4] Because of the profits they can make in the capital market, traders may sometimes accept FOB prices below what they would otherwise have accepted, and incur apparent trading losses.[5] These profits in the capital markets and the

TABLE 5.3   Breakdown of FOB-FAS Cost Differential for Wheat, Bahia Blanca, March, 1988[a]

	US$ per mt	Breakdown (%)
FOB Price:	110.00	100.00
	=====	=====
FOBBING COSTS:		
a. Taxes and Duties		
INTA Contribution, 1.5% of FOB	1.65	1.50
Stamps and Registration, 0.3% of FOB	0.33	0.30
Foreign Exchange Transfer, 0.6% of FOB	0.66	0.60
Merchant Marine Contribution, 0.36% of FOB	0.40	0.36
Sub-total Taxes and Duties:	3.04	2.76
	====	====
b. Handling Costs:		
Elevation, US$ 4.0 fixed[b]	4.00	3.64
Grain Board Inspection, US$ 1.00 fixed	1.00	0.91
Port Charges, US$ 1.00 fixed	1.00	0.91
Quality Differential (average), 0.3% of FOB	0.33	0.30
Standard Losses, 0.3% of FAS	0.03	0.03
Sub-total Handling Costs:	6.36	5.78
	====	=====
c. Miscellaneous Marketing Costs:		
Broker's Commission, 0.8% of FAS	0.08	0.07
Grain Analysis, 0.3% of FAS	0.03	0.03
Financing and guarantees, 0.2% of FAS	0.02	0.02
Sub-total Miscellaneous Marketing Costs:	0.12	0.11
	====	=====
Total Fobbing Costs in US$/mt:	9.52	8.65
	====	=====
FAS Cost in US$/mt	100.48	
	=====	
FAS Cost in Australes[c]	491.36	
	=====	

[a]FOB = Free On Board; FAS = Free Alongside Ship.
[b]A discount of 25 percent is given for ships topped up that have had their base cargo loaded at grain board elevators on the river Paraná.
[c]Average commercial buying rate for Australes during the month of March, 1988, was A$4.89/US$.
Source: AMEC, Inc.: *Argentina—Its Grain Trade,* (Consultant's Report to World Bank), Boston, June 1988.

higher shipping costs already referred to are probably the main reasons for the lower FOB prices in Argentina as compared with the Gulf ports (Table 7.20).

Because of the practice of trading prefinancing funds on the capital markets, it is virtually impossible to estimate the actual FOB prices of grains exported from Argentina. On average, actual FOB prices appear to be below listed and quoted FOB prices, primarily as a result of capital market operations with prefinancing funds by all major exporters. Consequently, any company unable to obtain prefinancing funds is at a serious disadvantage in exporting grains from Argentina.[6]

Farmgate Prices of Exports. To obtain the farmgate price from the FAS price, three main categories of cost have to be considered: (1) transport, (2) elevator costs, and (3) taxes and levies. A breakdown of these costs, as they obtained for wheat exports in March of 1988, is illustrated in Table 5.4.

Due to inefficiencies in the transportation system, costs per metric ton-km are high. However, because average transportation distances are small, transportation costs as a percentage of FOB are generally modest. According to a 1986 study, in 1985 it cost the equivalent of 2.8 cents per metric ton-km to transport grain by rail over the average distance of about 300 km, as compared with an average cost of 1.6 cents per metric ton-km in the United States.[7]

Transport, consisting of both short transport from farm to country elevator and long transport from elevator to shipping point, accounts for about 60 percent of the difference between FAS and farmgate prices. The country elevator can be located at the site of an intermediary trader, the country trader, or at the location of a primary cooperative.

Elevator costs incurred by country elevators or cooperatives include conditioning charges for operations such as drying, cleaning, and fumigation, and a handling and sales commission, generally from 3 to 6 percent of the value of the product. Levies and taxes are government charges to the grain trade.

Combining all costs, it is estimated that farmgate prices for wheat could be, on average, about 70 percent of FOB prices, if Argentina had no export taxes or differential exchange rates.

Although all costs are negotiable in the final instance, and thus are somewhat flexible, the largest degree of flexibility appears to be in the commissions charged by the country trader and in the conditioning costs incurred at the country elevator. Conditioning costs have some lack of transparency, since it is difficult for the producer to judge to what extent some or all operations he is charged for are necessary. Also, losses and shrinkage are calculated on the basis of government-distributed shrink tables, and a fixed percentage allowable for shrink as a result of cleaning, whether or not the grain is actually cleaned. Efficient country elevators may be able to reduce the losses and thus earn an extra margin. The flexibility of commission charges, conditioning costs, and, to a certain extent, transportation are among the cost-reducing opportunities exploited by newly emerging large-scale operators. These operators are integrating

TABLE 5.4   Conversion of FAS Cost of Wheat to Farmgate Cost, March, 1988

	Australes per mt	Breakdown (%)
Maximum Free Alongside Ship (FAS) Price [a]	491.36	100.0
COSTS BETWEEN FARMGATE AND FAS:		
a. Transport		
Short Haul (farm to country elevator)	12.00	2.4
Long Haul (country elevator to port) [b]	55.00	11.2
Sub-total Transport:	67.00	13.6
b. Country Elevator		
Grain Conditioning	9.00	1.8
Standard Losses	9.00	1.8
Commission Charges, at 4% of FAS	19.65	4.0
Sub-total Handling Costs:	37.65	7.6
c. Taxes:		
On Gross Receipts, at 0.9% of FAS	4.42	0.9
Stamp Duties, at 0.3% of FAS	1.47	0.3
Social Security Taxes, at 0.6% of FAS	2.95	0.6
Sub-total Taxes:	8.84	1.8
Total Costs FAS to Farmgate	113.50	23.1
Maximum Farmgate Price	377.86	
Farmgate Price as a % of FAS	77%	
Farmgate Price as a % of FOB [c]	70%	

[a] The price at which there is no profit, that is, the price is equal to the FOB price minus the Fobbing cost.
[b] Based on an estimated 30 percent of rail and 70 percent of truck transport.
[c] At the average commercial buying rate for Australes during March, 1988 (A$4.89 per US$).
Source: AMEC, Inc.: *Argentina—Its Grain Trade* (Consultant's Report to World Bank), Boston, June 1988.

from the shipping terminal backwards to having their own country elevators.

Prices in the domestic market are determined by the forces of supply and demand interacting within the three grain exchanges. Because such a large proportion of Argentina's grain production eventually moves to export markets, world prices become a major influence on domestic pricing.

Prices for cash sales are reflected through the Arbitration Chamber committee in daily prices announced to farmers and grain handlers throughout the country. The information appears to be widely distributed and easily accessible. Prices paid to farmers on the open market are generally equal to or close to the announced prices at the grain board. As described earlier, the first handler subtracts charges for services from the prices announced at the grain exchanges. The opportunity for manipulating prices at the country elevator level or throughout the marketing channel is quite limited.

As a result of the concentration of exports in the hands of a small number of multinational corporations, one might argue that prices do not reflect true competitive norms, but the competition among companies and the presence of cooperatives make it highly unlikely that farmers are not receiving the true market price. The charges for transportation, commissions, and service are subject to incorrect or inaccurate reporting in that each elevator is free to set its own charges. The protection against that form of price manipulation is competition, and the farmer's freedom to choose among elevators, processors, and export facilities. In the primary grain belt, farmers have easy access to several competing elevators. Some farmers deliver grain to two or more elevators, "just to keep them on their toes."

Financing of Grain Storage

Given the high real interest rates prevailing in Argentina—2 to 3 percent per month in real terms on occasion in recent years—private post-harvest financing of grain inventories would appear to be infeasible because product price increases after the harvest have been less than the cost of financing would have been. High financing costs have left most farmers heavily indebted. They have obtained production credit either in kind or in barter deals, and have had to sell at least part of their first crop immediately after harvest.[8] However, given the high risk involved in holding financial assets, some people discount the financial returns to such an extent that they prefer to hold their assets in the form of grain stocks. For a detailed review of stock financing, see the appendices to Chapter 13.

Notes

1. See: Bolsa de Cereales de Buenos Aires, *1986 Annual Report*, Buenos Aires, 1986.

2. S. Miró, "Las Limitantes de Infraestructura." In: Bolsa de Cereales de Buenos Aires, *Producción y Comercio de Granos—Las Limitantes de Infraestructura*, Separata Número Estadistico 1986.

3. Argentine grain circles have a confusing habit of calculating so-called effective exchange rates for each grain and milling by-product exported, taking into account all applicable taxes, including the export tax, if any, and the tax on the transfer of foreign exchange.

4. It also explains why half of the financial administrative personnel of one of the main exporting companies are engaged in managing funds, according to one of that company's directors.

5. Funds imported to pre-finance grain purchases have to be exported, in the form of grain, within the time period allowed under the law, or traders must pay a 30% penalty.

6. This is particularly true of the National Grain Board.

7. S. Miro, "Las Limitantes de Infraestructura," 1986.

8. In Argentina the harvest sequence is wheat, flax, corn, sorghum, sunflower seeds and soybeans. Harvests start in December and end in August.

6

Future Directions

Argentine grain producers operate in an economic environment which is much less favorable to production than that of their competitors in Brazil, Australia, Canada and the United States. Argentine farmers are substantially denied the incentives of world market prices by an expensive marketing system, cumbersome government regulations, and tax policies that discourage expansion of production. Table 6.1, which shows estimated producer subsidy equivalents (PSEs) for the major grains in five grain-producing countries, illustrates the extent of the discrimination against the grain sub-sector in Argentina compared with the more supportive environment within which the other major grain exporters operate.

Argentina is the only one of the five countries showing negative PSEs for the 1982–1986 period. Although one can debate the assumptions built into the calculations of PSE and might argue the accuracy of the exact values, the pattern of discrimination against Argentine producers is irrefutable, given the wide discrepancies shown in Table 6.1. If Argentine farmers

TABLE 6.1   Producer Subsidy Equivalents for the Major Grains in Selected Countries, Average for 1982–1986 (in percentages)

	Argentina	Australia	Brazil	Canada	U.S.A.
Soybeans	-76	n.a.	7	13.5	8.5
Maize	-65	n.a.	14	10.0	27.1
Wheat	-54	7.7	62	30.4	36.5
Rice	n.a.	13.0	49	n.a.	45.2
Sorghum	-111	n.a.	n.a.	n.a.	31.4

Source: USDA-ERS-Agriculture and Trade Analysis Division, "Estimates of Producer and Consumer Subsidy Equivalents," *Government Intervention in Agriculture, 1982–86*, Washington, D.C., April 1988

received a greater share of world market prices, they would respond with increased area and yields, export volume would increase, the trade balance would be improved, and economic activity in the private sector that uses grain as an input would be increased.

Argentina is the low-cost producer in inter-country comparisons, primarily because farmers are using limited amounts of purchased inputs such as fertilizers and chemicals. Use of fertilizers in Argentina is substantially lower than in Brazil, Australia, Canada, and the United States. Increased use of fertilizers and chemicals in Argentina would generate higher yields. However, encouragement of the use of these inputs should be done in the context of marginal costs and returns. Attempts to match Argentina's input use with that of the United States, for example, could result in Argentina becoming a high-cost producer in the world markets, and cause it to lose its competitive advantage. It is clear, however, that the political and economic environment within which Argentine farmers operate has discouraged the use of inputs and thereby artificially lowered Argentina's potential productivity. As an indication of Argentina's low use of fertilizers, Table 6.2 compares per-hectare consumption of fertilizers in Argentina with that in four competitor countries.

Should Argentina put in place the package of reforms suggested below, it could double the current value of its grain production in the ten years following implementation of the reform program provided that public sector investments in infrastructure are also made, particularly in transportation and ports. Yields of wheat, maize and sorghum, the three principal cereal crops could probably be doubled with the use of fertilizers. Argentina has the potential to expand the area under grain crops to upwards of 20 million hectares from the figure of about 15 million hectares common in the eighties. A combination of yield increases and area expansion could lead to production of upwards of 70 million metric tons of grain per annum within ten years of completing the implementation of the proposed reform program. However, a program to meet a fixed total production, regardless of the cost, would not be in the best interests of Argentina. Production expansion should be in response to inputs and output prices

TABLE 6.2   Consumption of Fertilizer per Hectare of Arable Land and Permanent Crops for Selected Countries, 1987 (kg N, P_2O_5, K_2O)

	Argentina	Australia	Brazil	Canada	U.S.A.
Nitrogen	2.6	7.9	12.4	25.8	50.0
Phosphate	1.4	17.6	21.0	13.8	19.6
Potash	0.4	3.1	19.9	8.6	23.6
Total	4.4	28.6	53.3	48.2	93.2

Source: FAO Fertilizer Yearbook, 1988.

established in world markets and competitively priced within Argentina. The most important policy changes required to improve Argentina's productivity are discussed below.

Macroeconomic Policies

The highest priority for the Argentine grain sub-sector is implementation of monetary and fiscal policies that will bring inflation to a level where it no longer plays a role in investment decisions. Putting in place a unified market-determined exchange rate system should be an integral part of the macroeconomic reforms. It is essential that stability be introduced into the economic and political environment in which farmers operate. Risk and uncertainty must be considered as farmers make plans for investment and production. Uncertainty adds to production cost as surely as if the government placed an even higher tax on farmers' use of inputs. Historically, the Argentine grain industry has concentrated more on export taxes and fluctuating and multiple exchange rates than on marketing and production efficiencies and cost reductions. Macro policies must be altered to create a favorable climate for production and marketing decisions. In such a changed macroeconomic environment, land would be held for agricultural production possibilities rather than as a hedge against inflation.

Stability in the macroeconomic environment, including control of inflation, is a necessary, though by no means a sufficient, precondition for the development of efficient financial markets, including those in rural areas, with the capacity and willingness to finance the short-, medium-, and long-term needs of the grain sub-sector. The availability of such financing will be essential if Argentina is to realize its full potential in grain production. Furthermore, macroeconomic stability, including control of inflation, would greatly facilitate the development of an organized private market for financing grain stocks. (For a detailed discussion of agricultural stock financing, see the appendices to Chapter 13).

Trade Policy

A neutral and liberal trade regime that removes quantity restrictions and export quotas and has a low and uniform level of tariffs would greatly enhance the prospects for Argentina's grain sub-sector. None of Argentina's major competitors in the world grain markets—Australia, Canada, the United States—handicaps its grain producers the way Argentina does.

Tax Policy

The historical policy of taxing exports has had damaging effects on the ability of Argentine farmers to compete in world markets. It has not only lowered the returns to farmers, and thereby their incentives to produce grains, but it has provided opportunities for manipulation by the grain traders, using tax laws and exchange rate regulations to their advantage. This income is achieved only through investing considerable time and resources in maintaining a high level of competence in manipulating those regulations. Removal of those taxes (underway in Argentina, but with no guarantees that they will stay at these lower levels) would go a long way towards reducing the aggregate cost of marketing, in addition to the direct effect that it would have on prices to farmers. A neutral tax system that relies on income taxes at the federal level and land taxes at the local level is suggested. This strategy has worked well in the United States, Canada and other developed countries. It taxes the income generated by the resources and capital, and provides neutrality in the tax regime, both within and between sectors. An income tax at the national level would result in far less distortion in the use of resources than do the current export taxes. In addition, it would have a positive effect on the public perception of fairness.

Transport Policy

The wide spread between farmgate prices and world export prices reflects, among other things, the high cost of marketing grain. One of these costs is an inefficient transportation system, where competition is limited or excluded. Government ownership of the railroad system, with no opportunity for more efficient firms to operate, leaves many shippers with no alternative but trucks over distances where rail would be more efficient. Public investment in road beds, with encouragement of more competition in service over those road beds, could decrease the cost of operation and reduce transportation rates. Increased public investment in waterways, particularly from Rosario through the Buenos Aires port, and rebuilding the facilities at Bahia Blanca, could also reduce transportation costs. Numerous regulations related to trucks apparently reduce competition and result in truck rates above competitive levels. Private traders operating in the domestic and export markets are providing vigorous competition for the grain board's facilities. However, the presence of the high-cost export elevators operated by the board (even though they may be subsidized) still adds to the aggregate cost of exports through their inefficiency.

Many of the major exporting countries provide subsidies for grain transportation. These subsidies can also result in distortions in terms of resource allocation, as has been frequently documented in the case of the

Crow rates in Canada (see Part Four on Canada). However, public investment in transportation facilities can increase economic growth in all sectors, and misallocation of resources is minimized if the subsidies are for the fixed cost rather than altering competitive rates within the system.

As the most important user of the national transport system, the grain sub-sector has a great need for liberalization and reform of government regulations that foster the creation and maintenance of rents in the transport sector, particularly in the rail and port systems, but also in the trucking industry. If the grain sub-sector is to realize its potential, it cannot continue to be burdened with a transportation system such as exists today in Argentina. In Australia and Canada, not only are grain producers not discriminated against in transport policies, they are in effect subsidized to some extent. For example, in Canada the Crow system, under which the Canadian government subsidized construction of a railroad in southeastern British Columbia in exchange for reduced freight rates for grain and flour moving eastward from the Prairie Region (see Part Four on Canada) served as a federal subsidy to grain producers, among others. Since 1983, under the Western Grains Transportation Act, the Canadian government has provided railways with an annual payment to cover the transportation of eligible grains east from Prairie shipping points.

Input Supplies for Agriculture

The high cost of inputs to Argentine farmers has resulted in less than optimal use of fertilizer and pesticides in the production of most grains. Prices for these inputs have been kept high through the imposition of import tariffs and quotas. The quotas have been used, in some cases, to protect small and inefficient producers of fertilizer within Argentina. Many advisors (including Dr. Norman Borlaug of CIMMYT) have recommended construction of fertilizer plants in Argentina and removal of the tariff that has held an umbrella over the high cost of the small fertilizer companies currently operating there. Argentina has the natural resources it needs to produce its own fertilizer. It lacks only the government decision to provide investors the necessary assurance that they will be allowed to profit from construction of a large-scale anhydrous ammonia plant. Instead of protecting high-cost fertilizer production, the government should be encouraging construction of plants to lower the cost of fertilizer to producers.

Targeted Bread Subsidies

Replacing the present generalized bread subsidy with one that is directly targeted to the poor, both in urban and rural areas, would better serve both the poor and the grain producers. If properly implemented, this one policy change could generate budgetary savings and lower the tax burden on grain producers, while simultaneously protecting or even enhancing the access of the poor to food. The current complex protective policies in the wheat and flour industries have provided little benefit to producers and consumers, but have become another opportunity for those marketing firms with sufficient expertise to manage the program to their advantage. The income transfers which were intended to be from producers to consumers (probably inappropriate even as a goal) have been more effective in transferring income from producers and taxpayers to grain handlers and processors.

Evaluation of Marketing Systems

The marketing board concept used by Canada and Australia for the marketing of wheat has occasionally been recommended as a substitute for the marketing systems operating in some other countries. At a time when Argentina may be considering a shift from the National Grain Board to an alternative form, it is important to evaluate the three alternatives of: (1) a regulatory body (the Grain Commission model); (2) a marketing board (the Canadian Wheat Board model); and (3) a grain board (the Argentine model, where the board combines regulation, marketing functions and consumer protection under one agency).

The performance of a marketing system can only be evaluated if performance norms are specified in qualitative, if not quantitative, terms. Performance norms in theory are generally broad and vague and intended to be comprehensive. When applied to individual industries, they must be made more explicit. The performance norms selected for this comparison include: (1) price level; (2) price stability; (3) bargaining power and pricing strategies; (4) marketing costs and efficiency; (5) entry and exit; and (6) resource allocation.

The level of prices in international markets is set by interaction of the buyers and sellers in the market. Although there are policies and restrictions that influence price in almost every country, these tend to counteract one another such that the overall world price level is set by forces of supply and demand. Therefore, any of the three forms of market organization (grain commission, marketing board or grain board) would face the same international price and presumably could have little influence on that price level.

However, the different forms of market organization can influence the extent to which the world price is transferred back to individual producers through the marketing chain. Unless the marketing board or grain board subsidizes an income transfer to producers, there is no a priori reason to believe it would give a larger share of the world food dollar to the farmer than would a free market system. In fact, most studies show that the marketing costs (including handling, distribution and risk shifting) are higher for centrally planned marketing systems than for a competitive industry consisting of many firms, each seeking its own best interest. Higher marketing costs generally mean lower returns to producers. One must, therefore, conclude that the grain commission model would be equal to or better than the marketing board or grain board with respect to world prices as well as farmgate prices.

Price Stability. Price stability (defined as long-run prices received) is similar to the discussion of price level in the preceding section. Supply and demand in world markets will move prices month by month and year by year. The exporting country has no alternative but to follow these prices if it is to continue to supply the world market. Only if it has sufficient volume and storage capacity to influence world price levels could it have any impact upon stability. A study by Hill and Mustard comparing the United States and Canada indicated that the Canadian Wheat Board had a larger share of the world market when prices declined and world supplies increased than when supply was low and prices high.[1] This suggests that the Canadian Wheat Board, by accentuating the swings, tended to destabilize rather than stabilize.

On the question of short-run price stability, the results are different. Fluctuations in a free market such as the Chicago Board of Trade occur day by day and even minute by minute. This type of instability is reduced or even eliminated by a marketing board or grain board system where prices are fixed by government decree in the domestic market. Prices at the local farm level under a free market system may also vary from firm to firm, region to region, and day to day. These fluctuations can be controlled with the authority of a marketing board, although at a cost of efficiency and responsiveness. Therefore, on the criterion of short-run price stability, the competitive model would rank below the performance of a controlled market.

Bargaining Power and Pricing Strategies. It has been argued that many centrally planned economies purchasing grain would prefer to purchase from single entities in the exporting country. However, there is no evidence to indicate that the U.S.S.R. or China have purchased preferentially on the basis of ideology or of the number of sellers in the exporting country. The U.S.S.R., in fact, has repeatedly demonstrated its ability to use the Chicago Board of Trade to its advantage. It has also been argued that a single committee will make better decisions as to when and where to sell than does the aggregated decisions of thousands of individual firms. The study by Hill and Mustard shows that Canada was a more aggressive seller in the market than the United States in periods of low prices than in periods of high prices.[2] Its market share was highest in years when prices were lowest, and lowest in years when prices were highest. This may be less the result of Canadian strategy than of United States actions to store grain across crop years during periods of excess supply in the world markets. Although causality is difficult to establish, the data do not support the hypothesis that the Canadian Wheat Board has exhibited superior marketing strategies or has generated consistently higher prices for its producers.

A study carried out by the United States General Accounting Office in 1976 identified some of the problems that the National Grain Board in Argentina has had in securing higher prices.[3] The report stated that in 1975 the board negotiated with Mexico to substitute 300,000 metric tons of sorghum for 200,000 metric tons of maize because the board thought Argentina was out of maize. The year ended with a surplus of 100,000 metric tons of maize, but insufficient sorghum to fill the Mexican order. The six month postponement of deliveries and logistics of substitution resulted in considerable losses to both buyer and seller. In 1976 there was evidence that the board restricted export sales of wheat, thinking that the downward price trend would have an upturn later in the year. The board guessed wrong. The continued price decline resulted in considerable losses and a large carryover into the next year.

Marketing Costs and Efficiency. Efficiency in this context should be defined to mean performance of all of the basic marketing functions at the lowest possible cost of resources. Costs of marketing can be minimized by not providing marketing services, so efficiency comparisons are preferable to comparison of marketing costs. Marketing services include: information on prices and quantities, transportation, risk shifting, and regulatory functions. A 1980 study by Hill and Dollinger showed that marketing costs for wheat (including transportation) were US$25.60 per metric ton in Canada and US$14.55 per metric ton in the United States.[4] When quality, geographical area, and export destination were matched for Canada and the United States, farm prices and export prices were equal for the two countries. This relationship was maintained in Canada only by subsidizing transportation to compensate for the extra US$11.02 per metric ton incurred in the marketing activity. McCalla and Schmitz also concluded that technological development in Canada was lagging behind that of the U.S. in terms of transportation, export facilities, and grain handling.[5]

Comparison with the Argentine system shows the same relationship. Technology used in transportation and handling is less under the Argentine Grain Board than is found in the free market sector of the grain industry. Cost of handling and marketing and the inefficiency of labor utilization has put the board's export elevators at a sufficient disadvantage to the free market to require heavy subsidies even to keep them operating.

The conclusion from the available studies is that the free market system generates a more efficient marketing system than any achieved by grain boards or marketing boards in the various countries. The severity of competition often results in capital losses to the inefficient firms. Many are forced out of the industry. However, the ones that remain are the most efficient and the process of "natural selection" results in a lower cost of marketing for the entire industry than has been achieved under any centrally planned marketing operation. The many daily decisions made by thousands of individual firms, each seeking its own individual gain, provides far more information and expertise than can be gained by a centralized committee of a few experts regardless of how well trained or well paid these commission members are.

Entry and Exit. By definition, the grain board and the marketing board systems restrict entry. Restriction is essential to exclude the possibility of lower-cost firms capturing market share. Exit of the less efficient firms is also precluded in the design of most board systems. Costs are often averaged or prices set to cover the highest-cost firm. There is little penalty for poor management or wrong decisions. The entry and exit of firms into the grain trade under a free market economy will often result in capital losses to individuals but a gain to the system in total. Under the proper regulatory environment, the ease of entry guarantees the absence of excess profits while encouraging maximization of efficiency. A grain board can eliminate all profits if that is the policy which it chooses to follow. However, the politicization of most government operations such as grain boards encourages the transfer of income from one sector to another for purposes deemed important in the overall economy. Grain boards such as that employed in Argentina seldom operate solely for the benefit of producers. Even the marketing boards used in Canada, South Africa and Australia are not always operated in the best interest of producers. Subsets of goals such as survival of the institution and positions for individuals within the organization sometimes comprise the criterion of "service to producers" . The competitive system also provides the necessary incentive and opportunity for exit of inefficient firms, while the marketing boards and grain boards are committed to maintaining their facilities and position regardless of relative costs or efficiencies of alternatives.

Resource Allocation. A free market price is the best allocator of resources and goods among alternative competing uses. It meets the economic criteria of perfect competition, although it is often criticized for inability to meet social and political objectives. However, the efficiency of the market-oriented system will generate more total income to be used to meet these social and political objectives than if these non-economic goals are used to distort the market system and resource allocation. A responsive and flexible system in terms of facilities for handling and shipment of grain is essential to minimizing costs and maximizing returns to producers. This flexibility, lacking in marketing boards and grain boards, is clearly evident in the ability of United States firms to make rapid changes in grain flows and mode of transport in response to every opportunity to arbitrage over time, form and space.

Social objectives such as equalizing producer returns often conflict with optimum resource allocation. Several studies have shown the less-than-optimal allocation of resources in Canada with grain being produced in areas in conflict with the principle of comparative advantage as a result of equalized prices (i.e., cross subsidization among crops and regions) throughout the growing area. Similar distortions have occurred in the U.S. production system as a result of price supports. To the extent that prices, or price supports, are established by government decisions, the United States has instituted a limited form of grain board. It is generally agreed that those price distortions created by government decisions have resulted in increased costs of production and decreased efficiency in production and marketing (see Part Five on the United States).

Based on the above analysis, we recommend that the trading functions and associated assets of the Argentine National Grain Board be returned to the private sector.

Many, if not all, of the regulatory functions now handled by the Argentine Grain Board are essential to the efficient functioning of private markets, and for that reason must continue to be discharged by a public sector institution.

Regulatory Environment—Establishment of an Argentine Grain Commission

All markets require regulatory controls in order to ensure efficient operation. Regulations to control and enforce contracts, to guarantee performance bonds, to set grades and standards, and to control predatory practices by firms are all examples of regulations that are essential to efficient market operation. These regulations can best be handled by an agency completely segregated and insulated from daily buying and selling activities.

In Canada, the Canadian Grain Commission, which is independent and separate from the Canadian Wheat Board, is responsible for control and supervision of grain quality and for all aspects of grain handling. It licenses grain handling facilities, sets grain standards, provides inspection, including of weighing equipment, handles complaints and ensures the quality of Canadian grains. It is supported by fees levied for its services.

The United States relies on several agencies to regulate the industry and to protect producers and consumers from unfair or non-competitive pricing practices. The Federal Grain Inspection Service of the USDA has responsibility for grading standards and inspection procedures for all official grades in domestic and export markets. Federal agencies also supervise weighing, especially in the export market, but also wherever official weights are requested on a grading certificate. The Food and Drug

Administration has responsibilities with respect to phyto-sanitary regulations, food safety and purity. Several other agencies are responsible for export licensing and reporting, and for the delivery of grain to developing countries. The Foreign Agricultural Service provides supervision of trading practices, records export sales, and supports staff in overseas markets. Although many of these agencies lack the coordination that might increase their efficiency, their role and their responsibility to a particular sub-sector is generally clear and explicit.

The federal government of the United States is responsible for quality control standards, for universal access to all navigable waterways, and for the enforcement of antitrust regulations. The Export Trading Act of 1982, however, limits the antitrust liability for grain export companies. Cooperatives have always had less stringent requirements with respect to antitrust regulation than other forms of corporation. The federal government is also responsible for maintaining waterways and the federal highway system. To the extent that these are heavily used by the grain industry, this provides an indirect subsidy to those firms operating in the transportation area. In both waterways and highway systems, the federal government provides the basic roadbed and charges user fees to recover a portion of the fixed cost. Operation on those roadbeds is by private companies competing without restrictions in terms of size and rates.

In addition, many of the regulations pertaining to grain are developed and implemented at the state level. Bonding, insurance and warehouse licensing are the responsibility of states. Some state governments also have regulations related to delayed pricing, and supervise the operation of country elevators with respect to speculation and to weights and measures.

To handle the regulatory functions now handled by the Argentine Grain Board, it is recommended that an Argentine Grain Commission be established, which would have at least the following regulatory functions:

- registration of all participants in the trade;
- standardization of all trading procedures through the design and administration of all trading, transport, and handling documentation;
- organization of forms and terms of contracts, as well as their control and regulation;
- quality control and issuance of Argentine quality certificates;
- maintenance of grain movement statistics; and
- maintenance of an export registry.

Argentina has already decided that some free market operations are beneficial to the economy as well as to producers. To hamper that decision by continuation of an inefficient, high-cost grain board operating alongside and parallel to the free market system is creating problems of high costs and unresponsiveness to world market conditions. Regulations which ensure freedom of entry and a "level playing field" for competitors whenever profit-making opportunities arise are the only guarantees required to avoid monopolies and the associated evils of excess profits and exploitation.

GATT Negotiations

Because Argentine agriculture is export-oriented, the outcome of GATT can have a significant impact if trade liberalization is achieved. The effect of trade barriers external to Argentina is almost as significant as are instruments, such as export taxes and distorted exchange rates, imposed by the Argentine government.

Development of Improved Production Technology

Although Argentine yields are low by international standards, substantial production increases have occurred, indicating that Argentine farmers are receptive to new technology. For example, in the 1970s soybean production grew at an annual rate exceeding 30 percent. While these growth rates declined to a more modest 9 percent in the 1980's, this is still an impressive performance. Wheat and soybean yields are second only to the United States among the group of countries considered in this study. Sunflower production has also increased rapidly since the 1970s. As indicated earlier, the National Institute of Agricultural Technology (INTA) made a major contribution to developing improved wheat varieties and incorporating soybeans into the Argentine grain sub-sector. In order to maintain Argentina's future competitiveness in grains, support for research, both public and private, should be given high priority by the government.

Notes

1. Lowell D. Hill and A. Mustard, "Economic Considerations in Industrial Utilization of Cereals," in Y. Pomeranz and Lars Munck, ed., *Cereals: A Renewable Resource*, St. Paul, Minn: American Association of Cereal Chemists, 1981, p. 25.
2. Hill and Mustard, "Economic Considerations."
3. See: U.S. General Accounting Office, *Grain Marketing Systems in Argentina, Australia, Canada, and the European Community; Soybean Marketing Systems in Brazil*, Washington, U.S. Government Printing Office, 1976.

4. Noreen L. Dollinger, "Comparative Analysis of Transport and Handling of Export Wheat in Canada and the U.S.," Unpublished M.S. Thesis, University of Illinois, 134 pp., November 1980.

5. Alex McCalla and Andrew Schmitz, "Grain Marketing Systems: The Case of the United States vs. Canada," *American Journal of Agricultural Economics*, 61:2 (May 1979) 199–212.

7

Statistical Annex

TABLE 7.1   Argentina—Gross Domestic Product, Agriculture's Share and Agriculture's Value Added, 1970–1987 (Australes of 1970)

	1970	1971	1972	1973	1974	1975	1976	1977	1978	1979	1980	1981	1982	1983	1984	1985	1986	1987
GDP at market prices	8,775	9,105	9,294	9,642	10,163	10,103	10,102	10,747	10,400	11,130	11,300	10,542	10,018	10,311	10,565	10,102	10,646	10,821
Net indirect taxes	1,000	1,038	1,061	1,100	1,159	1,152	1,152	1,226	1,186	1,269	1,288	1,201	1,142	1,175	1,204	1,152	1,214	1,234
GDP at factor cost	7,775	8,067	8,233	8,542	9,004	8,951	8,950	9,521	9,214	9,861	10,012	9,341	8,876	9,136	9,361	8,950	9,432	9,587
Agriculture	1,023	1,039	1,059	1,173	1,205	1,172	1,227	1,257	1,292	1,329	1,256	1,280	1,369	1,394	1,439	1,420	1,381	1,407
Agr. as a % of GDP (at factor cost)	13%	13%	13%	14%	13%	13%	14%	13%	14%	13%	13%	14%	15%	15%	15%	16%	15%	15%
Agriculture's Value Added	1,003	1,037	1,059	1,173	1,204	1,172	1,226	1,256	1,292	1,328	1,254	1,280	1,368	1,394	1,439	1,420	1,380	1,406
Cereals value added	214	204	194	242	229	214	241	244	249	233	231	292	314	314	321	296	262	240
Oil Seeds value added	51	47	43	55	52	58	72	107	137	170	153	164	186	200	264	286	267	285
Cereals and oil seeds as a % of agriculture's total value added	26%	24%	22%	25%	23%	23%	26%	28%	30%	30%	31%	36%	37%	37%	41%	41%	38%	37%

Source: Central Bank.

TABLE 7.2   Argentina—Area Harvested of Major Grains and Oilseeds, 1970–1988 ('000 ha)

	Wheat	Maize	Sorghum	Soybeans	Sunflower	Total
1970	5,191	4,017	1,872	26	1,347	12,453
1971	3,701	3,995	2,235	36	1,313	11,280
1972	4,295	3,147	1,420	68	1,286	10,216
1973	4,965	3,565	2,131	157	1,338	12,156
1974	3,958	3,486	2,324	344	1,190	11,302
1975	4,233	3,070	1,338	356	1,005	10,002
1976	5,271	2,766	1,834	434	1,258	11,563
1977	6,428	2,532	2,377	660	1,227	13,224
1978	3,910	2,660	2,254	1,150	2,000	11,974
1979	4,885	2,800	2,044	1,600	1,557	12,886
1980	4,787	2,490	1,279	2,030	1,855	12,441
1981	5,023	3,394	2,100	1,880	1,280	13,677
1982	5,926	3,170	2,510	1,386	1,673	14,665
1983	7,320	2,970	2,514	2,281	1,902	16,987
1984	7,073	3,025	2,370	2,910	1,389	16,767
1985	5,900	3,340	1,965	3,269	2,350	16,824
1986	5,382	3,231	1,280	3,316	3,046	16,255
1987	4,893	2,900	977	3,510	1,735	14,015
1988	4,875	2,400	965	4,270	1,992	14,502
Average annual growth rates [a]	1.67%	-1.28%	-1.92%	30.61%	3.45%	2.35%

[a] Calculated by solving for b in the equation $y=e^{(a+bt)}$

Source: Ministry of Agriculture, Livestock and Fisheries.

TABLE 7.3 Argentina—Production of Major Grains and Oilseeds, 1970–1988
('000 metric tons)

	Wheat	Maize	Sorghum	Soybeans	Sunflower	Total
1970	7,008	9,360	3,820	27	1,140	21,355
1971	4,920	9,814	4,660	59	830	20,283
1972	5,540	5,864	2,360	78	828	14,670
1973	7,900	9,700	4,960	272	880	23,712
1974	6,560	9,900	5,900	469	970	23,799
1975	5,970	7,700	4,830	485	732	19,717
1976	8,570	5,855	5,060	695	1,085	21,265
1977	11,000	8,300	6,600	1,400	900	28,200
1978	5,300	8,700	7,164	2,500	1,600	25,264
1979	8,100	8,700	6,200	3,700	1,430	28,130
1980	8,100	6,400	3,000	3,500	1,650	22,650
1981	7,780	12,900	7,550	3,769	1,260	33,259
1982	8,300	9,600	8,000	4,150	1,980	32,030
1983	15,000	9,000	8,100	4,000	2,400	38,500
1984	13,000	9,500	6,900	7,000	2,200	38,600
1985	13,600	11,900	6,200	6,500	3,251	41,451
1986	8,700	12,100	4,000	7,100	4,100	36,000
1987	8,700	9,250	3,000	7,000	2,200	30,150
1988	9,900	8,600	3,100	8,800	2,740	33,140
Average annual growth rates:[a]	3.72%	1.33%	0.57%	34.88%	8.42%	4.17%

[a] Calculated by solving for b in the equation $y=e^{(a+bt)}$

Source: Ministry of Agriculture, Livestock, and Fisheries.

TABLE 7.4 Argentina—Yields of Major Grains and Oilseeds, 1970–1988 (metric tons/ha)

	Wheat	Maize	Sorghum	Soybeans	Sunflower
1970	1.350	2.330	2.041	1.038	0.846
1971	1.329	2.457	2.085	1.639	0.632
1972	1.290	1.863	1.662	1.147	0.644
1973	1.591	2.721	2.328	1.732	0.658
1974	1.657	2.840	2.539	1.363	0.815
1975	1.410	2.508	3.610	1.362	0.728
1976	1.626	2.117	2.759	1.601	0.862
1977	1.711	3.278	2.777	2.121	0.733
1978	1.355	3.271	3.178	2.174	0.800
1979	1.658	3.107	3.033	2.313	0.918
1980	1.692	2.570	2.346	1.724	0.889
1981	1.549	3.801	3.595	2.005	0.984
1982	1.401	3.028	3.187	2.994	1.184
1983	2.049	3.030	3.222	1.754	1.262
1984	1.838	3.140	2.911	2.405	1.584
1985	2.305	3.563	3.155	1.988	1.383
1986	1.616	3.745	3.125	2.141	1.346
1987	1.778	3.190	3.071	1.994	1.268
1988	2.031	3.583	3.212	2.061	1.376
Average annual growth rates:[a]	2.02%	2.64%	2.54%	3.27%	4.8%

[a] Calculated by solving for b in the equation $y=e^{(a+bt)}$

Source: Ministry of Agriculture, Livestock, and Fisheries.

TABLE 7.5   Argentina—Grain Exports, 1974–1975 to 1987–1988 [a]

	Wheat	Maize	Sorghum	Soybeans
	(millions of tons)			
1974/75	2.2	5.8	2.5	0.0
1975/76	3.2	2.6	2.6	0.1
1976/77	5.6	4.4	4.6	0.5
1977/78	2.6	0.6	4.4	2.0
1978/79	3.3	6.7	4.3	2.8
1979/80	4.7	3.5	1.6	2.3
1980/81	3.9	0.9	4.9	2.7
1981/82	4.3	4.9	5.2	1.7
1982/83	7.5	6.5	4.9	1.4
1983/84	9.7	5.9	4.8	3.0
1984/85	0.8	0.7	3.4	3.3
1985/86	6.1	7.4	2.2	2.6
1986/87	4.3	0.4	1.0	1.3
1987/88	0.5	0.5	0.8	2.6
	(percent of world trade)			
1974/75	3.4	13.6	28.4	0.0
1975/76	4.8	4.9	25.7	0.3
1976/77	8.9	8.2	37.4	3.1
1977/78	3.6	10.4	40.0	8.6
1978/79	4.5	10.2	39.8	11.2
1979/80	5.5	4.7	13.8	8.1
1980/81	4.2	11.6	34.8	10.5
1981/82	4.2	7.3	38.2	5.9
1982/83	7.6	10.3	42.2	4.9
1983/84	9.5	9.7	36.9	11.9
1984/85	7.5	10.5	25.9	12.9
1985/86	7.2	13.6	25.3	9.9
1986/87	4.7	7.1	12.3	4.5
1987/88	5.0	8.7	9.8	9.0
Averages:				
'74 thru '88	4.9%	6.0%	30.4%	7.5%
'81 thru '88	4.8%	6.2%	29.2%	8.3%

[a] The commercial years run from July 1 through June of the next year.
Source: Junta Nacional de Granos, (National Grain Board) *Annual Report 1987*, Buenos Aires, 1988.

TABLE 7.6   Argentina—Amounts and Values of Grain Exports 1970–1987
(amounts in '000s of tons; values in millions of dollars)

	Cereals	Oilseeds	Sub-totals	Total Agric. Exports	Cereals and Oilseeds as a % of Total Agric. Exports
1970					
Amounts	10,105	9	10,114		
Values	509	3	512	1,333	38%
1975					
Amounts	8,025	16	8,041		
Values	1,067	7	1,074	2,059	52%
1980					
Amounts	9,889	2,825	12,714		
Values	1,631	671	2,302	5,277	44%
1981					
Amounts	18,360	2,298	20,658		
Values	2,830	648	3,478	6,168	56%
1982					
Amounts	14,615	1,965	16,580		
Values	1,822	460	2,282	4,806	47%
1983					
Amounts	22,339	1,539	23,878		
Values	2,894	366	3,260	5,907	55%
1984					
Amounts	17,249	3,370	20,619		
Values	2,240	952	3,192	5,979	53%
1985					
Amounts	20,316	3,454	23,770		
Values	2,265	735	3,000	5,579	54%
1986					
Amounts	13,563	3,221	16,784		
Values	1,245	647	1,892	4,591	41%
1987					
Amounts	9,065	1,531	10,596		
Values	718	321	1,039	3,375	31%

Source: World Bank, *Argentina—Agricultural Sector Review,* June 30, 1989.

TABLE 7.7 Argentina—The Value of Grain Exports as a Percent of Total Exports, 1980–1987 (values in billions of current US$)

	Total Exports	Grains and Derivatives	Grains etc. as a % of Total Exports
1980	8.021	3.084	38.4
1981	9.143	4.117	45.0
1982	7.624	3.049	40.0
1983	7.836	4.377	55.9
1984	8.107	4.723	58.3
1985	8.397	4.300	51.2
1986	6.852	3.230	47.1
1987	6.196	2.537	40.9

Source: Unpublished monograph by Dr. Marcelo Regunaga.

TABLE 7.8 Argentina—Grain Exports by Major Traders, 1980–1985 ('000 tons)

Traders	1980	1981	1982	1983	1984	1985
ACA	1,203	1,063	1,085	2,028	1,726	1,961
AFA	159	198	143	315	317	404
Bunge & Born	769	2,310	1,445	1,434	893	927
Cargill	1,226	2,011	1,894	2,336	1,820	2,638
Emiliana	265	266	446	721	771	1,121
Continental	736	1,312	705	1,162	1,812	1,488
FACA	693	1,530	1,050	2,149	2,157	2,261
Genaro Garcia	706	1,228	1,063	1,286	654	635
Italgrani	697	941	1,072	707	243	346
J.N.G. [a]	334	987	642	2,466	1,917	2,167
La Plata Cereal	902	1,241	1,221	1,697	704	888
Dreyfus	880	1,594	515	1,244	1,150	1,632
Nidera	1,106	1,419	1,691	2,228	2,543	2,521
P. Sudamericanos	698	919	468	876	851	995
Sasetru	668	–	–	–	–	–
Tradigrain	–	353	665	1,228	904	917
Totals :	12,603	20,487	16,623	23,726	20,640	23,723
	=====	=====	=====	=====	=====	=====

[a] National Grain Board.
Source: Interamerican Institute for Agricultural Sciences.

TABLE 7.9 Argentina—Export Taxes on Grains and Oilseeds [a] (% of FOB)

Date Of Effectiveness	Wheat	Maize	Sorghum	Soybeans Beans	Soybeans Oil	Soybeans By-Prod.	Sunflower Seed	Sunflower Oil	Sunflower By-Prod.	Linseed Seed	Linseed Oil	Linseed By-Prod.
08/04/81	0.0%	0.0%	0.0%	0.0%	-10.0%	-10.0%	0.0%	-10.0%	0.0%	0.0%	-10.0%	0.0%
08/05/81	0.0%	0.0%	0.0%	0.0%	-10.0%	-10.0%	0.0%	-10.0%	-50.0%	0.0%	-10.0%	-50.0%
12/24/81	10.0%	10.0%	10.0%	10.0%	-10.0%	-10.0%	10.0%	-10.0%	-50.0%	10.0%	-10.0%	-50.0%
05/05/82	10.0%	10.0%	10.0%	10.0%	0.0%	0.0%	10.0%	0.0%	50.0%	10.0%	0.0%	50.0%
07/05/82	25.0%	25.0%	25.0%	25.0%	10.0%	10.0%	25.0%	10.0%	15.0%	25.0%	10.0%	15.0%
12/16/83	18.0%	25.0%	25.0%	25.0%	10.0%	10.0%	25.0%	10.0%	15.0%	25.0%	10.0%	15.0%
01/04/84	18.0%	25.0%	25.0%	25.0%	13.0%	13.0%	25.0%	14.0%	17.0%	25.0%	10.0%	15.0%
05/24/84	18.0%	25.0%	25.0%	25.0%	13.0%	13.0%	33.0%	22.0%	17.0%	25.0%	10.0%	15.0%
06/02/84	18.0%	25.0%	25.0%	25.0%	19.0%	13.0%	33.0%	22.0%	17.0%	25.0%	10.0%	15.0%
07/02/84	18.0%	25.0%	25.0%	25.0%	19.0%	70.0%	25.0%	22.0%	17.0%	25.0%	10.0%	15.0%
08/24/84	18.0%	25.0%	25.0%	25.0%	10.0%	10.0%	25.0%	14.0%	15.0%	25.0%	10.0%	15.0%
10/29/84	24.0%	31.0%	31.0%	31.0%	16.0%	16.0%	31.0%	20.0%	21.0%	31.0%	16.0%	21.0%
11/23/84	24.0%	31.0%	31.0%	31.0%	25.0%	13.0%	31.0%	28.0%	23.0%	31.0%	16.0%	21.0%
12/07/84	18.0%	31.0%	31.0%	31.0%	16.0%	16.0%	31.0%	20.0%	21.0%	31.0%	16.0%	21.0%
01/09/85	18.0%	31.0%	31.0%	31.0%	16.0%	16.0%	26.0%	18.0%	0.0%	31.0%	16.0%	21.0%
02/07/85	18.0%	25.0%	20.0%	25.0%	10.0%	10.0%	25.0%	14.0%	15.0%	31.0%	16.0%	21.0%
03/04/85	18.0%	21.0%	20.0%	25.0%	10.0%	10.0%	25.0%	14.0%	15.0%	31.0%	16.0%	21.0%

06/11/85	28.5%	29.0%	28.0%	32.5%	19.0%	19.0%	32.5%	23.0%	23.5%	38.0%	24.5%	29.0%
10/10/85	15.0%	29.0%	18.0%	32.5%	19.0%	19.0%	32.5%	23.0%	23.5%	38.0%	24.5%	29.0%
11/07/85	15.0%	29.0%	28.0%	32.5%	19.0%	19.0%	32.5%	23.0%	23.5%	27.0%	13.5%	18.0%
01/10/86	15.0%	29.0%	28.0%	32.5%	19.0%	19.0%	27.0%	17.0%	22.0%	27.0%	13.5%	18.0%
02/22/86	15.0%	21.0%	20.0%	32.5%	19.0%	19.0%	27.0%	17.0%	22.0%	27.0%	13.5%	18.0%
03/20/86	15.0%	21.0%	20.0%	27.0%	15.0%	15.0%	24.0%	15.0%	20.0%	27.0%	13.5%	18.0%
07/08/86	5.0%	21.0%	20.0%	27.0%	15.0%	15.0%	24.0%	15.0%	20.0%	27.0%	13.5%	18.0%
09/17/86	5.0%	21.0%	20.0%	27.0%	15.0%	15.0%	24.0%	15.0%	20.0%	15.0%	1.5%	6.0%
01/05/87	5.0%	15.0%	15.0%	15.0%	3.0%	3.0%	15.0%	6.0%	11.0%	15.0%	1.5%	6.0%
Harvest 1987/88	0.0%	0.0%	0.0%	11.0%	0.0%	0.0%	10.0%	0.0%	0.0%	12.0%	0.0%	0.0%

[a] Negative figures denote subsidy; special levies are not included.
Source: Ministry of Agriculture, Livestock, and Fisheries.

TABLE 7.10 Argentina—Import Tariffs on Agricultural Inputs, (%), 1981–1988 [a]

	Tractors				Ag. Machinery			Fertilizer			Insecticides		
	15HP	15–35HP	35–140HP	140HP	P.D.	N.P.D.	Parts	Urea	N.P.D.	N.P.D.	N.P.D.	P.D.	N.P.D.
May/80	30	0	30	0	27	0	6/18	15	15/29	0	0	10/20/29/32	0
May/81	48	5	48	5	44	5	5/36	28	28/40	0	0	23/28/33/40/43	0
Nov/81	48	5	48	5	44	5	5/35	28	28/40	0	0	23/28/33/40/43	0
Mar/81	43	10	43	10	43	10	10/35	28	28/40	0	0	23/28/33/40/43	0
Dec/82	38	10	38	10	38	10	10/31	25	25/35	0	0	21/29/35/38	0
Apr/83	38	10	38	10	38	10	10/31	25	25/35	0	0	21/29/35/38	0
Apr/84	38	10	38	10	38	10	10/31	25	25/35	0	0	10/21/35/38	0
Apr/85	38	10	38	10	38	10	10/31	0	25/35	0	0	10/21/29/35/38	0

Dec/85	38–15	10–15	38–15	10–15	38–15	10–15 10/31–15	0 25/35–5	010/21/29/35/38–15	0–15
Apr/86	38–15	10–15	38–15	10–15	38–15	10–15 10/31–15	0 25/35–5	010/21/29/35/38–15	0–15
Dec/86	38–15	10–15	38–15	10–15	38–15	10–15 10/31–15	0 25/35–5	010/21/29/35/38–15	0–15
Apr.	38–15	10–15	38–15	10–15	38–15	0–15 10/31–15	0 25/35–5	010/21/29/35/38–15	0–15
May/88	38–15	10–15	38–15	10–15	38–15	0–15 10/31–15	0 25/35–5	010/21/29/35/38–15	0–15

aP.D. denotes produced domestically; N.P.D. denotes not produced domestically; /indicates range of tariffs;—indicates temporary surcharge under the Austral Plan.

Source: Ministry of Agriculture, Livestock, and Fisheries.

TABLE 7.11   Argentina—Production, Imports and Sales of Principal Farm Inputs, 1970–1988

| | '000 tons | | | | '000 units |
| | Seed Production | | | | |
	Maize	Soybeans	Fertilizer Consumption [a]	Pesticide Imports [b]	Tractor Sales
1970	-	-	-	7.4	11.3
1971	107.6	-	75.4	7.5	13.7
1972	73.0	0.2	76.1	7.9	14.2
1973	87.6	1.4	71.4	9.0	18.8
1974	78.6	9.0	62.6	13.5	20.7
1975	68.6	8.4	54.1	9.6	15.2
1976	66.0	11.4	34.7	9.0	21.0
1977	85.7	4.8	55.9	9.8	22.0
1978	87.1	18.5	63.3	8.2	6.5
1979	98.3	82.3	99.0	15.5	6.9
1980	102.3	82.4	94.4	10.7	5.0
1981	124.8	60.1	66.2	9.2	3.1
1982	87.4	94.4	76.2	13.3	4.4
1983	95.9	97.4	96.4	16.0	8.0
1984	80.6	123.3	127.8	17.8	12.4
1985	122.2	96.6	143.8	11.8	6.7
1986	94.7	109.6	101.2	15.8	7.5
1987	69.3	-	133.1	16.2	3.9
1988	56.6	131.3	140.9	15.7	5.3

[a] In plant nutrients of N, P and K.
[b] Insecticides, fungicides and herbicides.
Source: Ministry of Agriculture, SNESR.

TABLE 7.12   Argentina—Prices of Principal Agricultural Products
Relative to the Non-Agricultural Price Index

	Wheat	Maize	Sorghum	Soybeans	Sunflower
1960	0.85	0.77	0.60		1.64
1961	1.00	0.77	0.47		2.10
1962	0.86	0.93	0.74		1.37
1963	1.10	1.14	0.73		2.14
1964	1.13	0.86	0.51		1.98
1965	0.74	0.84	0.60	1.83	1.45
1966	0.66	0.73	0.48	1.18	1.17
1967	0.89	0.70	0.84	1.41	1.11
1968	0.81	0.69	0.58	1.67	1.09
1969	0.85	0.83	0.60	1.57	1.34
1970	0.77	0.73	0.49	1.47	1.53
1971	0.74	0.64	0.48	1.53	1.78
1972	0.63	0.75	0.52	2.16	2.21
1973	0.88	0.72	0.53	2.07	1.58
1974	0.88	0.79	0.66	2.00	1.69
1975	0.85	0.57	0.48	1.31	1.30
1976	0.90	0.63	0.54	1.90	1.66
1977	0.85	0.83	0.64	3.21	2.69
1978	0.74	0.73	0.50	1.63	1.39
1979	0.55	0.40	0.34	1.02	1.08
1980	0.61	0.45	0.33	0.61	0.51
1981	0.48	0.32	0.26	0.71	0.83
1982	0.74	0.50	0.36	1.15	1.22
1983	0.79	0.65	0.54	1.20	1.16
1984	0.65	0.61	0.42	1.38	1.98
1985	0.48	0.53	0.40	0.98	1.19

Source: Sturzenegger, A., *Trade, Exchange Rate, and Agricultural Pricing Policies in Argentina*, World Bank Comparative Studies, World Bank, Washington, D.C. 1990.

TABLE 7.13   Argentina—Farmgate and Export Prices for Cereals and Oilseeds: 1981–1987 (constant Australes of 1985 per ton)

	1981	1982	1983	1984	1985	1986	1987	1988[a]
CEREALS								
Wheat								
Farmgate	62.5	79.3	69.0	55.7	45.6	48.0	37.7	56.5
FAS	77.4	94.0	82.9	70.7	60.9	63.0	54.5	70.7
FOB	97.4	121.0	111.7	93.1	82.7	71.8	72.0	
Sorghum								
Farmgate	34.1	39.4	45.6	37.3	34.8	27.0	30.8	
FAS	45.9	50.9	60.2	51.2	47.0	39.6	40.6	48.0
FOB	61.9	64.6	80.9	75.6	67.9	56.2	54.4	
Maize								
Farmgate	40.8	49.3	56.1	54.1	48.6	36.6	40.8	
FAS	60.9	62.3	74.9	68.7	61.9	49.1	49.6	58.2
FOB	68.6	76.2	94.7	95.2	85.2	65.5	63.2	
OILSEEDS								
Sunflower								
Farmgate	90.3	114.0	99.1	153.1	108.4	69.6	110.3	
FAS	106.9	131.1	117.4	175.6	129.7	85.4	114.2	128.1
FOB	129.6	181.6	187.1	242.9	190.3	138.8	141.6	
Linseed								
Farmgate	91.3	143.8	113.9	117.9	103.0	101.1	83.7	106.4
FAS	111.9	159.5	127.7	136.1	123.3	119.6	100.9	119.6
FOB	140.6	212.2	170.2	196.0	195.8	149.0	0.0	
Soyabeans								
Farmgate	80.8	110.3	106.4	117.6	92.9	82.5	119.4	
FAS	96.5	127.2	125.6	137.2	110.6	98.8	131.9	160.1
FOB	125.8	153.0	172.5	190.4	158.3	146.6	157.5	
Exchange Rate:								
Australes of								
June 1985/US$0.5	0.7	0.8	0.7	0.8	0.8	0.8		

[a] Preliminary.
Notes:(1)   Farmgate and FAS prices are deflated by the wholesale price index for non-agricultural products (1985=100); current FOB prices have been converted to Australes using the exchange rate shown.
(2)   FAS = Free Alongside Ship; FOB = Free On Board (Ship).
Source: Ministry of Agriculture, Livestock, and Fisheries.

TABLE 7.14   Argentina—Estimated Gross Revenue per Hectare of Major Grains
and Oilseeds, 1970–1988 (Australes of 1988 per metric ton)

	Wheat	Maize	Sorghum	Soybeans	Sunflower
1970	5,308	15,729	9,271	14,191	12,014
1971	4,870	14,397	9,101	22,843	10,246
1972	3,907	11,839	7,339	21,035	12,118
1973	6,893	17,085	10,780	31,070	8,979
1974	7,097	19,956	14,876	24,095	12,192
1975	6,123	13,041	11,082	16,387	8,729
1976	6,939	10,624	11,921	24,425	11,439
1977	6,587	22,901	14,818	57,098	16,584
1978	5,044	20,652	14,770	32,825	10,301
1979	5,004	10,202	13,634	22,519	9,454
1980	5,614	11,431	8,099	10,883	4,694
1981	4,572	13,165	10,074	15,327	8,764
1982	5,469	13,744	10,584	21,593	13,129
1983	7,276	16,501	14,582	17,738	12,265
1984	5,623	15,991	10,220	27,925	18,344
1985	5,332	16,295	10,883	16,802	14,274
1986	4,303	12,807	7,015	16,827	8,569
1987	3,886	11,259	7,021	22,865	12,574
1988	5,599	12,116	5,915	22,615	11,501
Average annual growth rates [a]:	-0.70%	-0.90%	-1.55%	-.71%	0.59%

[a]Calculated by solving for b in the equation $y=e^{(a+bt)}$
Source: Ministry of Agriculture, Livestock, and Fisheries.

TABLE 7.15 Argentina—Marketing Margins for Wheat, December, 1986
(US$ per m. ton)

	Rosario	Buenos Aires	Bahia Blanca
FOB Prices	80.00	80.00	80.00
Port Related Charges	5.27	4.92	6.47
National Grain Board			
Elevation	3.32	2.86	4.52
Inspection	0.95	0.95	0.95
Losses (.3% of FAS)	0.18	0.18	0.18
Port Authority	0.82	0.93	0.82
Marketing Costs (incl. margins)	8.50	7.65	8.09
Analyses	0.04	0.04	0.04
Financing (1.25% of FAS)	0.72	0.74	0.71
Commissions (1% of FAS+US$0.15/ton)	0.75	0.76	0.74
Customs (1% of FOB)	0.80	0.80	0.80
Margins	6.19	5.31	5.80
Levies and Taxes	6.03	6.03	6.03
Export Tax (5% of FOB)	4.00	4.00	4.00
INTA Levie (1.5% of FOB)	1.20	1.20	1.20
Stamp Tax (0.15% of FAS)	0.09	0.09	0.09
Forex Transfer Tax (0.6% of FOB)	0.48	0.48	0.48
Merchant Marine Fund (2% of freight)	0.26	0.26	0.26
FAS Prices	60.20	61.40	59.41
Marketing and Conditioning	6.01	6.07	5.97
Conditioning			
Drying (Australes 0.8/ton)	0.67	0.67	0.67
Handling (Australes 0.5/ton)	0.42	0.42	0.42
Losses (Australes 2.3/ton)	1.92	1.92	1.92
Commission of acopiador	3.01	3.07	2.97
Charges and Taxes	1.14	1.15	1.13
Over Gross Income (1% of FAS)	0.60	0.61	0.59
Stamp and Registration (0.3% of FAS)	0.18	0.18	0.18
Social Security (0.6% of FAS)	0.36	0.36	0.36
Transport To Port	7.25	5.53	5.73
Transport To Country Elevator	3.38	3.38	3.38
Transport Proper (15 km at A.2/ton)	1.67	1.67	1.67
Handling (A.2.05/ton)	1.74	1.71	1.71
Farmgate Prices	42.42	45.27	43.20
Farmgate Prices as a % of FOB	53%	57%	54%
	===	===	===

Note: Exchange rate Austral 1.2 = US$1.0.
Source: Interamerican Institute for Agricultural Sciences.

TABLE 7.16 Argentina—FOB Buenos Aires Prices for Maize, 1960–1989 (current US$/m. ton)

	Jan.	Feb.	Mar.	Apr.	May	June	July	Aug.	Sept.	Oct.	Nov.	Dec.	Averages
1960	47.00	48.00	49.00	50.00	50.00	49.00	49.00	51.00	52.00	53.00	53.00	50.00	50.08
1961	49.00	49.00	48.00	48.00	47.00	48.00	50.00	52.00	0.00	0.00	0.00	0.00	48.88
1962	51.00	49.00	47.00	48.00	48.00	48.00	47.00	46.00	47.00	0.00	0.00	47.00	53.11
1963	49.00	49.00	49.00	50.00	52.00	56.00	56.00	56.00	57.00	56.00	56.00	57.00	53.58
1964	55.00	53.00	53.00	55.00	52.00	53.00	51.00	52.00	52.00	52.00	53.00	0.00	52.82
1965	56.00	57.00	58.00	56.00	57.00	51.00	53.00	54.00	54.00	57.00	54.00	54.00	55.08
1966	0.00	0.00	0.00	51.00	49.00	49.00	50.00	0.00	52.00	53.00	51.00	0.00	50.71
1967	0.00	0.00	55.00	0.00	0.00	0.00	52.00	53.00	54.00	57.00	60.00	60.00	55.86
1968	64.00	58.00	49.00	48.00	48.00	49.00	47.00	45.00	45.00	45.00	45.00	46.00	49.08
1969	48.00	46.00	45.00	45.00	50.00	51.00	53.00	56.00	59.00	59.00	60.00	61.00	52.75
1970	62.00	54.00	51.00	51.00	56.00	56.00	58.00	61.00	64.00	62.00	62.00	64.00	58.42
1971	63.00	61.00	59.00	56.00	57.00	61.00	61.00	57.00	55.00	52.00	53.00	52.00	57.25
1972	54.00	55.00	56.00	60.00	61.00	59.00	60.00	62.00	66.00	70.00	73.00	82.00	63.17
1973	87.00	76.00	74.00	73.00	90.00	110.00	118.00	119.00	108.00	104.00	109.00	114.00	98.50
1974	125.00	120.00	119.00	109.00	108.00	111.00	126.00	154.00	141.00	163.00	161.00	155.00	132.67
1975	0.00	124.00	126.00	130.00	128.00	126.00	0.00	0.00	0.00	130.00	121.00	122.00	125.88
1976	0.00	0.00	0.00	115.00	121.00	124.00	120.00	116.00	117.00	109.00	102.00	100.00	113.78
1977	0.00	0.00	105.00	102.00	97.00	91.00	85.00	79.00	77.00	89.00	98.00	105.00	92.80

(Continues)

TABLE 7.16 (Continued)

	Jan.	Feb.	Mar.	Apr.	May	June	July	Aug.	Sept.	Oct.	Nov.	Dec.	Averages
1978	95.00	97.00	102.00	109.00	105.00	111.00	99.00	95.00	97.00	105.00	115.00	100.00	102.50
1979	97.00	100.00	98.00	98.00	103.00	118.00	138.00	129.00	134.00	132.00	129.00	122.00	116.50
1980	136.00	149.00	155.00	158.00	144.00	147.00	155.00	169.00	175.00	181.00	183.00	174.00	160.50
1981	166.00	153.00	142.00	131.00	138.00	126.00	139.00	132.00	129.00	131.00	129.00	117.00	136.08
1982	117.00	114.00	110.00	111.00	110.00	113.00	119.00	115.00	105.00	93.00	100.00	104.00	109.25
1983	104.00	114.00	124.00	132.00	128.00	123.00	128.00	147.00	149.00	150.00	154.00	153.00	133.83
1984	136.00	130.00	134.00	142.00	139.00	140.00	138.00	138.00	140.00	141.00	141.00	0.00	138.09
1985	108.00	105.00	106.00	110.00	109.00	111.00	112.00	112.00	110.00	110.00	113.00	116.00	110.17
1986	99.00	88.00	84.00	84.00	89.00	89.00	82.00	80.00	90.00	69.00	71.00	67.00	82.67
1987	65.00	66.00	69.00	74.00	81.00	82.00	88.00	86.00	85.00	92.00	84.00	85.00	79.75
1988	86.00	87.00	85.00	81.00	76.00	115.00	130.00	117.00	119.00	119.00	114.00	115.00	103.67
1989	124.00	118.00	121.00	117.00	114.00	113.00	117.00	104.00	100.00				

Note: A value "0" denotes "not available".
Source: National Grain Board.

TABLE 7.17 Argentina—FOB Buenos Aires Prices for Soybeans, 1977–1989 (current US$/ m. ton)

	Jan.	Feb.	Mar.	Apr.	May	June	July	Aug.	Sept.	Oct.	Nov.	Dec.	Averages
1977	261	267	297	355	340	295	256	199	188	185	213	222	257
1978	210	205	246	252	252	240	231	224	242	251	245	246	237
1979	250	266	263	254	244	268	281	272	267	253	244	245	259
1980	238	245	228	210	209	207	250	262	291	303	321	311	256
1981	289	277	266	273	257	244	258	256	251	238	243	232	257
1982	233	232	225	234	235	227	227	219	220	200	212	214	223
1983	217	225	222	237	228	222	225	315	318	332	304	0	259
1984	284	265	282	280	300	271	237	229	219	224	230	222	254
1985	210	205	204	209	197	196	203	203	202	202	202	202	203
1986	203	194	190	187	190	188	186	190	194	179	181	176	188
1987	175	170	171	183	204	211	212	205	207	211	207	218	198
1988	226	223	221	237	253	328	317	313	314	287	276	280	273
1989	278	269	275	262	262	259	244	213	211				

Source: National Grain Board.

TABLE 7.18 Argentina—FOB Buenos Aires Prices for Sorghum, 1960–1989 (current US$/m. ton)

	Jan.	Feb.	Mar.	Apr.	May	June	July	Aug.	Sept.	Oct.	Nov.	Dec.	Averages
1960	39.00	40.00	39.00	38.00	40.00	40.00	40.00	41.00	41.00	41.00	40.00	39.00	39.83
1961	40.00	41.00	40.00	40.00	39.00	40.00	41.00	42.00	0.00	0.00	0.00	0.00	40.38
1962	41.00	39.00	37.00	37.00	37.00	38.00	37.00	36.00	36.00	36.00	0.00	36.00	37.27
1963	38.00	38.00	38.00	39.00	40.00	41.00	42.00	42.00	43.00	43.00	43.00	43.00	40.83
1964	43.00	41.00	41.00	42.00	41.00	41.00	40.00	41.00	41.00	41.00	42.00	0.00	41.27
1965	42.00	43.00	44.00	42.00	43.00	41.00	41.00	41.00	42.00	43.00	41.00	41.00	42.00
1966	0.00	0.00	46.00	0.00	0.00	0.00	0.00	0.00	0.00	0.00	0.00	0.00	46.00
1967	0.00	0.00	45.00	0.00	0.00	0.00	0.00	0.00	0.00	0.00	0.00	45.00	45.00
1968	0.00	45.00	41.00	0.00	0.00	0.00	41.00	41.00	0.00	39.00	0.00	0.00	41.00
1969	0.00	39.00	39.00	39.00	0.00	39.00	0.00	0.00	0.00	0.00	39.00	0.00	39.00
1970	0.00	0.00	0.00	0.00	0.00	0.00	0.00	50.00	53.00	52.00	51.00	53.00	51.80
1971	51.00	52.00	49.00	46.00	49.00	55.00	56.00	51.00	48.00	49.00	50.00	51.00	50.58
1972	51.00	52.00	51.00	50.00	50.00	48.00	50.00	52.00	56.00	58.00	60.00	73.00	54.25
1973	75.00	70.00	66.00	66.00	72.00	91.00	98.00	110.00	104.00	105.00	104.00	103.00	88.67
1974	104.00	107.00	106.00	92.00	84.00	82.00	95.00	0.00	0.00	0.00	0.00	130.00	100.00
1975	0.00	101.00	103.00	97.00	96.00	90.00	93.00	0.00	0.00	0.00	110.00	105.00	99.38
1976	102.00	105.00	104.00	96.00	95.00	99.00	100.00	97.00	0.00	0.00	85.00	89.00	97.20
1977	89.00	88.00	88.00	85.00	80.00	75.00	73.00	71.00	71.00	77.00	84.00	85.00	80.50
1978	77.00	76.00	82.00	89.00	85.00	84.00	77.00	75.00	79.00	86.00	90.00	80.00	81.67
1979	76.00	79.00	78.00	79.00	81.00	93.00	114.00	108.00	105.00	110.00	113.00	114.00	95.83

1980	112.00	123.00	140.00	152.00	134.00	138.00	149.00	167.00	173.00	170.00	172.00	165.00	149.58
1981	154.00	143.00	129.00	121.00	121.00	114.00	128.00	126.00	117.00	117.00	114.50	105.00	124.13
1982	105.00	99.00	98.00	99.00	86.00	93.00	100.00	99.00	94.00	93.00	101.00	104.00	97.58
1983	106.00	104.00	105.00	114.00	108.00	102.00	108.00	130.00	127.00	120.00	123.00	122.00	114.08
1984	116.00	109.00	109.00	113.00	110.00	109.00	103.00	102.00	105.00	106.00	103.00	103.00	107.33
1985	91.00	88.00	91.00	96.00	95.00	90.00	86.00	80.00	74.00	77.00	79.00	83.00	85.83
1986	90.00	81.00	77.00	76.00	78.00	74.00	65.00	69.00	69.00	62.00	65.00	62.00	72.33
1987	60.00	64.00	67.00	66.00	69.00	69.00	67.00	64.00	60.00	76.00	75.00	75.00	67.67
1988	74.00	76.00	72.00	66.00	64.00	100.00	105.00	94.00	93.00	104.00	98.00	101.60	87.30
1989	103.00	101.00	105.00	102.00	101.00	98.00	97.00	91.00	89.00				

Note: A value "0" denotes "not available".
Source: National Grain Board.

TABLE 7.19 Argentina—FOB Buenos Aires Prices for Wheat, 1960–1989 (current US$/m. ton)

	Jan.	Feb.	Mar.	Apr.	May	June	July	Aug.	Sept.	Oct.	Nov.	Dec.	Averages
1960	57.00	57.00	56.00	57.00	57.00	58.00	59.00	61.00	63.00	0.00	60.00	59.00	58.55
1961	60.00	60.00	60.00	61.00	60.00	59.00	59.00	65.00	65.00	65.00	63.00	62.00	61.58
1962	61.00	61.00	62.00	61.00	62.00	61.00	0.00	62.00	62.00	62.00	59.00	57.00	60.91
1963	58.00	58.00	58.00	59.00	0.00	0.00	0.00	57.00	61.00	64.00	64.00	65.00	60.44
1964	64.00	0.00	0.00	64.00	63.00	63.00	62.00	60.00	60.00	61.00	61.00	62.00	62.00
1965	58.00	56.00	56.00	54.00	55.00	49.00	50.00	49.00	49.00	48.00	49.00	49.00	51.83
1966	0.00	0.00	0.00	0.00	46.00	0.00	0.00	0.00	47.00	55.00	54.00	0.00	50.50
1967	0.00	0.00	59.00	0.00	0.00	0.00	0.00	0.00	0.00	0.00	0.00	59.00	59.00
1968	0.00	55.00	55.00	0.00	0.00	0.00	55.00	55.00	0.00	55.00	55.00	55.00	55.00
1969	0.00	59.00	59.00	59.00	0.00	59.00	59.00	58.00	59.00	57.00	57.00	55.00	58.10
1970	58.00	57.00	56.00	53.00	48.00	53.00	55.00	54.00	52.00	55.00	55.00	56.00	54.33
1971	58.00	59.00	59.00	60.00	58.00	58.00	58.00	57.00	59.00	58.00	63.00	62.00	59.08
1972	61.00	61.00	63.00	65.00	66.00	65.00	66.00	0.00	0.00	0.00	0.00	0.00	63.86
1973	0.00	0.00	0.00	0.00	0.00	0.00	0.00	0.00	0.00	0.00	0.00	0.00	0.00
1974	212.00	220.00	198.00	151.00	139.00	141.00	164.00	169.00	168.00	175.00	188.00	188.00	176.08
1975	0.00	0.00	0.00	144.00	0.00	128.00	132.00	166.00	165.00	161.00	145.00	136.00	147.13
1976	136.00	143.00	145.00	137.00	139.00	141.00	143.00	140.00	122.00	108.00	90.00	92.00	128.00
1977	94.00	97.00	98.00	95.00	96.00	96.00	96.00	97.00	99.00	106.00	113.00	114.00	100.08
1978	118.00	123.00	123.00	130.00	127.00	128.00	124.00	124.00	126.00	131.00	132.00	127.00	126.08
1979	125.00	127.00	128.00	129.00	137.00	166.00	189.00	179.00	175.00	174.00	175.00	185.00	157.42

Year													
1980	192.00	217.00	217.00	216.00	200.00	187.00	196.00	197.00	205.00	219.00	220.00	207.00	206.08
1981	213.00	213.00	212.00	202.00	185.00	178.00	177.00	182.00	180.00	181.00	183.00	179.00	190.42
1982	177.00	176.00	177.00	170.00	0.00	159.00	163.00	164.00	161.00	152.00	149.00	148.00	163.27
1983	148.00	147.00	146.00	134.00	127.00	129.00	139.00	141.00	146.00	140.00	134.00	130.00	138.42
1984	130.00	126.00	128.00	139.00	142.00	142.00	140.00	144.00	142.00	141.00	131.00	118.00	135.25
1985	109.00	111.00	113.00	113.00	113.00	107.00	107.00	98.00	94.00	92.00	99.00	112.00	105.67
1986	107.00	100.00	97.00	100.00	88.00	83.00	80.00	79.50	80.00	81.00	80.00	79.00	87.88
1987	84.00	90.00	92.00	92.00	92.00	85.00	83.00	83.00	89.00	94.00	95.00	96.00	89.58
1988	95.00	97.00	106.00	107.00	110.00	123.00	138.00	143.00	151.00	148.00	150.00	153.00	126.75
1989	161.00	160.00	156.00	156.00	160.00	158.00	156.00	154.00	148.00				156.56

Note: A value "0" denotes "not available."
Source: National Grain Board.

TABLE 7.20   Argentina—Differences in Price of Export Between FOB Argentina and FOB Gulf [a]

Year	Wheat	Maize	Sorghum	Soybeans
1980	27	38	22	-24
1981	24	-11	-21	-31
1982	4	-7	-21	-14
1983	-17	-4	-17	-14
1984	-18	-6	-18	-18
1985	-32	-10	-11	-17
1986	-27	-16	-14	-3
1987	-24	-2	-13	2

[a] In US$1980 per ton, at peak commercial periods.
Source: National Grain Board.

TABLE 7.21   Argentina—Difference in Shipping Costs of Grain Between Argentina and Other Countries, 1986

Origin	Destination	US$/ton	Differential Relative to Argentina (%)
US Gulf	USSR (Black Sea)	17.7	-35
San Lorenzo	USSR (Black Sea)	13.3	-51
Australia	USSR (Black Sea)	21.7	-21
Argentina	USSR (Black Sea)	27.4	0
US Gulf	Rotterdam	6.4	-57
San Lorenzo	Rotterdam	6.2	-59
Argentina	Rotterdam	15.0	0

Source: Interamerican Institute for Agricultural Sciences.

ARGENTINA
Grain Areas

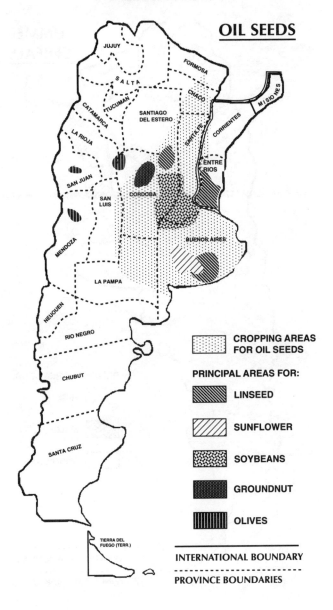

OIL SEEDS

CROPPING AREAS
FOR OIL SEEDS

PRINCIPAL AREAS FOR:

LINSEED

SUNFLOWER

SOYBEANS

GROUNDNUT

OLIVES

INTERNATIONAL BOUNDARY

PROVINCE BOUNDARIES

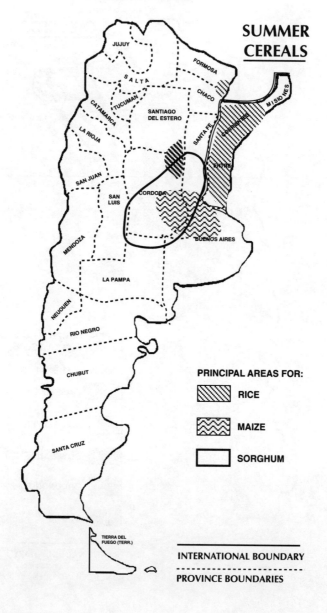

ARGENTINA
Grain Areas

SUMMER CEREALS

JUJUY
FORMOSA
SALTA
CATAMARCA
TUCUMAN
CHACO
SANTIAGO DEL ESTERO
SANTA FE
CORRIENTES
MISIONES
LA RIOJA
ENTRE RIOS
SAN JUAN
SAN LUIS
CORDOBA
MENDOZA
BUENOS AIRES
LA PAMPA
NEUQUEN
RIO NEGRO
CHUBUT
SANTA CRUZ
TIERRA DEL FUEGO (TERR.)

PRINCIPAL AREAS FOR:

RICE

MAIZE

SORGHUM

INTERNATIONAL BOUNDARY

PROVINCE BOUNDARIES

ARGENTINA
Grain Areas

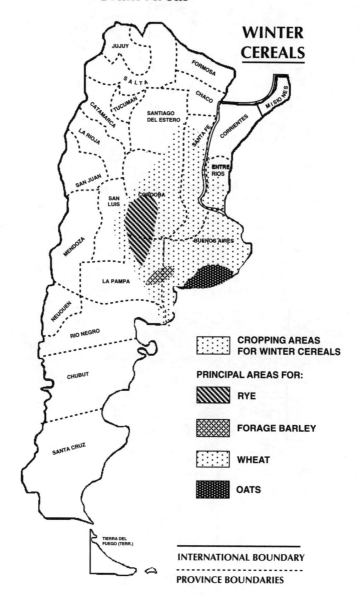

WINTER
CEREALS

CROPPING AREAS
FOR WINTER CEREALS

PRINCIPAL AREAS FOR:

RYE

FORAGE BARLEY

WHEAT

OATS

INTERNATIONAL BOUNDARY

PROVINCE BOUNDARIES

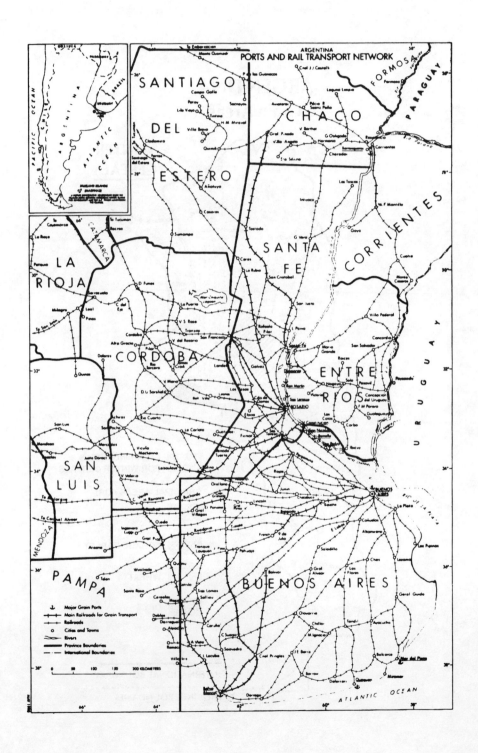

ARGENTINA
PORTS AND RAIL TRANSPORT NETWORK

Bibliography

AMEC, Inc. *Argentina—Its Grain Trade*. Consultant's report to World Bank. Boston; June 1988.

Bolsa de Cereales de Buenos Aires. *Producción y Comercio de Granos—Las Limitantes de Infraestructura*. Separata Numero Estadístico 1986. Buenos Aires, 1986.

_____. 1986 Annual Report. Buenos Aires, 1986.

_____. *Numero Estadístico 1987*. Buenos Aires, 1987.

Centro de Acopiadores de Cereales. *Deterioro Economico del Sector Acopiador Durante el Periodo 1984–1987*. Buenos Aires, undated.

Cirio, F. and Regunaga, M., *La Producción de Granos Argentina Frente a Las Condicionamientos del Mercado Mundial: Situación Actual y Perspectivas*. Presented at an IICA Seminar, Buenos Aires—November, 1986

Dollinger, Noreen L., "Comparative Analysis of Transport and Handling of Export Wheat in Canada and the U.S." Unpublished thesis. University of Illinois, 134 pp. November 1980.

Federacion de Centros y Entidades Gremiales de Acopiadores de Cereales. *El Acopio de Granos—Grain Companies Managing Country Elevators*. Buenos Aires, 1987.

Food and Agriculture Organization. *FAO Fertilizer Yearbook 1988*. Rome: FAO, 1988.

Hill, Lowell D. "Effects of Regulation on Efficiency of Grain Marketing," *Journal of International Law*, 17:31 (Summer 1985) 389–419.

Hill, Lowell D., and A. Mustard. "Economic Considerations in Industrial Utilization of Cereals," in Y. Pomeranz and L. Munck (ed.)., *Cereals: A Renewable Resource*, St. Paul, Minn: American Association of Cereal Chemists, 1981, pp. 25–53.

Hill, Lowell D., and Paulsen, "Maize Production and Marketing in Argentina," Bulletin 785, College of Agriculture, University of Illinois at Urbana-Champaign, July 1987.

Interamerican Institute for Agricultural Sciences (IICA). *Transporte de Granos por Ferrocarril—Tomo I: Situación Actual.* Informe 1.15. Buenos Aires, July 1987.

_____. *Diagnóstico de Situación del Sistema Comercial Argentino de Granos y Subproductos.* Buenos Aires.

_____. *Proyecto de Cooperación Para la Modernización del Sector Agropecuario Argentina—El Servicio Publico de Elevación y el Sistema Comercial.* Informe 2.5. Buenos Aires, June 1987.

_____. *Proyecto de Cooperación Para la Modernización del Sector Agropecuario Argentino—El Sistema Comercial Mixto y La Junta Nacional de Granos.* Buenos Aires, November 1987.

_____. *Proyecto de Cooperación Para la Modernización del Sector Agropecuario Argentino—Transporte de Granos por Automotor.* Informe 1.16. Buenos Aires, Julio de 1987.

_____. *Proyecto de Cooperación Para la Modernización del Sector Agropecuario Argentino—Transporte de Granos por Ferrocarril Tome II: Propuesta Para Su Desarrollo.* Informe 1.15. Buenos Aires, July 1987.

_____. *Proyecto de Cooperación Para la Modernización del Sector Agropecuario Argentino—Diagnostico Institucional de la Junta Nacional de Granos.* Informe 2.10. Buenos Aires, December 1987.

_____. *Proyecto de Comercialización de Granos, Documento Principal.* Buenos Aires, December 1988.

_____. *Proyecto de Cooperación Para la Modernización del Sector Agropecuario Argentino—Informe Final—Políticas de Producción,* Transporte, Almacenamiento y Embarque de Granos. Buenos Aires, Undated.

Junta Nacional de Granos (National Grain Board). *Annual Report 1987.* Buenos Aires, 1988.

_____. *Capacidad de Almacenaje—Instalaciones de Comerciantes; Instalaciones Oficales e Instalaciones de Productores.* Buenos Aires, 1986.

Lamarca, P., *Diagnostico de Situacion del Sistema Comercial Argentino de Granos y Subproductos,* Buenos Aires—Undated, probably 1987.

Lamarca, P., R. Lopez Guzman and Dr. O. Oszlak. *Privatización de la Gestión de Elevación Terminal Portuaria.* Buenos Aires, September 1988.

Leath, Mack N., and Lowell D. Hill. *U.S. Corn Industry.* National Economics Division, Economic Research Service, USDA, Agricultural Economic Report No. 479, Washington, D.C., February 1982.

Lewis, C. *Unpublished monograph on Argentinian Grain Trade.* Undated (probably 1987).

McCalla, Alex, and Andrew Schmitz. "Grain Marketing Systems: The Case of the United States vs. Canada," *American Journal of Agricultural Economics,* 61:2 (May 1979) 199–212.

Mercado a Termino de Rosario S.A./Bolsa de Comercio de Rosario *SOJA—Operaciones a Termino en Dolares*. Buenos Aires, April 1988.

Mielke, Myles J. *Argentina—Agricultural Policies in the Grain and Oilseed Sectors*. USDA Report Number 206. Washington, D.C., September 1984.

Miro, S. "Las Limitantes de Infraestructura." In Bolsa de Cereales de Buenos Aires, *Produccion y Comercio de Granos, Separata Numero Estadistico 1986*. Buenos Aires, 1986.

Obschatko de, Edith S. *La Transformación Economica y Tecnologica de la Agricultura Pampeana—1950–1984*. Buenos Aires, April 1988.

Office of Technology Assessment, Congress of the United States. *Grain— Quality in International Trade—A Comparison of Major U.S. Competitors*. Washington, D.C., February 1989.

Ortman, Gerald F. *Analysis of Production and Marketing Costs of Corn, Wheat and Soybeans among Exporting Countries*, Ohio State University, symposium, 1986.

Oszlak, Dr. Oscar. *Fortalecimiento Institucional de la Junta Nacional de Granos*. September 1988.

Piggott, Roley, Brian Fisher, Julian Alston, and Andrew Schmitz. "Grain Marketing and Policy in Australia." Consultant's report to the World Bank. University of California at Davis, January 1990.

Reca, Lucio G. *Argentina: Country Case Study of Agricultural Prices and Subsidies*. Working Paper No.386. Washington, D.C.: The World Bank, April 1980.

SAGyP—IICA. *Los Paises Productores de Cereales ante la Crisis Agricola Internacional*. Buenos Aires, 1987.

Storey, Gary G., Andrew Schmitz, and Alexander H. Sarris (eds). *International Agricultural Trade: Advanced Readings in Price Formation, Market Structure and Price Instability*. Boulder and London: Westview Press, 1988.

Sturzenegger, Adolfo C. *Trade, Exchange Rate, and Agricultural Pricing Policies in Argentina*, Washington, D.C.: The World Bank, 1990.

U.S. Department of Agriculture. *Agricultural Statistics 1988*. Washington, D.C.: USGPO, 1988.

USDA-ERS-Agriculture and Trade Analysis Division, "Estimates of Producer and Consumer Subsidy Equivalents," *Government Intervention in Agriculture, 1982-86*, Washington, D.C., April 1988.

U.S. General Accounting Office. *Grain Marketing Systems in Argentina, Australia, Canada, and the European Community; Soybean Marketing Systems in Brazil*, Washington, D.C.: USGPO, 1976.

Villegas, A. Walter. *Regimen Jurídico del Control del Comercio de Granos*. Segunda Edicion. Buenos Aires, 1982.

XX International Conference of Agricultural Economists. *The Argentine Agricultural Economy—An analysis of Its Evolution and Present Situation*. Buenos Aires, 1988.

Wheat Facts and Trends. Mexico: CIMMYT, 1988.

World Bank. *Argentina—Agricultural Sector Review.* Washington, D.C., June 30, 1989.

_____ . *Argentina: Reforms for Price Stability and Growth.* Washington, D.C., September 8, 1989.

_____ . *Brazil: Agricultural Sector Review,*Washington, D.C., July 26, 1990.

_____ . *The World Bank Atlas,* Washington, D.C., 1989.

PART TWO

Brazil

Grain Marketing, Institutions, and Policies

Michael J. McGarry, Richard Lacroix, Matthew McMahon,
Lowell D. Hill, Thomas Glaessner, and Joseph Reid

The views expressed here are solely those of the authors and should not be
attributed to the institutions with which they are affiliated.

8

Background

The Brazilian Economy

Brazil is the world's tenth largest economy, after the United States, Japan, Germany, France, Italy, the United Kingdom, the U.S.S.R., Canada and China. In 1988, its GNP was US$328.9 billion, behind that of China (US$356.5 billion), but ahead of that of Spain (US$301.9 billion). It has by far the largest GNP of the Latin American countries, and more than double that of its nearest competitor, Mexico (US$151.9 billion).

In 1988, Brazil's per capita GNP of US$2,280 placed it 33rd among the countries of the world (excluding those with a population of less than 3 million). However, it has the third-largest per capita GNP of the Latin American countries, behind that of Venezuela (US$3,170) and Argentina (US$2,640), but ahead of that of Mexico (US$1,820). In the period 1980–1988, the real growth rate of per capita GNP was 1.2 percent per annum.

Brazil has industrialized rapidly since the 1950s, and for 30 years experienced unprecedented economic growth. The industrial sector grew rapidly, with active government involvement, as did the service sector. Though agricultural production grew less fast, it exceeded population growth and contributed to substantial increases in per capita income (Table 8.1).

In the 1980s, Brazil's economy has been greatly weakened by inflation and the accumulation of long-term debt. Economic problems began after the oil shock of 1973. Massive investments in the public sector, backed substantially by foreign financing, led to a rise in inflation rates from 31.3 percent per annum in the period 1965–1980 to 166.3 percent per annum in the period 1980–1987. By early 1990, inflation was growing at about 80 percent per month. Efforts made to curb inflation through wage and price controls, and to restore the strength of the Brazilian currency, through the Cruzado

TABLE 8.1   Brazil—Average Annual Growth Rates of Real GDP, 1960–1987
(in percentages)

	1960–70	1970–80	1980–87
Industry	6.0	8.9	2.3
Services	5.1	8.7	4.0
Agriculture	4.0	4.2	2.8
Total Economy	5.3	8.2	3.3

Source: World Bank, *Brazil Agricultural Sector Review,* Washington, D.C., July 26,
1990.

Plan of 1985 and the Summer Plan of 1989, have been short-lived, and have
led to new rounds of inflation.

Brazil's population grew at an annual rate of 2.2 percent in the 1980s,
reaching 144.4 million by 1988. Population numbers have more than tri-
pled since 1940 and more than doubled since 1960. Life expectancy in 1988
was 65, compared with 75 in the United States.

There has been a strong trend to urbanization since the 1950s, when the
rise of manufacturing brought large numbers of people into the urban areas, principally located along the coast. Between 1960 and 1980, 28 million
Brazilians migrated from rural to urban areas. About 25 percent of these
migrants came from the Northeast, and 42 percent from the interior of the
Southeast (Minas Gerais, Espirito Santo, Rio de Janeiro and Sao Paulo).
The bulk of migration took place during the 1970s. Brazil's rural population decreased by almost two and a half million between 1970 and 1980. By
1980, 68 percent of the population were urban residents.

Table 8.2 shows the estimated geographic distribution, for selected
years, of the Brazilian population by region (see Map: Brazil Grain Production). In 1985, approximately 45 percent of the people lived in the Southeast.

TABLE 8.2   Brazil—Population Shares by Region, Selected Years (in percentages)

Regions	1950	1960	1970	1980	1985[a]
North	3.6	3.7	3.9	4.9	2.9
Northeast	34.6	31.6	30.2	29.3	29.6
Southeast	43.4	43.7	42.8	43.5	44.9
South	15.1	16.8	17.7	16.0	15.7
Center/West	3.3	4.2	5.4	6.3	6.9

[a] Excludes rural population in the North.
Source: World Bank, *Brazil Agricultural Sector Review,* Washington, D.C., July 26,
1990.

Changes in Employment Patterns

Dramatic changes have taken place in the structure of the work force since the 1950s (Table 8.3). During the period from 1950 to 1987, employment in the industrial and service sectors expanded rapidly, roughly doubling, while agriculture's share in total employment fell from about 61 percent to 25 percent.

From 1950 to 1987, few new permanent jobs were created in the agricultural sector. However, average labor productivity grew substantially, as evidenced by growth in agricultural output. Agricultural labor productivity increased at an average annual rate of 3.4 percent, while the rate for the industrial sector was 2.1 percent, and that for the services sector was 1.1 percent.

The pattern of change in employment structure has been similar for all the regions, with a fairly rapid increase in employment in the industrial and service sectors and a decline of employment in the agricultural sector. By 1987, the Southeast had the smallest proportion of its population engaged in agriculture (14 percent) while the Northeast had the largest (39 percent).

The Agricultural Sector

As already shown in Table 8.1, agriculture grew at over 4.0 percent per year in the period 1960–1980, but growth slowed to about 2.8 percent per year in the period 1980–1987. Agricultural growth in the eighties generally stagnated as a result of falling international commodity prices, poor weather, Brazil's increasingly difficult economic situation, and changes in the structure of incentives. Between 1980 and 1986, total planted area grew at an average annual rate of only 1.7 percent, as against the 3.3 percent growth rate of the 1970s. Annual agricultural exports of around US$9.0 to 10.0 billion fell to US$7.1 billion in 1986, primarily as a result of measures

TABLE 8.3 Brazil—Sectoral Shares of Employment, Selected Years (in percentges)

Year	Agriculture	Industry	Services	Total
1950	61	13	26	100
1960	55	12	33	100
1970	44	18	38	100
1980	30	25	45	100
1985	29	22	49	100
1987	25	24	51	100

Source: World Bank, *Brazil Agricultural Sector Review*, Washington, D.C., July 26, 1990.

taken under the Cruzado Plan, which brought about a sudden increase in domestic demand (Table 14.7).

A turnaround occurred after 1986, due to rising international prices, good weather and stabilization of the economy. Brazil had a record grain harvest of about 60 million metric tons in 1987. Similar high production levels were achieved in 1988 and 1989.

Agriculture has played a central role in Brazil's economic development. Major crops such as sugar, rubber, coffee and soybeans have supplied the wealth to create the country's infrastructure and fund its industrial expansion.

Although the rate of structural change has varied between regions, the pattern has been similar throughout Brazil, with industry and services gaining at the expense of agriculture. The Southeast shows the smallest share of agriculture in total regional production, at 5.6 percent in 1980 (down from 21 percent in 1949). The Center/West, Brazil's agricultural frontier up to about 1985, shows the highest share for agriculture in total regional output, at 21 percent in 1980 (down from 40 percent in 1949). Even in Brazil's poorest region, the Northeast, agriculture's share in the regional economy has dropped—from 37 percent in 1949 to 16 percent in 1980.

Brazil is self-sufficient in most foodstuffs, with the notable exception of wheat, and produces a wide range of raw materials for a diversified agro-industrial sector. It is the world's leading producer of coffee, oranges, sugarcane, sisal, cassava and bananas. Also, it is the world's leading exporter of orange juice, soybean oil, soybean meal, and pepper, and the second-largest supplier of soybeans and soybean products to the world's markets after the United States. Until 1985, it was the world's second-largest exporter of poultry meat. It has a strong horticultural production sector, led by the large immigrant community of farmers of Japanese descent. Increasingly, exports of counter-seasonal fruits and vegetables are made to the European Community winter market.

The composition of exports is changing, moving increasingly to semi-processed and processed goods. There is increasing diversification–the newer exports such as orange juice and poultry are expanding rapidly, and the older staple exports (coffee and sugar) are declining in importance. However, Brazil still has a 25 percent share of the world's coffee market, 8 percent of the sugar market (much of the sugarcane produced in Brazil is used domestically to make ethanol), and a 15–18 percent share of the world's cocoa market.

Table 8.4 compares the importance of the agricultural sector to the overall economy in five of the world's leading grain exporters—Argentina, Brazil, Australia, Canada, and the United States. One of the most striking features of the table is the relative importance of agriculture to the economies of Argentina and Brazil. For example, in Brazil over 40 percent of export earnings are derived from agricultural sources.

TABLE 8.4   The Agricultural Sector in Five Grain-Exporting Countries
(in percentages)

	Gross Domestic Product [a]	Total Export Earnings	Employment
Argentina	13	70 [b]	13 [b]
Brazil	12	44	25
Australia	4	29	n.a.
Canada	3	18	4
USA	2	10	3

Sources: [a] The World Bank Atlas, 1989.
          [b] 1987 data, World Bank, Brazil Agricultural Sector Review, July 26, 1990.

Geography and Land Use

Brazil's five major ecological zones are: the humid tropics in the North, the semi-arid tropics in the Northeast, the woodland savannah, the humid temperate zone and the grassland zone. Distinct climate zones, from tropical to temperate, allow for the production of a wide range of agricultural commodities.

The total land area of Brazil is 846 million hectares. In 1985, 44 percent of this area was devoted to agriculture—an increase of 17 percent since 1950. Most of the expansion in agricultural use occurred after 1970. Much of the new agricultural land has been developed in the Center/West, but new areas have also been opened in the North and Northeast.

Much of the farmland in the Center/West region of Brazil was previously virgin land, with limited rainfall, acid soils and sparse vegetation. Some areas were used for extensive cattle holdings. Research performed by the federal Brazilian Company for Agricultural Research (EMBRAPA) on soil treatments for the acidic soils of the cerrados revealed the potential of this frontier area for mechanized annual cropping. Farmers tried new crops and many adopted higher-yielding varieties of soybeans, cotton, maize, wheat, fruits and vegetables.

The pace of development in the Center/West has slowed almost to a halt since 1985, as most of the arable land in the area is now occupied by farmland. In the South, Southeast and part of the Northeast practically all arable land is under production. Brazil's agricultural frontier is pushing further north, occupying increasing areas of what used to be tropical rain forest, of varying quality for agricultural production.

Modest changes in land use nationwide since 1950 mask substantial differences in certain areas, particularly at the agricultural frontier. While the proportion of utilized farm land under crops nationwide rose from 15 percent in 1950 to 22 percent in 1985, in the South it almost doubled, rising from 19 percent to 39 percent. In the Center/West, the increase in cropland was even more pronounced, rising from 1.7 percent to 9 percent.

Land Occupancy

The productive agricultural land is occupied by about 6 million holdings of various sizes. Nearly 50 percent of these farms are in the Northeast and 38 percent are in the South and Southeast.

Although the total area farmed increased by well over 60 percent in the last 35 to 40 years, concentration of land ownership has not changed fundamentally (Table 8.5). However, very small holdings have increased from about 34 percent to 53 percent of all farms, and the area they occupy has doubled, to 2.7 percent of all farmland. Very large holdings have decreased somewhat in relative importance, occupying about 44 percent of all land in 1985, as against 51 percent a generation ago.

Crop Production

About 9 percent of the total land area of Brazil (75 million hectares) is devoted to crop agriculture. Twelve major crops account for 98 percent of the total crop area, and for more than 90 percent of total crop value. Seven of these (coffee, sugar cane, cocoa, cotton, citrus, tobacco and soybeans) are primarily industrial feed stocks or are intended for export, often after primary processing. The other five (beans, rice, maize, cassava, and wheat) are domestic food crops. Fifteen percent of the total crop area is devoted to soybean production, and 30 percent to wheat, maize and rice. Table 8.6 shows, for selected years, the areas under production along with the production of grains, food crops, and industrial raw materials. The areas under grains and food crops for domestic use have increased almost 60 percent since 1965, keeping pace with population growth. The area devoted to export/industrial crops has more than doubled.

A structural shift in crop production began in the late 1960s. Growth rates of the basic staple grains (maize and rice), of other basic staple crops (principally beans and cassava) and of the traditional export crops (notably coffee and cotton) slowed. Much of the new growth came from five crops: soybeans, sugar cane (mainly to support the ethanol fuel program), cocoa, citrus and wheat. Soybeans, cocoa and citrus are major export crops. Wheat is an import substitution product, and as such it benefited from high protection in the form of support prices. The structural change in production was accompanied by shifts in patterns of regional production. Part of the traditionally coffee-growing South shifted to soybeans and wheat. In the Southeast, there was a shift from coffee and cotton to soybeans, sug-

TABLE 8.5 Brazil—Size Distribution of Land Holdings, 1950–1985 (in percentages)

Farm Size (ha)	1950 No.	1950 Area	1960 No.	1960 Area	1970 No.	1970 Area	1980 No.	1980 Area	1985 No.	1985 Area
0—10	34.4	1.3	44.8	2.4	51.3	3.1	50.3	2.4	52.9	2.7
10—20	16.7	2.1	16.4	3.1	15.6	3.6	14.9	2.9	14.0	3.0
20—50	23.6	6.6	20.2	8.3	16.7	8.6	16.5	7.2	15.6	7.5
50—100	10.6	6.6	8.2	7.6	6.9	8.1	7.6	7.5	7.5	8.0
100—1000	13.0	32.5	9.4	34.4	8.4	37.0	9.5	34.8	8.9	35.0
above 1000	1.6	50.9	1.0	44.1	0.7	39.5	0.9	45.1	0.9	43.8

Total Numbers and Areas, ('000s)

	1950 No.	1950 Area	1960 No.	1960 Area	1970 No.	1970 Area	1980 No.	1980 Area	1985 No.	1985 Area
	2,065	232,211	3,338	249,862	4,924	294,145	5,160	364,854	5,835	376,287

Source: World Bank, Brazil Agricultural Sector Review, July 26, 1990.

TABLE 8.6 Brazil—Crop Cultivation, 1965–1988 (area in '000 ha.) (production in '000 metric tons)

	1965	1970	1975	1980	1985	1988	1988 Breakdown
Grains (wheat, maize and rice)							
area	14,157	16,733	19,092	20,817	19,211	22,620	30%
production	19,692	21,769	24,116	30,147	31,036	42,301	10%
Foodcrops (beans and cassava)							
area	19,179	22,242	25,280	27,476	26,394	30,287	40%
production	46,974	53,445	52,516	55,581	56,656	66,804	16%
Ind.+ Export Crops (coffee, sugar cane, cocoa, cotton, citrus, tobacco, soybeans)							
area	10,721	10,637	14,995	18,889	21,660	22,136	30%
production	82,433	85,137	106,278	168,329	271,631	299,887	73%

Source: World Bank, *Brazil Agricultural Sector Review,* July 26, 1990.

ar cane and citrus, and in the Center/West a shift from cattle breeding and maize production to soybeans.

A marked shift to substitution of capital for labor over the past few decades is evidenced by the increase in mechanization (Table 8.7). For example, between 1950 and 1985 tractor use, per 1,000 persons employed, increased more than fifty fold. This shift in the use of production factors has slowed since 1980, in part as a result of an economy-wide recession.

TABLE 8.7 Brazil—Indicators of Mechanization in Agriculture, 1950–1985

	Tractors per 1,000 ha of Cultivated Land	Persons Employed per 1,000 ha of Cultivated Land	Tractors per 1,000 Persons Employed
1950	0.44	543	0.81
1960	2.14	432	4.94
1970	5.35	422	12.67
1980	11.08	257	43.06
1985	12.45	290	42.93

Source: World Bank, *Brazil Agricultural Sector Review,* July 26, 1990.

TABLE 8.8   Brazil—Livestock Inventory—Cattle, Hogs and Poultry, Selected Years *(in '000 head)*

Year	Cattle	Hogs	Poultry
1950	44,562	22,887	213,622
1975	101,673	35,152	286,810
1980	118,086	32,629	413,180
1985	127,643	30,067	429,732

Source: World Bank, *Brazil Agricultural Sector Review,* July 26, 1990.

Livestock Production

Though reliable statistics on the sector are lacking, Brazilian estimates are that livestock production may account for between 20 and 25 percent of agricultural GDP. Table 8.8 gives national figures for the stock holdings of cattle, hogs, and poultry for 1980–1985, based on the Agricultural Census performed by the Brazilian Institute of Geography and Statistics, IBGE.

The cattle herd has grown at an average annual rate of about 3 percent since 1950. Since the early 1970s most of this growth has occurred in the Center/West and the North—much of the new land opened up to agricultural development in the North is grazed to cattle. In spite of its large cattle population, Brazil is not an important dairy producer. More than half of its modest production originates in the Southeast. Retail prices of dairy products are controlled.

Poultry production has been stimulated by export demand and by the availability of domestic feedstuffs (maize and soybean meal). By 1984, Brazil had become the world's second largest exporter of poultry meat. As of 1985, Brazil lost that position, partly due to United States farm programs. Domestic demand for poultry meat has also increased rapidly, and poultry meat has displaced some of the pork in the Brazilian diet.

9

The Grain Sub-Sector

Sub-Sector Performance

The grain sub-sector—wheat, maize, rice and soybeans—makes a significant contribution to Brazil's agriculture. It accounts for about 30 percent of the gross agricultural product, just under 50 percent of the total value of crop production, and about 45 percent of the area under crops. Between 1975 and 1986, it also accounted for 13 percent of total export earnings, and 27 percent of agricultural exports (Table 14.7).

Since the mid-1970s, exports from the grain sub-sector have consisted almost entirely of soybeans and soybean derivatives. Brazil's other major grain crops—wheat, maize and rice—have been essentially reserved for the domestic market. For 1989, soybean-related exports generated US$3.3 billion, about 10 percent of total export receipts. These exports were mostly in the form of processed products (oil and meal).

Production

The past 25 years have seen a dramatic expansion in soybean production, and a large increase in the production of wheat. The production of maize and rice has increased, but at a much slower rate. Production of these two staples slowed in the late 1960s. During the 1970s, there was a major expansion of soybean production, and a large increase in that of wheat. From 1980 to 1988, production of maize, rice and soybeans grew modestly, while wheat production sustained an annual average growth rate of 15.8 percent (Table 9.1). A combined record total of almost 61 million metric tons of wheat, maize, rice and soybeans was produced in 1988, compared with just over 48 million metric tons in 1980 (Table 14.5).

The main changes accompanying the growth in Brazil's grain production in the last 25 years have been:

- Government support for wheat production, which led to an increased area devoted to wheat in Rio Grande do Sul;
- Introduction of soybeans as a complementary crop to wheat;
- Spread of soybeans into the agricultural frontier areas such as Paraná and Mato Grosso do Sul, to be followed by wheat;
- Soybeans/wheat is now the most common crop sequence in the South, and there have been large production increases in both crops;
- Yield increases for soybeans and wheat, due to introduction of improved varieties and use of improved soil management techniques;
- Increased technical sophistication, contributing to increased yields of grains and oilseeds, but also to shifts to other products;
- In the main grain-producing areas, production is now generally mechanized.

Table 9.1 gives the key indicators (production, exports, area and yield) for the principal grain crops in Brazil. In terms of area and physical production, the most important grain crops are maize, soybeans, rice and wheat.

TABLE 9.1 Brazil—Key Indicators for the Principal Crops, Selected Years

	Wheat	Maize	Rice	Soybeans
Production ('000 mt) (1984–88)	4,698	23,140	10,150	16,464
% of world production (1984–88)	0.9	5.1	2.2	16.4
Prod. growth % p.a. (1970–79)	6.0	n.a.	n.a.	15.5
Prod. growth % p.a. (1980–88)	15.8	2.7	3.0	2.4
Exports ('000 mt) (1984–88)	0.0	0.0	0.0	3,230
Exports as % of national production (1984–88)	0.0	0.0	0.0	17.4
Exports as % of world trade (1984–88)	0.0	0.0	0.0	12.1
Area ('000 has) (1984–88)	3,042	13,652	5,533	9,690
Growth in area % p.a. (1970–79)	6.6	n.a.	n.a.	16.4
Growth in area % p.a. (1980–88)	5.9	2.0	-0.8	2.4
Yield (kg/ha) (1984–88)	1,510	1,830	1,840	1,700
% Growth yield % p.a. (1970–79)	-0.7	n.a.	n.a.	2.2
% Growth yield % p.a. (1980–88)	9.9	0.8	3.8	0.0

Source: USDA, *Agricultural Statistics*, Washington, D.C., various issues.

Main Production Areas

The South (the states of Paraná, Santa Catarina and Rio Grande do Sul) is the dominant grain-producing area; in 1988, it accounted for 48 percent of total production. However, the annual rate of growth of production in the South is levelling off—between 1980 and 1988, it was less than 1 percent, compared with a 3.6 percent national average growth rate.

The Center/West, comprising the states of Mato Grosso do Sul, Mato Grosso, Goiás and the Federal District, is the next-largest grain-producing area; in 1988, it accounted for over 24 percent of total production. Production in this area is increasing rapidly—its annual average growth rate, over the period 1980 to 1988, was over 10 percent.

The third most important grain-producing region is the Southeast; in 1988, it accounted for about 18.4 percent of total production. Its annual average growth rate over the 1980–1988 period was 3.5 percent, about equal to the national average growth rate of 3.6 percent.

Changes in the geographic distribution of grain production are the result of opening up new lands and of technical innovation. Frontier development has been a major contributor to the increase in total production and to a shift in the distribution of production nationwide. Increased technical sophistication has been responsible for yield increases and, in part, for shifts to other products (Tables 14.1 to 14.6).

Changes in the relative profitability of certain crops (directly linked with government pricing policy for agricultural commodities) have also led to shifts in cropping patterns. The greatest changes in cropping patterns have taken place in the South (Paraná and Rio Grande do Sul) and Center/West (Mato Grosso).

Between 1980 and 1988, Paraná and Rio Grande do Sul both had a negative annual average growth rate of soybean production, -3.6 percent and -4.6 percent, respectively. During this eight-year period, some soybean production yielded to maize in Paraná and to rice in Rio Grande do Sul. The annual growth rate for maize production in Paraná was 1.4 percent, while the growth rate for rice in Rio Grande do Sul was 6.7 percent.

In Mato Grosso, soybean production has increased over the 1980–1988 period by almost 50 percent per year. This one state accounted for over 15 percent of total national production of soybeans in 1988 (Table 14.4).

Yields and Productivity

Growth rates in production have outstripped the growth rates of yields for all grains except rice. The substantial increases in production shown for maize and soybeans since 1980 have been achieved primarily through expansion of the area dedicated to these crops. For wheat, about one third of the increase in production is attributable to expansion of the area dedicated to wheat; the remaining two thirds is due to increases in yields. For rice, increases in production are fully attributable to increases in yields—the rice area has declined since 1980.

The Principal Cereals and Oilseeds

During the past 25 years, the main focus of growth in the grain sector has been on soybeans, followed by wheat. Maize and rice production has grown unevenly, and at a much slower rate. Each crop will be examined separately. Wherever possible, changes in production and (in the case of soybeans) export patterns are compared to changes occurring in four other major grain-producing countries—Argentina, Australia, Canada, and the United States.

Wheat

Wheat is grown primarily in Brazil's southern states of Paraná, Rio Grande do Sul, Mato Grosso do Sul, Santa Catarina and Sao Paulo, in rotation with soybeans. Acid soils and numerous diseases are the major constraints in this region. Rio Grande do Sul was Brazil's traditional leader in wheat production until the late 1970s. Only 20 years ago, Brazil's current leader in wheat production, Paraná, was still mostly forested in the west and planted to coffee in the north. Following a frost in the mid- 1970's that killed many of Paraná's coffee plantations, a search was made for other crops to diversify the state's agricultural economy. Two complementary crops, soybeans and wheat, both became important within a few years. In 1988 Paraná produced about 55 percent of the country's wheat on about 1.7 million hectares (Table 14.1).

The national harvested wheat area has averaged about 3 million hectares over the past five years, about 30 percent of the soybean area. Most of the wheat area is double-cropped with soybeans. The area harvested to wheat has grown steadily at about 6 percent over the past 20 years.

Production has averaged almost 5 million metric tons since 1985 (Table 9.1). A high of 6.2 million metric tons was reached in 1987. This record harvest supplied 87 percent of domestic needs for that year. Production in 1989 was about 5.5 to 6.0 million metric tons and current consumption is close to 7.5 million metric tons. Thus, Brazil is close to self-sufficiency in wheat. It may well achieve self-sufficiency soon, due to decreased demand as well as increases in production—in 1989, consumption of wheat was expected to decrease or at least to stabilize as a result of reduced subsidies for wheat-based products and because of the general malaise in the Brazilian economy.

Brazil has traditionally been a wheat importer, although the volume of imports has been decreasing. Canada used to be a major supplier of wheat to Brazil, but a recent bilateral trade agreement with Argentina has displaced Canada from this position.

Brazil has no comparative advantage for wheat production. To the contrary, domestic resource cost ratios for wheat production in Brazil were as high as 2.5 in 1968, in the early days of production, and fell to a still uneconomic level of 1.35 in 1976/77.[1] With the high nominal rates of protection shown in Table 10.1, wheat is likely still an uneconomic crop for Brazil, although the gap between the domestic price and border prices is narrowing.

Increases in production of wheat during the 1970s are totally attributable to increased area seeded—yields were actually decreasing during this expansion phase (see Table 9.1).

Wheat has always been a difficult crop to produce in Brazil, because of acid soils and diseases. For a long while after wheat was introduced, yields remained stagnant at about 800 kg/ha. A modern phase in wheat research began in the early 1970s, with the establishment of the farmer-supported research center, FECOTRIGO, in Cruz Alta, Rio Grande do Sul, the EMBRAPA National Wheat Research Center (CNPT) in Passo Fundo (also in Rio Grande do Sul), and somewhat later, a farmer-supported institution, the Eloy Gomes Research Center, in Cascavel, Paraná. The work of these research centers has been supported by other state institutions such as the Agronomic Institute of Paraná (IAPAR); the Agricultural Institute of Campinas (IAC) in Sao Paulo; the State Agricultural Research Company of Minas Gerais (EPAMIG); and the State Agricultural Research Company of Rio Grande do Sul (IPAGRO).

The research undertaken in the various centers has paid handsome dividends, both in terms of crop management and plant breeding. Over the past decade, the centers have released high-yielding varieties resistant to disease and to the aluminum toxicity associated with acid soils. These new varieties have been adopted by farmers all across Brazil's wheat belt. The centers have also developed crop rotations that reduce soil-borne diseases and increase fertility—these have made the wheat-soybean rotation sustainable.

Adoption of new varieties and improved crop management have led to a 10 percent growth rate in wheat yields in the eighties. Compared to an average world growth rate of 2.8 percent (see Table 9.2), this is extremely high. Further yield increases can be expected as the new technology spreads. Brazil's yield level of 1,510 kg/ha for wheat compares favorably with the levels attained in Argentina (1,900 kg/ha), Australia (1,470 kg/

TABLE 9.2   Key Indicators for Wheat, Major Grain Producers, Selected Years

	Argentina	Brazil[a]	Australia	Canada	U.S.A.	World
Production ('000 mt) (1985–89)	10,220	4,698	14,600	24,420	57,000	513,800
% of world production (1985–89)	2.0	0.9	2.8	4.8	11.1	100
Yield kg/ha (1985–89)	1,900	1,510	1,470	1,780	2,370	2,290
Growth in yield % p.a. (1980–89)	1.7	9.9	2.8	-1.7[b]	0.6	2.8
Exports ('000 mt) (1985–89)	4,740	0.0	13,000	18,240	34,000	95,300
Exports as % of production (1985–89)	46.3	0.0	89.0	74.7	59.7	18.6
Exports as % of world trade (1985–89)	5.0	0.0	13.6	19.1	35.7	100

[a]1985–1989
[b]1980–1988
Sources: USDA, *Agricultural Statistics*, Washington, D.C., various issues; Canadian Wheat Board.

ha), and Canada (1,780 kg/ha); but it is still well below the average world yield of 2,290 kg/ha and the yield in the United States of 2,370 kg/ha.

Maize

In the period 1985–1989, Brazilian maize production averaged 24.6 million metric tons, 5.4 percent of total world production (Table 9.3). None of the maize produced is now exported. While modest quantities of maize were exported until the late 1970s, increased domestic demand and a support price structure that has not favored exports made this trade disappear. Brazil is now importing increasing quantities of maize, particularly for the Northeast.

Although maize is produced throughout the country, about 70 percent of the annual crop is harvested in the southern states of Paraná, Rio Grande do Sul and Santa Catarina, and in the states of Minas Gerais and Sao Paulo (Table 14.2). Domestic annual consumption of maize was about 22 to 25 million metric tons on average from 1986 to 1988. About 30 percent of this amount is consumed, in various forms, on the producers' farms. Another 10 percent is lost in storage and transport. This leaves an average of 13 to 15 million metric tons of marketed maize. The poultry industry uses about 40 percent of this amount (Table 9.4).

TABLE 9.3   Key Indicators for Maize, Major Producers, Selected Years

	Argentina	*Brazil*[a]	*U.S.A.*	*World*
Production ('000 mt (1985–89)	9,425	24,580	186,360	452,120
% of world production (1985–89)	2.1	5.4	41.2	100
Yield kg/ha (1985–89)	3,400	1,830[b]	6,880	3,640
Growth in yield % p.a. (1980–89)	1.0	0.8[c]	0.4	n.a
Exports ('000 mt (1985–89)	4,000	0.0	44,940	60,729
Exports as % of production (1985–89)	42.4	0.0	24.1	13.4
Exports as % of world trade (1985–89)	6.6	0.0	74.0	100

[a] 1985–1989
[b] 1984–1988
[c] 1980–1988

Sources: USDA, *Agricultural Statistics*, Washington, D.C., various issues;
Canadian Wheat Board.

Not enough is known about the costs of growing maize in Brazil to enable us to make any statements about its comparative advantage or disadvantage. Nevertheless, in 1988, the nominal rate of protection for maize grown in the Northeast was estimated at 80.1. Estimates for maize grown in other areas are not available.

TABLE 9.4   Brazil—Consumption of Maize, 1986–1988 ('000 metric tons)

	1986	*1987*	*1988 (est.)*	*Difference Between 1988 and 1986 (%)*
Main Categories				
Industry				
Poultry	5,370	6,300	5,790	7.8
Hogs	3,960	4,360	3,960	0.0
Other Animals	1,300	1,450	1,320	1.5
Milling	2,750	3,200	3,200	16.4
Seed	160	140	140	-12.5
Subtotal Industries	13,540	15,450	14,410	6.4
Rural Consumption	6,840	7,740	6,970	1.9
Losses	1,820	2,620	2,350	29.1
Total Consumption	22,200	25,810	23,730	6.9

Source: Production Financing Company (CFP)

Soybeans

Brazil is the world's second-largest producer of soybeans after the United States. Between 1984 and 1988, it produced over 16 million metric tons of beans per year, 16.4 percent of world production. Following explosive growth in the 1970s, a steady growth rate of 2.4 percent per year was achieved throughout the 1980s. All of the increase in production has come from an increase in the area harvested (Table 9.5).

The largest areas devoted to soybeans are in the South (states of Rio Grande do Sul, Paraná), and in the Center/West (states of Mato Grosso, Mato Grosso do Sul, and Goias). Soybeans are also grown in the Southeast (Table 14.4).

Soybeans were established as a crop in Rio Grande do Sul during the 1950s and 1960s. In the initial phase, the driving force for soybean production was the wheat price support program. As farmers opened land for wheat, they began to look for a complementary crop; soybeans were the crop of choice. Soybeans could be grown in the opposing season (summer vs. winter) and could make use of the same capital plant as that used for wheat, thereby reducing the fixed unit costs for the two crops. The huge infrastructure that had been constructed under the wheat support scheme was available to switch directly into soybeans.

The soybean production technology that led to the initial expansion of soybean production was imported directly from the southern United States, and adapted to Brazilian conditions by local research institutes, by commercial firms, and by the farmers themselves. Technology in terms of equipment, fertilizers, seed, and proven cultivation practices was brought directly to Brazil from other developed countries, including the United

TABLE 9.5   Key Indicators for Soybeans, Major Producers, Selected Years

	Argentina	Brazil [a]	U.S.A.	World
Production ('000 mt) (1985–89)	7,160	16,464	51,470	100,200
% world production (1985–89)	7.2	16.4	51.3	100
Yield (kg/ha) (1985–89)	1,960	1,700	2,160	1,800
Growth in yield % p.a. (1980–89)	-1.5	0.0 [b]	1.4	n.a.
Exports ('000 mt) (1985–89)	1,780	3,286	18,608	26,800
Export as % of production (1985–89)	24.9	17.4	36.2	26.6
Export as % of world trade (1985–89)	6.6	12.1	69.4	100

[a] 1984–1988
[b] 1980–1988
Source: USDA, *Agricultural Statistics*, Washington, D.C., various issues.

States, by the commercial firms. By 1963, the Brazilian soybean industry was firmly established.

As soybeans became more attractive as a crop in their own right, they soon replaced wheat as the lead crop. As the crop expanded northwards into Paraná and Mato Grosso do Sul, soybeans were used to open the land; wheat followed as the secondary crop, the reverse of the pattern initially established in Rio Grande do Sul.

By the mid-1970s, Paraná (where much of the rapid expansion of soybean production was a result of the Brazilian coffee eradication scheme) had overtaken Rio Grande do Sul as the main producer of soybeans. These two states produce about 46 percent of Brazil's soybeans.

While soybean yields in Brazil are quite respectable, averaging 1,700 kg/ha over the past five years, they are inferior to the yields obtained by Brazil's main competitors, Argentina and the United States. These latter two countries have yields of 1,960 kg/ha and 2,160 kg/ha, respectively (Tables 9.5 and 9.6). The lower yields in Brazil are related to double cropping. It must be recognized that the yield of Brazil's crop is an average of many areas where soybeans are being double cropped, requiring producers to sacrifice some timeliness of cultural operations and shorten the growing season in order to accommodate the double cropping. Double cropping is increasing in Argentina and the United States, but is not yet a major factor in national average yields. Yields in Brazil are higher in those areas where soybeans are the primary crop rather than a complementary crop (for example, in the state of Mato Grosso).

During the 1970s, yields grew at the impressive rate of 2.2 percent per year. This growth in yields occurred at a time of rapid expansion into the more fertile areas of Paraná and also coincided with an increased impact of new technology—introduction of better-adapted varieties developed within Brazil itself. During the 1980s, the growth rate in yields has slowed to zero as soybean production has expanded into areas less suited to beans. The last decade has also seen a buildup of diseases and pests, particularly in Rio Grande do Sul.

TABLE 9.6 Average Soybean Yields in Selected Countries, 1985–1989

Country	Yield (mt/ha)
Argentina	2.0
Brazil	1.7
China	1.4
Paraguay	1.6
U.S.A.	2.2

Source: USDA, *Agricultural Statistics*, Washington, D.C., various issues.

   In the southern states, yield declines are due to the increased double-cropping with wheat, which requires shorter-season varieties and less than optimum timing in planting. Currently, yields in Rio Grande do Sul are about 70 percent of those in Paraná. Double-cropping in Rio Grande do Sul is at the margin for sufficient season length to mature both crops.

   Further north, the longer season in Mato Grosso do Sul and Mato Grosso allows more double-cropping and produces wheat yields of about 3,000 kg/ha under irrigation. The returns justify installation of irrigation systems for wheat. At the end of the dry season, the land is prepared for soybeans. If the rain is delayed, the irrigation system can give the soybeans a good start at marginal cost. For these reasons, the northern expanding areas have a potential for increasing average yields and total production.

   Brazil exports only 17 percent of its national production of beans, compared to 36 percent for the United States. The country has a well-developed crushing industry with capacity that has often exceeded national production. This excess capacity was generated under government policies to encourage investment in crushing capacity. Brazil has frequently imported soybeans to better utilize the fixed investment in capacity and in response to government policies and financing schemes that made it profitable to import beans and export meal. Most of the oil produced by this crushing industry is consumed internally, leaving only 27 percent for export. While Brazil is not a major player in international oil markets, it is a major player in the world soybean meal trade, with a market share of 33 percent, equivalent to 76 percent of its national production (Table 9.7).

   The data in Table 9.8 indicate that the average production cost for soybeans in Brazil is below that of the United States, its principal competitor. Soybean producers have been discriminated against in Brazil as a result of taxes, unfavorable exchange rate policies, and protection for local processing and input supply industries. Low production costs and rapidly ex-

TABLE 9.7   Production and Exports of Soybeans and Soybean Products, World and Brazil, 1985–1989

	Soybeans		Oil		Meal	
	World	Brazil	World	Brazil	World	Brazil
Production (million mt)	100.2	18.6	14.9	2.7	65.8	11.1
Percent of world production	100.0	18.6	100.0	17.9	100.0	16.8
Exports (million mt)	26.8	3.2	3.7	0.7	25.7	8.4
Percent of world production	26.8	3.2	24.9	4.9	39.0	12.8
Percent of national production	0.0	17.4	-	27.4	-	76.3
Percent of world exports	100.0	12.1	100.0	19.6	100.0	32.8

Source: USDA, *Agricultural Statistics*, Washington, D.C., various issues.

panding exports in spite of such discrimination indicate that Brazilian soybean production is internationally competitive.

Rice

Rice is produced in all states of Brazil. The rice grown is either dryland or irrigated. In the South, long-grain and medium-grain rice are produced under irrigation. Dryland rice is an important first crop on newly cultivated lands in the Center/West. These new lands, and the established lands in the Northeast, produce a rice based on traditional varieties, and have low yields (Table 9.1) and high production costs.

TABLE 9.8   Comparative Soybean Production Costs, 1986 (US$/ton)

| | United States | | Brazil | |
Production Costs	Overall	Maize belt [a]	Double crop with wheat	Soybeans alone
Variable:				
Seed	12.87	11.30	14.57	14.57
Fertilizer/lime	13.04	8.33	50.90	55.04
Chemicals	24.53	20.04	14.82	14.82
Custom operations	5.08	3.56	n.a.	n.a.
Fuel and lubricants	16.26	12.98	20.76	20.85
Repairs	10.22	8.22	6.55	6.58
Hired labor	1.93	1.62	n.a.	n.a.
Miscellaneous	0.37	0.29	5.89	6.09
Interest on var. exp.	4.06	3.01	3.86	4.01
Sub-total var. costs:	88.36	69.35	117.35	121.96
Fixed:				
Overhead	14.61	14.93	2.59	2.59
Taxes/insurance	15.96	18.08	3.27	4.67
Capital replacement	33.07	30.15	13.43	13.49
Labor	16.68	13.79	6.45	6.48
Interest on capital exp	11.51	10.59	6.46	6.48
Land cost (1985)	62.95	67.06	35.25	42.74
Sub-total fixed costs	154.78	154.60	67.45	76.45
Total production costs	243.14	223.95	184.80	198.41

[a] Includes Great Lakes region.

Source: Norman Rask, Gerald Ortmann, and Walter Stulp. Comparative Costs Among Major Exporting Countries, Ohio State University, Dept. of Ag. Economics, Occasional Paper, Columbus, Ohio, Jan. 1987, app. 3.

Brazil's annual rice harvest of approximately 10 million metric tons is about 2 percent of world rice production. The Center/West and the Northeast combined typically account for about 40 percent of the total harvest. Rio Grande do Sul, the main producer of long-grain rice, typically contributes about another 33 percent. The irrigated varieties grown are medium- and long-grain rice strains, developed from United States seed stock (Table 14.3).

Brazil consistently exported rice through 1981, though quantities exported varied from year to year. Record exports took place in 1974 when about US$1.0 billion dollars of rice was shipped abroad. Exports have shown a diminishing trend since the early 1980s. Two factors account for this loss of export marketing. First, in most years, government's support price for rice has been higher than export parity, at the official exchange rate. Second, the support price policy has emphasized quantity (yields) over quality. As a result, Brazilian rice has lost the quality image it once had in the international trade.[2]

Notes

1. For 1968 data, see: Peter Knight, Brazilian Agricultural Technology and Trade—A study of Five Commodities, New York: Praeger, 1971. For 1976/77 data, see: Ricardo Pereira Soares, Avaliaç&o Econômica da Política Triticola de 1967 a 1977, Brasilia-CFP, Coleçaô Análise e Pesquisa, vol. 20, 1980.

2. The rice referred to here is the long-grain rice produced under irrigation in the South (locally called "argulinho").

10

Policies Affecting the Grain Sub-Sector

Policy Background

The government's main objectives since the 1950s have been to build an industrial society and to develop a consumer durable goods industry in as short a time as possible. These objectives have been supported with a strategy of import substitution, including high tariffs on imports of manufactured goods to protect "infant" industries, and the control of trade in basic foodstuffs (including at one time or another all the major grains) to ensure adequate supplies of low-cost food for the urban work force.

Except for wheat, cereals and oilseeds have been discriminated against both as non-industrial products—subject to taxes, export restrictions, and high import tariffs on inputs—and as basic foodstuffs, whose prices must be kept low to make them readily available to urban consumers. Wheat production and consumption have been encouraged through subsidies, though wheat marketing has been controlled by a government monopoly. Maize and rice have been reserved for the domestic market in the past few years. Soybeans have received mixed treatment. Interventions to ensure adequate supplies of soybean oil (considered a staple foodstuff) for the domestic market have generally harmed Brazilian producers in the international markets, and shifted benefits from soybean producers to processors.[1] However, government protection to industry has helped create a flourishing export trade in soybean derivatives.

Although the main thrust to industrialization has continued over the past 40 years, policies and policy instruments have changed frequently. Critical factors for the grain markets have been government control of trade; stockholding of grains; supply of credit that has generally gone to the larger producers of export and industrial crops; and the lack of maintenance of the transportation network since the 1970s, which lead to the high cost of transporting grains.

From the 1940s through the early 1960s, government efforts were directed at extracting rents from the agricultural economy. From the mid-1960s to the early 1970s, there was a general liberalization of the economy, with active government support for exports, including agricultural commodities. An attempt was made to reform trade policies through tariff reform (1967) and to lower the real value of the cruzeiro through mini-devaluations. By the beginning of the 1970s, the terms of trade began to shift in favor of agriculture, although some implicit taxation of agriculture remained.

The liberalization period ended following the world oil crisis of 1973. To sustain export growth and maximize domestic value added, the government introduced large subsidies for manufactured exports. New limitations were placed on the export of unprocessed agricultural products. In an effort to ensure domestic supplies and curb inflation, price controls on basic foodstuffs were reinforced. Direct and indirect taxation on agriculture increased substantially.

Since the 1970s, government credit and subsidies to agriculture have been used to counter the effects of discrimination against the sector. The producers of industrial crops have benefited the most from this support. To compensate for the high rate of taxation on agriculture and the damage to agricultural incentives caused by its import substitution and price control policies, the government provided large amounts of credit to the sector at preferential rates for production, investment, and marketing; and reinforced its pricing policy. By 1980, the major policy instrument for managing the sector was massive credit subsidies. The government also increased resources devoted to agricultural research and extension and to infrastructure.

After the second major oil price increase in 1979, but particularly after 1982, Brazil faced an escalating debt crisis. Agricultural credit subsidies were cut back sharply. To lessen the impact of this policy change, the government increased its intervention in the commodity markets, mainly through its Minimum Price Program, a program intended to stabilize prices and assist farmers by providing loans on post-harvest stocks of grains, and purchases of grains at the guaranteed minimum price.

In recent years, the government has cut back on interest rate subsidies. The remaining major intervention in the grain sub-sector is through the Minimum Price Program (MPP). Through the MPP, the government has become a major stockholder of maize and rice. Its stockholding policies, together with restrictions on exports and a policy of selective imports to ensure adequate domestic supply, have depressed prices for maize and rice and effectively excluded Brazil from the export markets in these two commodities.

Two summary measures of the effects of government policies on the grain sub-sector are nominal protection rates at the official exchange rate, and producer subsidy equivalents. Nominal protection rates are equal to domestic producer prices minus border price equivalents, divided by the border price equivalents, expressed as percentages. A negative or positive value (taxation or protection, respectively) is shown only when the domestic price is either below FOB (taxation) or above CIF (protection). When the domestic price lies between the two, this indicates neither taxation, nor protection, hence a zero value in those cases. Data for the period from 1970 to 1988 are shown in Table 10.1.

The Producer Subsidy Equivalents (PSEs) shown in Table 10.2 attempt to cover all direct and indirect payments to grain farmers, as well as the rents extracted from the grain economy. As such, they are a broader measure of public policy impact on the grain economy than the nominal protection coefficients shown in Table 10.1. However, because of data problems, the following were not included: crop insurance, fuel, and fertilizers. Also, the assumption is made that subsidies on credit are a subsidy to the crop, ignoring the problem of credit leakages. But the more funda-

TABLE 10.1   Brazil—Nominal Protection Rates at the Official Exchange Rate for the Major Grains, 1970–1988 (in percentages)

Year	Wheat (South)	Maize (South)	Rice (South)	Soybean (South)
1970	79.8	-30.2	–	-9.0
1971	50.1	-20.6	–	-23.6
1972	51.4	0.0	–	-27.6
1973	28.3	-27.8	-51.9	-52.6
1974	0.0	-17.1	-43.0	-13.9
1975	24.7	-14.2	0.0	-12.1
1976	21.9	-13.4	0.0	-17.0
1977	81.9	-2.3	0.0	-23.3
1978	58.6	0.0	0.0	-15.2
1979	20.1	0.0	0.0	-8.5
1980	25.3	0.0	0.0	-10.0
1981	89.0	0.0	-8.1	-9.5
1982	65.6	0.0	0.0	-14.5
1983	21.4	-27.7	0.0	-24.6
1984	30.2	-17.2	-0.5	-22.0
1985	61.2	0.0	0.0	-18.3
1986	0.0	0.0	0.0	-11.2
1987	24.3	0.0	0.0	0.0
1988	60.7	0.0	0.0	-15.9

Source: World Bank—*Brazil Agricultural Sector Review*, Washington, D.C., July 26, 1990.

TABLE 10.2   Brazil—Producer Subsidy Equivalents for Major Grains, 1982–1986 (in percentages)

	1982	1983	1984	1985	1986	Average 1982–86
Wheat	77	55	63	64	52	56
Soybeans	2	13	-13	3	28	7
Maize	17	5	-24	12	59	14
Rice	40	53	34	56	64	49

Source: USDA-ERS-Agriculture and Trade Analysis Division, "Estimates of Producer and Consumer Subsidy Equivalents, "Government Intervention in Agriculture, 1982-86, Washington, D.C., April, 1988.

mental problem is that one distortion (tax) cannot be cancelled by another distortion (subsidy); yet this is precisely what the PSE methodology assumes.

Policy Instruments

The main policy instruments used by the various governments to pursue their objectives vis-a-vis the grain sub-sector have been: (1) exchange rates; (2) trade controls; (3) import tariffs; (4) value-added (ICM) taxes; (5) other taxes and levies; (6) price controls; (7) credit policies; (8) price and income stabilization through the Minimum Price Program; (9) buffer stocks; and (10) the wheat program. Each of these policy instruments is discussed below.

Exchange Rates

Probably the most important policy for the grain sub-sector has been that governing exchange rates, where there has been systematic overvaluation of the currency. Currency overvaluation has led to scarcity, and de facto rationing, of foreign exchange. Table 10.3 shows official and parallel exchange rates between 1970 and 1988. Although the parallel rate is not a true free-market rate by virtue of the distortions resulting from the overvalued official rate, it does give a likely upper limit to the market rate, and is at least indicative of the extent of distortion and of the losses in revenue, in local currency terms, suffered by the grain sub-sector.

TABLE 10.3   Brazil—Official and Parallel Market Exchange Rates, 1970–1988 [a] (in percentages)

Year	Official	Parallel Market	Official/Parallel Minus 1
1970	4.59	5.13	-11
1971	5.29	6.03	-12
1972	5.93	6.77	-12
1973	6.13	6.77	-10
1974	6.79	7.79	-13
1975	8.13	10.18	-20
1976	10.67	14.04	-24
1977	14.14	17.68	-20
1978	18.08	22.51	-20
1979	26.82	33.34	-20
1980	52.81	59.08	-11
1981	93.35	111.10	-16
1982	180.37	274.83	-34
1983	576.94	910.00	-36
1984	1,846.98	2,255.83	-18
1985	6,200.00	8,685.33	-29
1986	13.66	22.00	-38
1987	39.23	52.72	-26
1988	225.26	319.10	-29

[a]*Cruzeiros ([Cr$]/US$ through 1985; Cruzados [Cz$]/US$ thereafter)*
Source: World Bank, *Brazil Agricultural Sector Review,* July 26, 1990.

Trade Controls

The export of agricultural products, particularly grains, has been guided by the notion of "exportable surplus" , treating exports as a residual after domestic demand is met. To ensure supplies at home, outright bans, export quotas, licensing and other restrictive measures have been applied at one time or another. The "surplus" rationale has recently become more dominant in trade policy because of the need to reduce inflation and the effect of some of the tradeable commodities on the consumer price index.

Three of the five key commodities that have been at the center of agricultural trade policy and controls are soybean products (beans, meal and oil), maize and rice. These grains and their derivatives are targeted because they figure prominently in both domestic food consumption and industry. They have all been important exports at various points. Only soybean products are now consistently a major export.

Trade policy for soybeans is based on two objectives: (1) to assure domestic supplies of oil and (to a lesser extent) meal; and (2) to maximize value added to soybean exports, consistent with Brazil's emphasis on industrial exports. Trade controls on soybean exports have taken various forms: quotas, preferential taxation for meal and oil exports over bean exports, and outright bans at various points on bean and oil exports. The variability of this trade policy, especially as concerns the export of the raw beans, has led to much uncertainty and risk. Table 10.4 illustrates the effects of the policy on soybean exports over the period 1970–1988.

To encourage domestic processing and exports of processed products instead of beans, government has applied lower ICM taxes (taxes on the circulation of goods) for meal and oil exports, concessionary export credits for oil and meal, various tax exemptions, and the drawback scheme (allowing imports of beans for processing into export products).

Trade policy for maize and rice is based on assuring domestic supplies of these basic foodstuffs. Exports have periodically been restricted or banned. Imports are equally restricted, based on the need to hold down inflation and to economize on the use of foreign exchange. Import restriction is a primary tool for domestic protection. During the Cruzado Plan, import restrictions were temporarily lifted, and foreign exchange was made available. As a result, over 2 million metric tons of rice and an equal amount of maize were imported in 1986.

Trade policy reforms instituted in 1988 were aimed at reducing trade barriers for soybean products (beans, meal and oil), rice and maize. Under this policy, exports and imports of these commodities were free from quantitative restrictions, but were subject to continued licensing by CACEX, the trade control agency of the Central Bank, and to a variable tar-

TABLE 10.4   Brazil Soybean Exports, 1970—1988 ('000 metric tons)

	Net Exports [a]				Total	Trade as a % of Production	
	Beans	Oil	Meal	Total	Production	Beans	Total
1970–73 [b]	830	40	1,110	1,980	2,960	28.0	66.9
1974–77	3,070	320	3,720	7,110	10,380	29.6	68.5
1978–81	650	760	6,510	7,920	12,480	5.2	63.5
1982	-880	850	7,720	7,690	12,890	-6.8	59.7
1983	1,290	1,080	8,500	10,870	14,530	8.9	74.8
1984	1,580	910	7,370	9,860	15,340	10.3	63.4
1985	3,220	820	8,630	12,670	18,210	17.7	69.6
1986	940	410	6,930	8,280	14,190	6.6	58.4
1987	2,540	910	8,020	11,470	17,070	14.9	67.2
1988	2,900	750	7,840	11,490	18,170	16.0	63.2

[a]   As of 1978, soybeans were imported for processing in the off-season; 1982 imports exceeded exports by 880,000 metric tons.
[b]   Annual averages for periods indicated, through 1981.
Source: World Bank, Brazil Agricultural Incentives and Marketing Review, 1989.

iff. The tariff was intended to raise the CIF import price to a level equal to a floor price, derived from the trigger price formulas introduced to guide the operations of the Production Financing Company (CFP), the public company that purchases grains under the Minimum Price Program, in the disposal of stocks in the domestic market.

Import Tariffs

Policies toward the manufacturing sector have been highly protection-ist, designed to protect "infant" industries from foreign competition. Man-ufacturing industries have been allowed to import raw materials and some intermediate capital goods duty free. Other sectors, such as agricul-ture, face discriminatory tariffs. Import tariffs for inputs needed by the grain sub-sector range from about 20 percent for certain fertilizers to 45 percent for tractors and 50 percent for selected pesticides. The trade also faces many non-tariff barriers such as import licenses, formal and informal quotas, outright trade bans, and foreign exchange restrictions. The end re-sult is that the sub-sector is implicitly taxed through the higher prices it pays for the domestically produced industrial inputs it needs.

Value-Added (ICM) Taxes

The value-added taxes, called taxes on the circulation of goods (ICM), are meant to stimulate state industrial development. The ICM is levied on the sale of goods at all stages of production. Changes made in 1989 mean that the federal government retains authority for setting rates on interstate and international sales, but the states have the authority to set rates and determine exemptions on intra-state sales. The ICM is not applied to in-dustrial exports and goods subject to specific excise taxes, but it does apply to many services. Some processed agricultural exports are considered in-dustrial, and are exempt under tax law if the cost of the agricultural input is less than 50 percent of the final price of the product. Soybeans are subject to the ICM when exported, as are all unprocessed agricultural exports. Ag-ricultural inputs (fertilizers, chemicals and seeds) are exempt, but capital inputs (machinery) are not; the tax paid on capital inputs is not subject to a rebate. For maize, the ICM is paid when it is sold to produce flour for hu-man feed, but not when it is sold to produce animal feed.

The ICM is a major source of revenue for the states. It accounts for near-ly 5 percent of GDP. Nominal tax rates are 17 percent for intra-state trans-actions and final consumption; 12 percent for most interstate transactions, except for shipments from the South and Southeast to the Northeast, Cen-ter/West and Espirito Santo, which are taxed at 9 percent; and 13 percent for exports (mostly agricultural).

ICM's burden on agriculture is not as great as might be expected. It is estimated that the agricultural sector accounts for about 6 percent of total ICM revenues, although agriculture accounts for about 12 percent of GDP. The heaviest implicit tax burden (greater than 10 percent) falls on coffee, wheat, milk and vegetable oils; and on the processing of wheat, soy products, poultry and vegetable oil refining (6–10 percent). Analysis of marketing margins has shown that taxes are the single largest item, as high as 20 percent for rice.[2] For export crops, the ICM falls most heavily on unprocessed commodities, and thus favors the processed commodities. This is clearly evident in the case of soybeans.

Other Taxes and Levies

The government imposes a number of other taxes and levies for specific purposes. Other taxes levied on marketing transactions in addition to the ICM are: FUNRURAL, the rural development fund (rate = 2.5 percent); FINSOCIAL, the social investment fund (rate = 0.6 percent); PIS, the mandatory private savings program (rate = 0.75 percent); IPI, the industrial products tax; and IOF, the tax on financial transactions. These taxes generally fall less heavily on agriculture (except for agro-industry) than on the other sectors. Goods transported by road are also subject to a road tax.

Table 10.5 shows the direct taxes levied on the products of the soybean industry in 1988. The domestic bean trade is taxed slightly below the export trade. However, the domestic meal and oil trade is taxed well above exports.

Price Controls

Direct and indirect controls on consumer prices have been a longstanding feature of Brazilian economic policy. Beginning in the 1970s, use of direct price controls became an increasingly popular tool in attempts to restrain inflation. Price controls have included: select controls for basic commodities; economy-wide price freezes; control of processing and marketing margins; and "gentlemen's agreements" between government and supermarkets to restrain prices of key consumer commodities, including rice, maize and soybean oil. Price controls and freezes have generally been arbitrary and unpredictable, increasing uncertainty for producers and discouraging stockholding and investment. The price freezes have not been successful in slowing Brazil's inflation. To the contrary, they have destabilized agricultural markets and, by increasing government crop purchases and financing, contributed to inflation.

TABLE 10.5   Brazil—Taxes Levied on Soybeans, 1988 (in percentages)

	Beans		Meal		Oil	
	Domestic	*Exports*	*Domestic*	*Exports*	*Domestic*	*Exports*
FUNRURAL	2.50	2.50				
ICM [a]		13.00	12.75	11.10	17.00	8.00
PIS	0.75	0.75	0.75	0.75	0.75	0.75
FINSOCIAL	0.60		0.60		0.60	
Road Tax	5.00		5.00		5.00	
Classification						
Tax	6.00		7.00		13.00	
Total Taxes	14.85	16.25	26.10	11.85	36.35	8.75

[a]   May vary by state.
Source: The Vegetable Oil Industry Association (ABIOVE).

Credit Policies

The government has been supplying funds for rural credit since the mid-1960s. Interest rates on rural credit have been subsidized. Up to 1973, real interest rates were moderately negative, but from 1974 to 1983, they became strongly negative. In 1982, agricultural credit subsidies were cut back sharply, but negative rates continued. From 1984 to 1988, the government indexed credit to the price of government bonds (ORTN). As a result, real interest rates on official credit came close to being positive, but were still well below market rates. In January 1989, as part of the Summer Plan, the government increased interest rates on government-funded rural credit to 12 percent plus monetary correction.

Credit volume grew at an average annual rate of 17.5 percent in real terms during the 1970s. By the mid-1970s, the volume of credit had reached 92 percent of agricultural GDP. By the end of the decade, it peaked at about US$16.7 billion equivalent, over four times its 1970 volume. In the 1980s, rural credit, and the demand for it, have both been greatly reduced.

Relatively large producers of export crops in the South, Southeast, and Center/West have been the principal beneficiaries of rural credit. Money creation to pay for the subsidies on rural credit contributed to inflation, a fundamental Brazilian problem of long standing.

Price and Income Stabilization

The government has attempted to stabilize prices, improve producer incomes, and reduce risk through its Minimum Price Program (MPP).

The MPP was first introduced in 1943, and remained relatively small until the late 1970s. Its role was to absorb market surpluses. With the cuts in credit subsidies in 1982, the MPP became the major form of government intervention in agriculture. Through its purchases of large stocks of key commodities at the guaranteed minimum price, the government has acquired large inventories of key commodities, including maize and rice, and has become a major holder of grain stocks.

The MPP covers 25 commodities, including rice, maize and soybeans. It is the most encompassing price-support program for the agricultural sector.

The MPP has multiple objectives: (1) to stabilize domestic prices, both between seasons and between years; (2) to insulate the domestic economy from external price variations; (3) to improve producer incomes by encouraging storage and off-season sales, and by establishing a floor price for the commodities covered; (4) to provide liquidity to agricultural markets, particularly to smaller producers and disadvantaged regions and, through subsidized interest rates, to reduce marketing and storage costs; (5) to reduce producer and processor risk; and (6) to compensate (especially through subsidized marketing credit) for other distortions in the agricultural economy.

The MPP works as follows: each year, the government posts floor prices for the major grains, and, in theory, the government has an obligation to purchase all amounts offered at the posted floor price. Its executing agency is the Production Financing Company (CFP), a semi-autonomous agency under the Ministry of Agriculture modeled after the U.S. Commodity Credit Corporation. CFP calculates and recommends minimum prices, buys and sells commodities under the price support program, contracts for storage and transportation, and monitors market and price information. The National Monetary Council makes the final decisions on minimum prices.

The principal instruments for the MPP are the government storage loans (EGF) and the government stock purchase program (AGF). Under the EGF, the government makes short-term credit available to producers, processors, traders and cooperatives, often at subsidized rates, with agricultural products as collateral, as an inducement for them to hold stocks after harvest. Under the AGF, the government guarantees to purchase any quantity of product at its minimum guaranteed price, established and announced well before the harvest, indexed for any producer until harvest time, and indexed for holders of EGF loans for the full period of the loan.

Within a specified time frame, those to whom EGF is available with a purchase option may convert to the AGF and may sell their product to the government. As a general rule, EGF with a purchase option is available only to farmers and cooperatives. Traders and agro-industry have access to EGF for most products, but without the option of subsequent sale to the government.

Those who hold an option for conversion of EGF to AGF will take this option if the market price falls below the guaranteed minimum price. For persons eligible for AGF, the guaranteed minimum price is indexed to inflation up to the time of eventual conversion of their EGF loans into government AGF purchases. As an added advantage, when the conversion is made the government is not only liable to pay the guaranteed and indexed minimum price, but also part, or all, of the cost of storage incurred for the entire period of validity of the EGF loan.

Minimum prices are set in August for the main commodities and are indexed for inflation. Recommendations on price levels made by the CFP to the National Monetary Council are based on a combination of analysis of production costs, domestic and international price trends, and production trends. The weight given to each of these factors is unknown, but it is clear that the government has often sought to influence the mix. Allocations of funds between AGF (purchases) and EGF (loans) also varies. CFP buys all commodities offered under AGF at the minimum price and sells them through the grain exchanges (through auctions), and to government institutions. Producers, cooperatives, traders and processors are eligible to apply for EGF at differing terms and in differing amounts depending on the crop. Interest rates are set by the government as part of official rural credit interest rates plus monetary correction.

The subsidized interest rates of the EGF loans have made them very popular. In 1970, when it began, the EGF program accounted for 6 percent of all rural credit; by the early 1980s, it accounted for 22 percent.

Since 1979, the MPP has grown greatly in importance, as a result of several factors. First, in 1979/80, the basis for annual crop credit was changed from prices to production costs. Before then, minimum price levels were restrained to prevent escalation of crop credit, which would have led to excessive monetary creation. Second, in 1981/82, minimum prices began to be indexed for inflation up to harvest, but not afterwards. This was a change from the previous system, under which minimum prices were announced in advance of the planting season, with an estimate of future inflation incorporated into the price. In 1984, indexing was extended to the harvest period. Third, with sharp cuts in subsidies for production credit in 1982, the MPP assumed increasing importance as the major form of government intervention for agriculture. Real resources spent on the MPP more than doubled between 1979 and 1986, and in particular, resources spent on AGF relative to EGF increased dramatically. Government stockholding and direct buying operations escalated.

The Price Band System. In 1987, in order to regulate stock sales, minimize market disruptions from government sales, and avoid political pressures on sales of AGF commodities, CFP introduced a price band system, which governed the release of its stocks to the market for major commodities, including soybeans, maize and rice. Intervention prices are calculated as 60-month averages of the wholesale price in Sao Paulo, with an upper price 12 percent above the target (17 percent for beans) and the floor price at the minimum price plus 5 percent (to compensate for differences between farmgate and wholesale prices). Government AGF stocks must be sold at or above the floor price to prevent dumping of stocks, provided that sales do not entail a direct subsidy, i.e., that sales revenue covers carrying costs. The bands are also intended to serve as reference prices for compensating tariffs for imports and exports.

As shown in Table 10.6, year-to-year variations in allocated funds to EGF and AGF have been large.

As a consequence of the Summer Plan of January 1989, the subsidized interest rates on CFP's credits to finance the AGF program were abolished. Henceforth, CFP has to pay market rates. These rates are strongly positive with respect to inflation (inflation was running at about 40 percent per month by mid-1989). As a result of these changes, the carrying costs of the government's grain inventories may increase more rapidly than the market prices of the same grains.

Funds available for the MPP programs are increasingly limited as a result of Brazil's economic crisis. CFP has calculated that NCr. (Novo Cruzados) 13 billion (of mid-August 1989) would be required to finance all the forecast purchases under the AGF program for 1990. The government has announced that only 60 percent of this amount will be available under its program. It is unclear how this money will be rationed.

TABLE 10.6   Brazil—Yearly Allocations to Government Price Support Operations, EGF & AGF, 1980—1987 (millions of constant Cruzados of 1987)

Year	EGF		AGF		Totals
	Value	% of Total	Value	% of Total	
1979	50,876	91.68	4,616	8.32	55,492
1980	70,450	96.00	2,932	4.00	73,382
1981	86,796	88.93	10,800	11.07	97,596
1982	91,439	69.36	40,389	30.64	131,828
1983	55,107	84.32	10,247	15.68	65,354
1984	24,158	72.25	9,277	27.75	33,435
1985	51,908	41.28	73,840	58.72	125,748
1986	76,595	60.46	50,095	39.54	126,690
1987	49,944	61.32	31,506	38.68	81,450
Average annual growth rate :	-3.57%		36.28%		3.92%

Source: Production Financing Company (CFP)

The ability of the MPP to meet its many, and sometimes contradictory, goals has been mixed for several reasons: shifts in government perceptions of the use of the MPP; the inconsistency of the program's goals; important changes in design; variations in real resources assigned to the MPP; and the impact of trade and price control policies outside the MPP.

Impact on Small Producers. One of the justifications for the MPP, particularly for subsidized credit, has been to improve access to marketing finance for smaller producers, especially those in remote or disadvantaged regions. The MPP has not fully met this goal. The commercial farmers, traders and processors of soybeans and cotton in the South, Southeast, and Center/West of the country have been its chief beneficiaries. Between 1970 and 1988, these areas accounted for 85 to 97 percent of EGF operations, and for 81 to 96 percent of the AGF. Nor was the EGF/AGF distributed evenly according to crop production shares. Although the Center/West accounts for 24 percent of national rice production, it received 37 percent of total EGF/AGF financing for that crop. The Center/West (the frontier), considered disadvantaged relative to the South, has benefited from the MPP in recent years, whereas the Northeast, clearly a disadvantaged area, has not. Although the Northeast has about 18 percent of national rice production, it received only 3 percent of EGF/AGF resources going to rice. Most of EGF financing has gone to the industrial crops, soybeans and cotton, not to maize and rice, although soybeans have traditionally had the greatest access to overseas and domestic marketing finance. For example, between 1980 and 1987, soybeans and cotton accounted for 50 percent of EGF financing, while corn and rice accounted for only 29 percent.

Price Stabilization. The MPP aimed to stabilize prices, both between seasons (by encouraging storage and thereby smoothing out price peaks through EGF), and between years (by guaranteeing floor prices to producers through AGF). The record of the MPP has been mixed in this area as well, in part because of the large variability from year to year in price levels and in resources devoted to the program.

Overall, minimum prices do not appear to have stabilized market prices throughout the year. This was particularly true before minimum prices were fully indexed to inflation; without full indexation, they deteriorated in real terms after harvest. Ratios of market prices in the off-season to those at harvest time show great variation between years as well as between commodities (Table 10.7). In many years, mid-season prices for maize, rice and soybeans, in real terms, are less than harvest prices, indicating that returns to stockholding would have been negative. It could be said that this variability of market prices is a strong argument for the MPP. However, the MPP existed throughout the 1970–1988 period and seems to have had little impact. Particularly in the years between 1984 and 1987, when official government purchases under the MPP were at their greatest and official minimum prices at their highest, the market ratios were almost consistent-ly below 1.00.

Price stabilization between years has also not been achieved. Between 1972 and 1988, real minimum prices for maize and soybeans fluctuated more than farmgate prices (Table 10.8). Farmgate prices can be expected to fluctuate from year to year, but the inconsistency of minimum prices has introduced additional uncertainty. However, minimum prices for rice have been more stable than market prices.

Changing objectives for the MPP over the last two decades have con-tributed in a major way to the variability in real minimum prices. Before 1979, the objective was to restrain credit subsidies and domestic food pric-es; thus, minimum prices were kept low. In 1984 and 1985, by contrast, real minimum prices were increased significantly to compensate the agricul-tural sector for the withdrawal of credit subsidies and to encourage do-mestic food production, which had stagnated in the early 1980s.

Apparently, there is no consistent policy on setting minimum prices. Comparisons of minimum prices with costs of production and with border prices show significant instability among years.[3] Although CFP's technical analysis and resulting recommended prices are based on estimated pro-duction costs and forecasted international prices, a range of other factors

TABLE 10.7   Brazil—Maize, Rice, and Soybean Ratios of Off-Season to Harvest Prices, 1974–1987 [a]

Year	Maize Farmgate Paraná	Maize Wholesale Sao Paulo	Rice Farmgate RGS	Soybeans Farmgate Paraná
1974	0.97	1.06	1.19	1.10
1975	1.21	1.01	1.14	0.95
1976	0.93	0.93	0.77	1.34
1977	1.01	1.00	0.98	0.70
1978	1.10	0.90	1.06	0.99
1979	1.32	0.99	1.33	1.20
1980	1.81	1.72	0.95	1.23
1981	0.81	0.76	n.a.	0.93
1982	0.85	0.97	0.78	0.87
1983	2.52	2.08	1.16	1.97
1984	1.04	1.02	1.17	0.87
1985	0.91	0.99	0.83	1.04
1986	0.88	0.97	0.91	0.99
1987	1.04	1.36	0.97	1.42

a   Off-season prices are averages for October and November.
Harvest prices are averages for March and April.
All prices were corrected to constant terms of November 1984.
Source: World Bank, Brazil Agricultural Incentives and Marketing Review, 1989.

TABLE 10.8   Brazil—Indices of Price Instability for Major Grains, 1972–1988 (standard deviation as a percentage of the mean)

	International Prices [a]	Official Minimum Prices [b]	Farmgate Prices
Irrigated rice	38.1	12.8	20.7
Upland rice		13.7	22.8
Wheat [c]	27.5	20.1	20.1
Maize	20.6	16.9	15.8
Soybeans	23.0	22.0	17.5

[a]   Rice: Thai white 5 percent brokens, FOB Bangkok.
Wheat: Canadian no.1 Western Red Spring.
Maize: US no.2 Yellow, FOB Gulf ports.
Soybeans: US, CIF Rotterdam.
[b]   Irrigated rice, soybeans and wheat: Rio Grande do Sul; upland rice: Maranháo; maize: Paraná.
[c]   Minimum prices are the same as farmgate prices due to government monopoly on marketing.
Source: World Bank, *Brazil Agricultural Incentives and Marketing Review,* 1989.

comes into play at the decision-making level. Overall, the evidence suggests that the MPP's impact on domestic price stabilization has been limited.

Impact on Producer Incomes. An analysis of the MPP's impact on producer income is more elusive since product price is only one of several variables that contribute to income. Systematic information on other determinants of total income—such as inputs used, and their costs, yields and labor productivity, access to credit and availability of risk management instruments—is not available. Nevertheless, when price stabilization of the MPP is looked at alone, regional distribution patterns of EGF and AGF show that these programs have primarily benefited producers in the most productive areas of the country (Table 14.8).

To a large extent, small farmers are the major producers of traditional foodcrops; certain regions, such as the Northeast, mainly produce traditional foodcrops. The MPP is not, and never has been, designed to meet the needs of small subsistence farmers; it is fundamentally a program oriented toward commercial farmers. In recent years, the MPP has clearly played an important role in opening up the frontier areas and has doubtless contributed to higher producer incomes in those regions, primarily through subsidizing transport costs. Minimum prices are set on a panterritorial basis, making them extremely attractive in the interior, where underdeveloped and costly transport decreases producer returns. The greatest proportion of AGF purchases, particularly for soybeans, maize and rice, has been in the Center/West, where distance from export and processing facilities result in lower market prices at the farm gate (Table 14.8). Consequently, minimum price schemes are much more attractive to farmers located a greater distance from export and processing points.

Effects on the Markets. Since financial costs, including those of post-harvest stock financing, are a major part of food marketing expenses in Brazil, EGF may have restrained marketing costs somewhat. However, another major part of marketing costs are taxes, primarily the ICM.

The major impact of CFP stocks on the markets has been in the uncertainty engendered by large inventories overhanging the market, with unclear rules about their disposal. In 1986, CFP had nearly 2 million metric tons of rice (20 percent of domestic consumption) and 1.5 million metric tons of maize in stock. However, in 1986 the government banned external trade in rice and maize, except for official imports; in 1987, government stocks swelled to 3.6 million metric tons of rice (40 percent of domestic consumption) and nearly 4 million metric tons of maize (15 percent of domestic consumption). The government became the largest single stockholder in the markets for these two commodities. The government's overwhelming weight in domestic rice and maize markets, coupled with the 1986 price freeze and bumper crops, served to depress domestic prices.

Buffer Stocks

Brazil's buffer stock (or regulatory stock) program has 3 major goals: to assure supply of food to areas where transportation and marketing channels are poor, including provisioning of some social programs, such as school lunches; to regulate prices through sales from government stocks; and to facilitate the supply of low-cost inputs to food processing industries. The major grain products covered under the program are milled rice and maize, and soybean oil.

The current buffer stock program originated in the Cruzado Plan period, 1985, when government developed a policy to dampen urban consumer price increases through selective sales of commodities. Stocks are stored by COBAL, a government-owned food store chain. The program is not large by Brazilian standards, representing less than 500,000 metric tons of stock by the end of 1986 and 136,000 metric tons by end 1987, but it requires subsidies for its operation.

The Wheat Program

The government's monopoly of domestic wheat marketing dates from 1962. In developing policies for wheat, the government has tried to subsidize both producers and consumers. For producers, the aims have been: to support farm incomes, to reduce the foreign exchange costs of wheat imports by increasing domestic production, and to compensate for subsidies to farmers in other countries. For consumers, subsidies were introduced to offset high world wheat prices in 1972. They have been retained for two reasons: to fight inflation (wheat products weigh heavily in the consumer price index) and to provide food security for low-income groups. Though consumer subsidies were first introduced as a short-term measure, they have become entrenched.

Since the introduction of a consumer subsidy for wheat in 1972, costs of the program have risen drastically. In 1980, government receipts covered only 20 percent of outlays for the program. In 1982, when domestic prices were US$322/mt, import prices (CIF port) were only US$170/mt. At that point, disease problems and other factors had caused domestic wheat production to stagnate. The government, in the face of a budget deficit, tried to reduce producer and consumer subsidies, but its proposals were not fully implemented.

To help the program cover its costs, in 1987 the government raised the consumer price of flour by over 500 percent, reduced the producer price from US$240/mt to US$188/mt, and raised the price to millers. However, further price freezes and lags in price adjustment led to more subsidies.[4]

Combined Cost of the Support Programs

The average yearly costs to the Brazilian treasury of the MPP, the buffer stock program, and the wheat subsidy program between 1984 and 1988 has been estimated as follows: wheat US$1.5 billion; MPP US$0.6 billion; and buffer stocks US$0.3 billion.[5]

Other Policies Relevant to the Grain Sub-Sector

Public Storage

The government's focus on storage investment dampens private initiative. On-farm storage, in particular, has not been well developed.

Stocks acquired under AGF, as well as buffer stocks, are kept in government storage. The public storage companies own about 20 percent of Brazil's estimated 60 million metric tons of storage capacity and rent additional storage, thus making use of about 50 percent of the country's total storage capacity. CIBRAZEM, the federal storage company, concentrates its activities in the Center/West, North and Northeast, and has tended to discourage private investment in those areas due to its domi-nance and its subsidized rates (Tables 14.13 and 14.14).

Transportation

Large investments made in road transport, particularly in the late 1960s and early 1970s, helped stimulate agricultural development. Since the mid-1970s, public investment in roads has been greatly reduced, and the road network built earlier has deteriorated rapidly. The density of state roads (0.17 km per square km) is low. These roads are important for the transport of agricultural inputs and products. As a consequence of high-way neglect, the cost of transporting grains is high. High costs particularly affect the export competitiveness of soybeans.

Seed Production and Distribution

Brazil seed legislation is supported by Act No. 6507 (1977), Decree No. 81771 (1978) and several (more than eighty) Federal Ministerial Resolu-tions. Most of the Brazilian states have their own state legislation—State Resolutions; some of them, such as Sao Paulo, pioneered seed legislation in the country during the 1920s. Nevertheless, Brazil has no plant breed-ers' rights enactment, and there are high concerns raised in the country on that matter between groups favoring and opposing proprietary rights on plant varieties.

Breeding institutions are mostly supported by public funds. The private sector is scarcely involved in soybean variety development. EMBRAPA, the national agricultural research institute, and some state and university breeding institutions are the main developers of soybean varieties. Coop-erative organizations multiply and market public soybean varieties. There are some private soybean varieties developed by private companies but international firms are reluctant to release open pollinated cultivars in the Brazilian market due to lack of protection. Enactment of plant breeders' rights in the future could emphasize Distinctiveness, Uniformity and Sta-bility (DUS) for variety registration.

Seed firms and farmers seldom rely on seed certification. Even PLA-NASEM, a national plan for seed politics and improvement of seed provi-sion enforced some years ago, did not foresee a generalized use of certified seed. Seed-related agents do not trust government-guaranteed seed and prefer more emphasis on seed controls than on seed production.[6]

New soybean varieties in Brazil must be approved by a commission ap-pointed by the Minister of Agriculture. There are, in fact, two commis-sions: one for the southern part of Brazil and the second for the remainder of the country. The two commissions test and approve varieties for release in each of these regions. The Commission is composed of one representative from each of the following groups: the Ministry of Agriculture, EMBRAPA, the State Research Organization, the State Extension Service, and the seed producers of the country.

The procedure for testing includes two years of preliminary testing performed by the breeder, followed by one year of intermediate testing at five locations in Brazil. The best lines from these five locations are sent for final testing at 10 locations over a two-year period. The Commission then meets to discuss the characteristics of each variety and decides which ones will be released. The decision is then published in the official newspaper. The Commission reviews the criteria of yield, stability, disease resistance and agronomic characteristics. It will not release a variety unless it is equal to or better than the two varieties selected as the standard.

The two varieties for the standard are selected to represent four maturity groups. The two best varieties in each maturity group become the standards. One variety is selected for its highest yield, the second because it is the most popular currently being planted in the region. Oil and protein content are identified, but release of new varieties has not been restricted for lack of higher oil and protein. Brazil has the potential for controlling varieties to meet a gradually rising standard of quality with respect to oil and protein, but in practice this criterion is not being applied.

Notes

1. See The World Bank, Brazil Agricultural Incentives and Marketing Review, December 1, 1989, p.77.

2. See The World Bank, Brazil Agricultural Incentives and Marketing Review, December 1, 1989, p. 56.

3. See The World Bank, Brazil Agricultural Incentives and Marketing Review, December 1, 1989.

4. This analysis of the wheat program is drawn from the World Bank, Brazil Agricultural Incentives and Marketing Review, December 1, 1989.

5. World Bank, Brazil Public Expenditures, Subsidy Policies and Budgetary Reform, June, 1989.

6. J.M.F.J. da Silveira, Progreso Tecnico e Oligopolio : as especificidades da Industria de Sementes no Brasil, [Mimeo], UNICAMP, Campinas, Brazil, 1985.

11

Institutions Affecting the Grain Sub-Sector

Public Institutions

Many public-sector institutions, from federal ministries to regional and state agencies and grassroots organizations, are involved in the grain sub-sector. Policies and funding tend to come from the central government, but programs are frequently implemented by state and local authorities. Because of the large number of institutions, each with its own administration and operating procedures, coordination of agricultural programs is cumbersome and complicated.

In 1988, at least 8 institutions at the ministerial level dealt with agriculture, including grains—the Ministries of Agriculture, Finance, Planning, Industry and Trade, Interior, Irrigation, Justice and Agrarian Reform.

Two ministries heavily involved in the grain sub-sector are Finance and Agriculture. The Ministry of Finance is the largest government spender on agriculture: in 1986 and 1987, its credit and subsidy programs were responsible for two thirds of federal spending on agriculture. This ministry also sets producer prices for wheat, and consumer prices for bread and flour. The Ministry of Agriculture is administratively responsible for the Production Financing Company (CFP); this ministry has several specialized agencies, including the research agency, the Brazilian Enterprise for Agriculture and Livestock Research (EMBRAPA), and the extension agency, the Brazilian Enterprise for Technical Assistance and Rural Extension (EMBRATER). In 1986 and 1987, it accounted for 16 percent of government spending on agriculture.

A mixed public/private sector commission, the National Monetary Council and the National Council for External Commerce (CONCEX), makes formal decisions about trade policy. However, many important trade policy decisions, e.g., imposing bans or quotas, are made at the administrative level. This situation may change under Brazil's new constitution, as there may be a movement away from administrative law, favoring a greater role for the legislature in creating trade legislation.

Because the Federal Trade Agency (CACEX) has responsibility for the foreign trade portfolio of the Central Bank, it controls the trade of commodities of primary domestic importance. CACEX is responsible for administering nearly all of Brazil's trade in the agricultural sector, except for coffee and sugar. Its responsibilities include issuing import and export licenses, registering imports and exports, and price checking. All exporters and importers must register with CACEX.

The National Monetary Council formulates rural credit policy. Until 1987, the Central Bank administered all official rural credit. Administration of official credit was transferred to the National Treasury in 1988 to enable tighter budget control over official rural credit and credit subsidies. In the same year, an inter-agency Credit Limits Committee (GLC), made up of representatives from the Central Bank, the National Treasury and the Secretariat of Special Economic Affairs of the Ministry of Finance, was established to determine the allocations of official credit to the banks.

In 1979, the Federal Price and Supply Secretariat (SEAP) was created in the Ministry of Planning to centralize authority for determining prices. SEAP plays an important role in price control. However, in the 1980s this was mostly done through "gentlemen's agreements" and buying and selling of stocks by SEAP.

In early 1986, under the Cruzado Plan, an Interministerial Price Council (CIP) was established in the Ministry of Finance. CIP administers price controls throughout the economy. In early 1987, price controls were mostly lifted, but were reimposed on a short-term basis in June 1987. CIP continues to watch over price increases and has intervened from time to time.

The National Wheat Buying Commission (CTRIN) has a monopoly over the domestic marketing and storage of wheat, and over wheat imports. Import contracts for wheat are made by the National Superintendency of Supply (SUNAB) in the Ministry of Finance. Imports come primarily from Argentina under a five-year trade agreement.

CTRIN was created in 1962 and given monopoly status for handling domestic and imported wheat to reduce abuses in wheat trading. CTRIN is responsible for the storage of wheat, primarily through cooperatives and state storage companies and for transfer of wheat to millers.

Wheat millers need quotas for supplies from CTRIN, which allocates the quotas on the basis of the mill's capacity and the availability of storage. CTRIN delivers the wheat to the mill. The wheat milling industry is concentrated—20 percent of the 179 wheat mills account for 80 percent of the throughput—and is sheltered from competition. The industry has sufficient capacity but needs modernization.

The Brazilian Federal Storage Company (CIBRAZEM) stores grain stocks acquired under AGF purchases, as well as buffer stocks. CIBRAZEM concentrates its activities in the Center/West, North and Northeast. CIBRAZEM and the state storage companies own about 20 percent of Brazil's estimated 60 million metric tons of storage capacity and rent additional storage, thus making use of about 50 percent of the country's total storage capacity. CIBRAZEM was created by a merger of two storage regulatory agencies in 1972. Its role was significantly expanded in 1975 under the National Storage Program, PRONAZEM. Its market share in the South is very small (between .04 and 1.8 percent of capacity). However, in the North it has the lion's share, 76.5 percent in Rondonia and 91 percent in Roraima. Its policy is to concentrate its activities in deficit regions, considered unattractive to private investors. It has discouraged investment in storage in those areas due to its market dominance and subsidized rates. But in the past few years, it has divested some of its storage capacity to state storage companies.

CIBRAZEM's responsibilities are: (1) to participate directly in planning and executing government storage programs; (2) to regulate agricultural markets; (3) to provide storage services in deficit areas; and (4) to manage the federal storage system. It may legally function as a general warehouse company; and can issue warrants, participate in and coordinate state warehouse companies, and collect data.

A number of weaknesses, including poor management, underpaid staff, and lack of well-defined policies, prevent CIBRAZEM from providing efficient service. In 1986, it had a total deficit equivalent to about US$4 million, mostly due to operational losses.

Agencies of the Ministry of Agriculture

The National Supply Secretariat (SNAB) is responsible for inspection, supervision, classification and grading of grains, and for technical studies.

The Production Financing Company (CFP). CFP, a semi-autonomous agency under the Ministry of Agriculture, modeled after the United States Commodity Credit Corporation, is the executing agency for stock loans (EGF) and stock purchases (AGF) made under the Government's Minimum Price Program. CFP calculates and recommends minimum prices, buys and sells commodities under the price support program, contracts for storage and transportation, monitors and publishes market and price information, and does ad hoc studies about topical issues related to the marketing of agricultural products. It does not own or operate grain handling facilities.

The Brazilian Enterprise for Agriculture and Livestock Research (EMBRAPA). EMBRAPA, founded in 1973, is a public company attached to the Ministry of Agriculture. Its principal objectives are to improve the quantity and quality of food supply and the production of raw materials for agro-industries and to improve the productivity and management of all factors of agricultural production. Its new master plan gives high priority to basic foods—rice, maize, wheat and soybeans among others.

EMBRAPA operates 44 national commodity and regional centers as well as state and territory-level research units. It coordinates a cooperative research system that involves universities, producer cooperatives, scientific and technical organizations, and private industry. EMBRAPA is funded from the federal budget but has considerable financial and administrative autonomy. It supplements its income by doing contract research for other public and private bodies and by the sale of improved plant material and livestock. Its staff of over 21,000 includes over 4,000 technical, 11,000 support and 6,000 administrative staff. It is regarded as one of the best-endowed agricultural research systems in the developing world. In 1987, 6.8 percent of its research expenditures went to maize/sorghum, 4.9 percent to soybeans, 6.2 percent to rice and beans, and 4.6 percent to wheat.

EMBRAPA's major achievements include soil treatment for the cerrados, improved yields for wheat and improved soybean varieties. The development of tropical soybean varieties has enabled this crop to be grown within seven degrees of the equator and has led to greatly increased production. Weaknesses of the agency are related to political pressures, shifts in programs due to late releases of funding, and failure to develop technologies to benefit resource-poor and subsistence farmers.

Agencies Handling Market Information

Statistics are gathered by a wide number of agencies. The agencies that gather market information include the Brazilian Institute of Geography and Statistics (IBGE), which collects mainly production and storage data; FGV, which tracks consumer and producer price movements; CFP, which mainly collects market data for its own use; and CIBRAZEM, which collects information on storage.

The Brazilian Institute of Space Research (INPE) is developing ways of improving crop forecasting through remote sensing, beginning with a pilot project in Parana. IBGE, which will participate jointly in these projects, is also improving its own system of crop data collection.

Private Institutions

Cooperatives

The cooperatives play an important role in grain purchasing, storage, and marketing. Several successful cooperatives have been established by groups of Japanese and European immigrants. The largest of these is COTIA, a cooperative founded by Japanese immigrants. COTIA's headquarters are located close to Sao Paulo.

Cooperativism is furthest developed in the traditional grain-producing areas of the South. Pioneer farmers on the new lands in the Center/West and the North, many of whom come from the South, have often started their ventures within a cooperative structure. Founding members are often sons and daughters of leaders of the Southern cooperatives, who have moved to new areas as they have no access to land in the area of their parents' farms.

Cooperatives are most important at the primary level of marketing, where they receive product directly from the farmer. They are less important in the subsequent phases of commercialization, particularly exports. However, cooperatives are increasingly integrating forward, entering into rice milling and soybean crushing. The largest cooperatives market end products under their own brand names. Several cooperatives are organized into regional and national entities, which provide coordination from input supplies to final products in many agriculture-related products.

The availability of storage facilities at the cooperatives and the Government requirement that grain must be supplied to accredited warehouses for eligibility under the EGF and AGF schemes have been driving forces in the development of cooperatives. On-farm storage for grains is limited in Brazil to an estimated 3 percent of all storage, as compared with about 60 percent in the United States.

The government has supported cooperative development in several ways. It has: (1) supplied credit to cooperatives on a preferential basis; (2) failed to accredit on-farm storage for eligibility under government credit schemes; and (3) bowed to the political clout of large cooperatives when scarce resources, for instance, under the EGF/AGF programs, have to be rationed. Cooperatives have become so strong that in Rio Grande do Sul, for instance, rice marketing is considered to be a cooperative's monopoly. In Mato Grosso, recent settlers have been unable to start rice production because the local cooperative does not admit new members as it cannot handle (store) their products. The MPP program has thus had an important indirect influence on the development of cooperatives in Brazil.

Interest Groups

There are a number of trade associations for the various agricultural inputs, including ABRASEM (seeds); ANDA and SIAGESP (fertilizers); ANDEF (chemicals); and ANFAVEA, SINDIMAQ/ABIMAQ and SINFAVEA (farm machinery).

The Brazilian vegetable oil industry, dominated by soybeans, has an association, the Vegetable Oil Industry Association (ABIOVE). The association is headed by a president who is usually an upper echelon manager of one of the major crushers. ABIOVE has an economic research staff of about 20 professionals who collect and disseminate performance data on the industry. Its research office analyzes topical issues of importance to the industry and provides the background information needed for presentations by ABIOVE's president and for the association's lobbying efforts. ABIOVE also provides information to its members through a monthly newsletter, in which statistical data on production, exports, and crush are published. Its lobbying efforts include development of production costs, which are used in establishing minimum prices under the government price support program for soybeans. The combination of quantitative analyses done by ABIOVE's staff and the aggressive defense of the industry by its president appear to result in effective lobbying. For instance, soybean products have been excluded from Brazil's trade agreement with Argentina, ostensibly as a result of ABIOVE's intervention

Grain Exchanges

Grain exchanges have existed for many years in Brazil, but the present institutions deal with only a fraction of the grain trade. Almost every grain-producing state has an exchange. The largest is in Sao Paulo. Most of the grain trade at these exchanges is Government auctions of stocks held by the CFP.

Numerous brokers operate individually in an unorganized market, particularly in Sao Paulo. Though many have offices in the Sao Paulo Exchange, they mostly operate outside its structure.

There are several reasons for lack of public interest in the exchanges. A major one is that it is easier to avoid ICM taxes by by-passing the exchanges with their open bidding system. It also makes it easier to conceal third-party payments and other irregularities.

The exchanges appear to be run by a small côterie of friends and acquaintances, mostly brokers. This group does not attract people outside their inner circle. Rivalry between exchanges inhibits cooperation or the exchange of information.

The exchanges may soon lose the sale of CFP-owned stocks, the only substantial activity remaining to them, as the Government is considering withdrawing from the grain trade. In an effort to revitalize its functions, the Sao Paulo Exchange is proposing: (1) installation of a computerized trading system, to be operated by the Exchange, with fees charged for real time access; and (2) institutional changes that would channel the grain stocks under EGF agreements through the exchanges for sale.

Under the first proposal, the Exchange would set norms, establish guidelines, and act as the clearing house for all transactions. Automation of sales would capitalize on the recently installed nationwide phone and telex communication system.

The second proposal would allow sales of EGF-financed stock through the exchanges before the ending date of the EGF contract. At present, EGF financing essentially freezes a substantial inventory of grains out of the market for 3 to 8 months after harvest, until the due date of the loan. If a producer or cooperative wishes to sell grains with an EGF lien, the EGF loan has to be paid off first. Few farmers have enough borrowing capacity to repay their EGF loans in full with other loans. If the holders of inventories under EGF convert to AGF, or default on their loans, it may take another six months before stocks are freed from sales on CFP-organized auctions. Thus, under the prevailing EGF/AGF system substantial parts of each grain harvest are withheld from the market for long periods.

The Exchange proposes to allow sales of EGF commodities through the exchanges, which would act as clearinghouses, with responsibility for repayment of the loan and remittance of the balance to the owner of the grain. This proposal would strengthen the exchanges by channelling more business through them. At the same time, it would alleviate the serious market distortion that results from the freezing of stocks under the EGF. An issue as yet unresolved is the handling of any subsidy, in cases where the market price is below the minimum guaranteed price that underlies the EGF loan. However, the question of a subsidy may be academic in this context since, presumably, the grain will only be put up for sale if the owner can expect to receive a positive balance after the transaction.

Soybean Processors and Exporters

The soybean processing and exporting firms are a mixture of cooperatives, independent private firms, and multinationals. All the major multinational grain companies are involved in some phase of assembly, processing, and marketing of the grain. The cooperatives handle a large share of the processing and local assembly of soybeans.

12

Grain Marketing

Most agricultural commodities in Brazil are marketed by the private sector. However, there is substantial government intervention in the marketing of grains, particularly in the marketing of wheat, maize, rice and soybeans. Wheat marketing is controlled by a government monopoly. For maize and rice (considered major domestic food items), government price supports, marketing credit, management of stocks, trade restrictions, and consumer price controls are key determinants of market structure and price. Soybeans are subject to government intervention as domestic food items (soybean oil), industrial raw materials, and exports. Circumvention of export controls is sometimes extremely profitable, and unusual marketing structures have developed as a result.

Marketing Channels

The marketing channels used vary by region and by type of grain. Grains are generally marketed through the private sector in the South and Southeast. In the frontier areas, the government has become the major buyer of maize, rice and soybeans, through its AGF program. In all regions, the government is the buyer of last resort for corn, rice and soybeans, and the only buyer for wheat.

Wheat

CTRIN, the government wheat monopoly, purchases all domestically produced wheat.

Maize

Most of the maize grown in Brazil (90 percent) is for animal consumption. There are many maize producers, both large and small, and many processors of maize for feed or food. Producers sell their maize to cooperatives or traders, directly to processors, or to the CFP.

In the South and Southeast, which account for about 70 percent of production, the maize is sold directly to a cooperative or to industrial users such as the poultry industry or feedstock manufacturers. If the minimum price is above the market price at the time that the producer wants to sell, the CFP may purchase a large part of the harvest.

In the Northeast, which accounts for about 10 percent of production, there are few large institutional buyers. Itinerant trucker traders purchase the maize in bags at the farm gate, for cash. In the state of Pernambuco (Northeast), most of the maize produced goes to the local poultry and pig industries; the rest goes to the animal feedstock industry, or is milled for human consumption. Even in months of peak harvest, only 80 percent of the state's demand is covered by local supplies. The states of Bahia (Northeast) and Goias (Center/West) supply Pernambuco with maize, which is brought in by truck.

Rice

Rice producers in Rio Grande do Sul sell directly to cooperatives, which usually have their own mills, or to private millers. The state has 34 cooperatives and more than 400 millers. Farmers rarely store rice at the cooperative for future sales. Most of the larger cooperatives and independent millers sell the milled rice directly to supermarkets or institutional users, either under their own or under the customer's brand name. The wholesale market has many sellers and only a few buyers. Price information is poor, and the cash markets (bolsas) are rarely used, except for auctions of government rice stocks.

In other areas of the country, the government buys most of the rice crop under the AGF program. Short-grain rice is often used to open up new lands in the Center/West. There is only a limited demand for this rice. Urban consumers have a marked preference for long-grain rice.

Brazil last exported substantial amounts of rice in 1977 (almost 410,000 metric tons) and a smaller amount in 1978 (about 180,000 metric tons). Small, but not significant, exports have been made every year since then. At present, Brazil is an unknown quantity in the international rice market and has no established trading channels for rice exports.

A trade agreement signed with Argentina and Uruguay during the period of the Cruzado Plan in 1986 is interpreted as implying a Brazilian obligation to import 400,000 metric tons of rice yearly from these two countries. Imports of agricultural products are to be offset by exports of manufactured goods.

Soybeans

The soybean market is one of the most open and competitive commodity markets in Brazil. Much of the crop is marketed through cooperatives, which are very strong in the South. Soybeans are also marketed to processors or traders. Nearly all of the soybeans produced are delivered at harvest directly from the farm to either country elevators or processors. Almost the entire soybean crop (except for a portion sold to CFP) is sold directly to industry and exporters. In rare cases, they may be delivered direct from farm to port. However, the majority of the soybeans delivered to ports are sold through cooperatives or private elevators and are delivered in the elevator's name. Some large producers sell beans directly in their own name.

Since over 80 percent of annual production goes into local processing plants, the marketing channels are directed toward supplying these plants with their monthly crush requirements. About 17 percent of the crop is exported in the form of beans. About 76 percent of the soybean meal, and 27 percent of the oil produced from the remaining 80 percent of the crop, are also exported (Table 9.7). Exports are seasonal—the harvest-time surplus moves into the export channel—and are therefore concentrated in a relatively short period. Over 75 percent of the beans exported move into world markets between April and August, and over 90 percent by the September following spring harvest. Meal is produced and exported year-round.

Trading and crushing of domestic beans starts in early April and lasts through September. Exports of cake and oil last until December. Brazil usually imports beans in October, between harvest periods, mainly to maintain supply for the local oil market. The beans are imported under drawback arrangements to be crushed from October through March of the subsequent year. Drawback arrangements typically allow for 180 days to re-export the cake produced, although the terms for this trade have varied as a result of domestic economic conditions. Imports are usually ordered in October, November and December, while re-exports are made in January and February. A small amount of beans, estimated at 300,000 metric tons, is used for domestic consumption in raw form, mainly by the Japanese community.

Grain Handling and Transportation

Drying

In most of the grain-growing regions, high humidity is a problem at harvest time. The grains must be cleaned and dried before storing, both on the farm and in storage warehouses. Drying helps reduce transport charges, which are very high, particularly in some of the more inaccessible regions. Many warehouses provide cleaning and drying services.

A high proportion of the soybean crop must be artificially dried. Since few farmers have either drying or storage facilities, the country elevators and processors provide the capacity for drying as well as storage. Fuel for drying is primarily wood or coal, while some firms in areas where sugar cane is produced use bagasse for fuel. Soybeans are normally handled in bulk and moved directly from farms into storage or processing during the harvest period.

High humidity and high temperatures at harvest time and the high moisture content of maize makes drying essential to ensure an acceptable quality. The standard for maize for humidity in the market channel is 13 to 14 percent. Maize that is used on the farm where it is grown, as feed for livestock, is generally dried on the ear prior to shelling and use. Rice also requires careful drying, with special attention to breakage associated with high drying temperatures and rapid removal of moisture.

Storage

The storage capacity of soybeans is provided primarily at the cooperatives or processors. There is a shortage of adequate storage space for maize, which often presents a problem, particularly in years of high production, even though a relatively small proportion of the total crop is moved into the market channel. Rice farmers have an undetermined amount of on-farm storage, built in the 1970s using funds from PRONAZEM, a World Bank-funded line of credit for the construction of on-farm storage. Apparently, these storage facilities have been certified by CIBRAZEM, and rice stored in these facilities is eligible for EGF financing.

Brazil has an estimated 60 million metric tons of usable storage capacity. In years of bumper grain harvests, storage capacity is insufficient. In 1986/87, for example, grain production reached a record 65 million metric tons Dynamic agricultural growth in the frontier areas and shifts to new crops (soybeans and maize) have contributed to the storage shortage. The ratio of storage capacity to production is estimated at 0.8, including marginally acceptable storage facilities. By contrast, the ratio in the United States is 1.5. FAO recommends a minimum ratio of 1.25 in order to provide adequate storage for fluctuating harvest volumes. Averaged over several crop years, a ratio of one is conceivably acceptable, since that would provide one metric ton of storage capacity for each metric ton of production. However, this would be acceptable only if storage and production were proportionately distributed throughout the entire geographical region. Variations in crop production from year to year with fixed storage capacities in their locations means that storage supply never matches exactly with storage demands.

The main shortage is of bulk-storage facilities. In 1986, this was estimated at about 29 million metric tons (Table 14.13). For bag facilities, there is a total excess supply of about 15 million metric tons, mostly in the South and Southeast.

The South and Southeast have the best overall storage situation. Together they account for 70 percent of agricultural production and 80 percent of storage capacity. The Northeast accounts for 14 percent of production, and has only 4 percent of capacity.

The private sector owns about 80 percent of the storage facilities in Brazil. The public storage companies own about 20 percent, but the government rents additional storage, making use of about 50 percent of the country's total storage capacity (Table 14.14). About half of all storage capacity is reserved by agro-industries, marketing companies, and the Brazilian coffee and sugar institutes for their own use.

General warehouse companies have traditionally been strong in the port areas and adjoining cities. The warehouse companies and the public storage companies are the major suppliers of storage in the frontier areas, particularly the Center/West. In the South, cooperatives and agroprocessing facilities provide much of the storage.

On-farm storage is inadequate. Brazil has no legal standards for warehouse construction. Because of the lack of standards for on-farm storage, it is difficult for farmers to qualify for government crop-financing unless they store their grain off the farm, usually at a cooperative or general warehouse. Large quantities of grain are stored under government EGF loan, or by the government under AGF purchases. In peak years, and in certain regions, government agencies are forced to accept substandard or emergency storage. Some large storage losses are attributed to storage in open-air "swimming pools" made with walls of bagged grain, with bulk grain in the center.

Transportation

In 1987, road haulage accounted for 74 percent of all movements of soybeans and soybean products (compared with 16 percent in the United States), rail transport for 23 percent, and river transport for 3 percent. The average transportation costs for moving soybeans to export ports in 1989 were about US$30/metric ton, about 20 percent of ex-farmgate prices. Due to increased production on the frontier lands, the average transportation distance for moving soybeans to port or to processor increased from about 660 km in 1980 to almost 900 km in 1989. Nationwide, the average freight cost to export ports is about US$30/metric ton, but soybeans transported from Mato Grosso face costs of over US$60/metric ton, which is more than 50 percent of farmgate prices (Table 12.1). Average port charges in Brazil for soybean exports are now the equivalent of US$10 per metric ton, as compared to US$3/metric ton in the United States.

Transportation costs are very high in Brazil. For example, it costs about US$45 to ship one metric ton of soybeans from a producer in Mato Grosso to Paranagua, but only US$7 to ship the same amount from Paranagua to Rotterdam. It is also cheaper to ship one metric ton of maize from the United States to the Northeast of Brazil than to ship the same amount to the Northeast from the South. High costs of transport increase domestic marketing costs and, thus, partly offset Brazil's farmgate cost advantage in soybean production (Table 12.2).

In the South and Southeast, grains are trucked or shipped by rail to cooperatives, mills, or ports. Most of the grain produced in the Center/West is transported by truck to the agro-industrial centers near Sao Paulo and Ponta Grossa (Parana), and to ports for export, or to be shipped to other regions of Brazil.

Due partly to problems of geography, and to physical problems, Brazil's waterways have not been developed to serve as low-cost transport links.

Table 12.3 compares Brazil's road and rail networks and those of 18 other countries, all of which have more railroad per square kilometer of land area than Brazil, ranging from about 15 times as much in the United States,

TABLE 12.1 Comparative Transportation Costs for Soybeans in Brazil, 1989 (US$ per metric ton)

	Average Brazil	*Mato Grosso*	*USA*
FOB Export Port	220	220	220
Freight to Port	30	62	15
Port Charges	10	10	3
Taxes	34	34	
Net Farmgate	146	114	202
Inland freight as a % of farmgate price:	21%	54%	7%

Source: Association of Brazilian Vegetable Oil Industries (ABIOVE)

TABLE 12.2. Comparative Soybean Marketing Costs, United States and Brazil, 1986 (US$/metric ton)

	United States		Brazil	
	Overall	Maize Belt [a]	Double Crop With Wheat	Soybeans Alone
Total production costs	243.14	223.95	184.80	198.41
Marketing costs	24.60	24.60	43.50	43.50
Grand total costs	267.74	248.55	228.30	241.91

[a] Includes the Great Lakes region.
Source: Norman Rask, Gerald Ortmann, and Walter Stulp, *Comparative Costs Among Major Exporting Countries*, Ohio State University, Dept. of Ag. Economics, Occasional Paper, Columbus, Ohio, Jan. 1987, app. 3.

to over 3 times as much in Argentina and Canada, and about 1.5 times as much in Australia.

Truck Transport

Road haulage is the dominant means of transport for grains. Brazil's network of paved highways is heavily concentrated in the coastal states, particularly in the South and Southeast. The frontier areas, which are a long distance away from major consumption and export centers, lack adequate transportation infrastructure. Existing roads are often in need of maintenance.

In years when grain surpluses are produced, such as 1987, the government purchases almost all the soybeans and maize in many of the states of the Center/West, due to an inadequate supply of truck transportation from this region and the subsequent high cost of transport to ports or processing plants.

During the harvest season, truck transport becomes erratic. Seasonal demand raises prices, trucks are not always available, and handling of grain in smaller truck lots leads to inefficiencies, including higher losses. Trucking demand is highly seasonal in Brazil. Truckers move around the country, or into Paraguay, supplying services wherever they can earn the highest returns. In many cases, state storage companies or other agencies provide transit services to storage centers.

Rail Transport

Only about 15 percent of the grain crop is transported by rail, amounting to about 6 percent of total rail traffic. The rail infrastructure, of which

TABLE 12.3 A Comparison Between Road and Rail Transport in Selected Countries

Country	Roads		Rail		kms. of rail/ km of road	
	Km of Road per sq.km. of Area	Ratio Country x/ Brazil	Km of Rail per sq.km. of Area	Ratio Country x/ Brazil	Values	Ratio Country x/ Brazil
Brazil	0.17	1.00	0.004	1.00	0.02	1.00
South Africa	0.15	0.90	0.019	5.57	0.13	6.18
West Germany	1.97	11.81	0.111	31.81	0.06	2.69
Argentina	0.08	0.46	0.013	3.73	0.17	8.14
Australia	0.10	0.62	0.005	1.52	0.05	2.44
Belgium	0.91	5.48	0.125	35.57	0.14	6.49
Canada	0.04	0.24	0.012	3.44	0.31	14.60
China	0.10	0.59	0.005	1.54	0.06	2.62
South Korea	0.51	3.08	0.062	17.58	0.12	5.70
Cuba	0.31	1.84	0.131	37.36	0.43	20.32
USA	0.68	4.09	0.052	14.85	0.08	3.63
France	0.71	4.24	0.064	18.20	0.09	4.29
Great Britain	1.44	8.63	0.067	19.24	0.05	2.23
India	0.49	2.92	0.019	5.34	0.04	1.83
Italy	1.00	5.99	0.053	15.26	0.05	2.55
Japan	2.96	17.77	0.056	15.88	0.02	0.89
Mexico	0.11	0.66	0.008	2.28	0.07	3.47
Portugal	0.56	3.38	0.038	10.86	0.07	3.21
USSR	0.07	0.40	0.006	1.85	0.10	4.61

Source: Adapted from SETA (a Cargill—Brazil publication), Year 8; no. 45, Jan/Feb 1988.

about 30,000 km is operational, is mostly located in the South and along the coast. Major problems that constrain use of the railroads include poor track conditions, inadeqate management, and general ineffectiveness in meeting customer needs.

Brazil's rail system, built by private railroad companies as a means to facilitate exports at a time when roads were underdeveloped, is old and in need of upgrading. Early agricultural development of export crops in Rio Grande do Sul, Paraná and Sao Paulo was directly linked with the development of local railroads.

Several plans for new railroads are being discussed, and some are in advanced stages of planning. The potentially most important lines for grain transport are the "East-West" line and the "Production Railroad" (*Ferrovia da Produçao*).

The proposed East-West line would start in Cuiabá, in Mato Grosso, and would link with the existing rail network in Sao Paulo, providing a direct link between Cuiaba and the port of Santos. In subsequent stages it might be extended into Rondonia. Reportedly, detailed studies have been made, including demand projections, while cost estimates of about US$1.2 billion, equivalent to about US$1.2 million/km for construction, have been discussed. Construction costs are relatively low due to a favorable terrain for most of the projected route. Only one major bridge would need to be built, over the Paraná river.

The Production Railroad would link the southern part of Mato Grosso do Sul, the area of Dourados, with the existing network in Paraná, and thus with the port of Paranaguá. About 420 km of new railroad would have to be built. Mato Grosso do Sul is Brazil's third largest producer of wheat and soybeans. The Dourados region produces 60 percent of these crops. Total planted area in the region of Dourados is already about 1.7 million hectares.

River Transport

The location of Brazil's navigable rivers limits the opportunity for major volumes of grains to be moved by water. Although some interesting possibilities for developing water transport exist, river transport and the cabotage trade are both insignificant, with river transport of grains in its infancy.[1]

In 1987, a local grain trading company exported, on an experimental basis, about 3,000 metric tons of soybeans from Caceres in Mato Grosso do Sul down the Paraguay river to the port of Nueva Palmyra in Uruguay, a distance of about 3,200 km. If feasible, this would be an alternate outlet for the grain production also targeted by the Production Railroad. The beans were shipped in small barges from Caceres to Corumba, then transferred to larger barges for the remainder of the trip. Though it took about 5 months to get the grain from Caceres to Nueva Palmyra due to numerous difficulties, including strikes of marine personnel, equipment failure, groundings, and problems with export documentation, the trading company which organized the experiment believes that this river transport for grains may be competitive with road transport to Paranaguá.

In April 1988, there was an international meeting in Campo Grande (Mato Grosso do Sul) to discuss the development of the river transport potential of the Paraguay and Paraná rivers. Representatives of all countries involved (Bolivia, Paraguay, Uruguay, Argentina and Brazil) took part in the deliberations. Transportation demand estimates made in preparation for the meeting predicted that, through the year 2000, about 30 percent of the expected traffic would be grains, second only to minerals. The remainder would be mostly petroleum products, alcohol, cement and fertilizer.

Methods of Sale and Payment for Grains

Wheat is purchased for cash by CTRIN, the wheat monopoly. In the case of maize, in the South and Southeast, which account for about 70 percent of production, it is sold directly to cooperatives or to industrial users such as the poultry industry or feedstock manufacturers, usually for cash upon delivery. If the minimum price is above the market price at the time that the producer wants to sell, the CFP may purchase a large part of the harvest. In the Northeast, which accounts for about 10 percent of production, there are few large institutional buyers. Itinerant trucker traders purchase the maize in bags at the farm gate, for cash. Each bag is sampled individually and tested for product integrity and general cleanliness. There are no other quality considerations. Humidity is sometimes determined by hand.

Rice producers in Rio Grande do Sul generally sell to cooperatives or to rice millers for cash. Farmers rarely store rice at the cooperative for future sale. Most of the rice grown in Rio Grande do Sul is long-grain rice, which urban consumers prefer. In other parts of the country, the government buys most of the rice produced under the AGF program.

For milled rice, which millers or cooperatives sell to supermarkets or institutions, payments are received in 30 days or more. An average of 45 to 60 days elapses between costs incurred at the mill or cooperative and receipt of payment. In Brazil's inflationary economy, such delays entail substantial nominal interest costs. The long, average-payment period indicates a buyer's market. Although it is likely that millers' margins are being squeezed as a result, the producer takes the major brunt of the depressed market. Producer prices are set fortnightly by the cooperatives and are based on actual prices received for milled rice, mostly in the Sao

Paulo market, during the previous fourteen days. Cooperatives, private millers, and supermarkets keep inventories to an absolute minimum.

In general, soybean producers have five sales options.

- *Cash Payment on Delivery.* Producers sell at spot or current bid prices and receive cash payment usually within 48 hours.
- *Delayed Payment with Cash Advance.* The producer receives an advance, typically from 40 to 70 percent of the price of the beans and pays interest on it until, at a day the producer selects, the soybeans are priced at the elevator. Usually, the farmer has from 6 to 12 months in which to price his beans. Different buyers set different time limits, and most agree that the limit is negotiable.
- *Deferred Price and Deferred Payment.* Farmers may sell on a deferred price agreed upon at the time of delivery, with payment to be made at the deferred date. For example, the soybeans might be delivered in April with price set according to a July price and payment made in July.
- *Storage at the Country Elevator for Later Sale.* Farmers may store the grain at an agreed-upon storage cost at the country elevator and sell at their option at a later time.
- *Sale to the Government at Minimum Price.* Farmers may sell to the government at the established minimum price, with delivery to a country elevator or a CFP-designated warehouse. Payment is made at time of delivery. Lately, the government has purchased, through its AGF program, most of the crop produced on the new lands in the Center/West. The crop is later sold to crushers and exporters.

The delayed payment option is used most extensively. It can be compared to the Argentine system of "precio à fijar" (delivery, with price to be fixed at a later date: see Chapter 5).

Producers' marketing strategies are heavily influenced by the economic situation in Brazil and by each producer's personal financial situation. In most cases, an immediate sale is necessary or, at least, an advance against a future payment.

Price Determination in the Brazilian Markets

Both government purchases of grains under the Minimum Price Program and government actions to control exports and imports of grains greatly affect prices.

Wheat

Prices are set by SEAP, in the Ministry of Planning. The wheat is purchased by CTRIN, a subsidiary of the Bank of Brazil.

Yearly negotiations take place between the Treasury and the Ministry of Agriculture about the resources to be made available for wheat production and purchase. Due to the deteriorating fiscal situation, resources have been reduced in real terms. In 1988, it was decided to freeze available funds at the 1987 level. The purchasing price was maintained at the 1987 level of US$186/metric ton, down from well over US$200/metric ton earlier. Table 12.4 shows production, import and price data for wheat for the period 1982–1988. Production has increased significantly; imports have fallen as have producer prices. In 1989, only 80 percent of the resources estimated to be necessary to purchase the total wheat crop were made available.

The government has reduced the producer price for wheat in real terms. The purchasing price of wheat from the 1988 harvest was equivalent to Cz$18,437/metric ton on March 1, 1988, or about US$185/metric ton at the official exchange rate and less than US$140/metric ton at the parallel market rate. The Chicago price was just over US$120/metric ton. CFP calculated 1986 cost differences for wheat between FOB Chicago and delivered-Brazilian-mill at about US$42/metric ton. This resulted in a border price of about US$162/metric ton, between a domestic price based on the official and the parallel exchange rate, respectively.

The subsidies that have been an integral part of Brazil's wheat policy since 1938 have led to a substantial increase in per-capita wheat consumption. In the early 1960s, per-capita wheat consumption averaged 30 kilograms. This average increased to 49 kilograms per capita by the late 1970s. The per-capita consumption of more traditional foods—beans, rice, cassava, and maize—decreased. The government's wheat policies, including its substantial consumer subsidies, have led to a partial substitution of traditional foodstuffs by wheat-based products over the last 20 years. By 1980, the total subsidy on wheat products had reached an historic high of 82 percent. The subsidy was later reduced, but increased sharply again as a result of the Cruzado Plan in 1986 when the sales price of wheat to the mills was frozen. The average subsidy in 1986 was calculated at 70 percent.

Maize

The official minimum price for maize under the MPP largely determines the market price. The market price varies according to season, location, and other market factors, such as the cost of obtaining loans for

TABLE 12.4 Brazil—Wheat Production, Imports and Prices, 1982–1988
(amounts in '000 of metric tons, prices in US$/metric ton)

	Production	Imports	Domestic Producer	Price Average Import	Sale to Mills
1982/3	1,827	4,224	275	164	136
1983/4	2,237	4,709	204	157	110
1984/5	1,983	4,621	226	152	114
1985/6	4,247	4,552	248	137	121
1986/7	5,638	2,166	241	97	78
1987/8	5,847	2,687	186	94	105

Sources: World Bank, *Brazil Agricultural Incentives and Marketing Review*, 1989.
ETAC, *Weekly Newsletter on Brazilian Agricultural Trade*, Vol. 9, no. 405, Curitiba—
PR, Aug. 7, 1989.

purchasing maize at government auction, but these variations occur
around the minimum price.

Under the MPP, the government posts two floor prices for maize, one
for the South and one for the North. The prices are set in August, at least
two months before the earliest planting—the corn season in Brazil is from
November to March in the South and from March to July in the North. Under the AGF stock-purchase program, the government has an obligation to
purchase all maize offered to it at these prices, which are periodically adjusted for inflation.

To limit its purchasing obligation, and to aid in the operation of the
maize market by inducing farmers to hold maize inventories, the government extends credit to producers and cooperatives under the EGF storage
loan program at below-market interest rates. The maize stocks the borrower has stored serve as collateral for the loans. The stocks must be kept in
government-accredited storage.

If at the end of the loan period, usually 5 to 8 months after harvest, the
market price is below the minimum price, producers and cooperatives can
cede their maize stocks to government in lieu of paying off the loan. Once
the government has accepted the maize, it becomes responsible for payment of storage costs over the entire period that the loan was outstanding.
During the period that the loan is in effect, the underlying stocks cannot
be sold unless the loan is paid off first.

Resources made available for the combined purchase and loan programs are usually well below demand. The government uses the need to
ration resources as a policy instrument to influence the maize market both
over time and according to region. Ostensibly, the government aims to
help the small farmer in marginal areas.

Financing limits on EGF loans are set as a percentage of the value of the stock, based on the minimum price. The limits set vary by region and by size of producer. The total EGF financing available can be as low as 40 percent of the value of the stock for large farmers in the South, or up to 100 percent of the value of a small farmer's maize crop in the Northeast.

The maize ceded to the government at the end of the loan period is released gradually into the market as the loan is liquidated, and the balance paid out in 3 to 7 installments. Payment may take up to 6 months. Thus, the combined loan and purchase programs withhold a substantial amount of each maize harvest from the market for a period of 12 to 16 months.

From 1970 to 1981, government purchases of maize under the AGF program averaged slightly over 1 percent of the yearly maize crop (Table 12.5 and Figure 12.1). During that period, the government's main support for agriculture was through subsidized production credit. However, real resources spent on the MPP more than doubled between 1979 and 1986, and the funds available for maize increased more than fourfold. By 1987, almost 50 percent of all funds allocated to crop purchases (AGF) were spent on maize, representing almost a quarter of that year's crop. Almost 10 percent of all funds available for stock financing (EGF) were also spent on maize (Table 12.6 and Figure 12.2). Particularly noticeable is the shift of resources from lending (EGF) to purchasing (AGF) since the early 1980s (Figure 12.1).

Because the government set excessively high floor prices for maize, its AGF purchases increased greatly (Table 12.5). By 1987, CFP had nearly 4 million metric tons of maize in stock, equal to about 15 percent of that year's total domestic maize consumption. At the same time, the government banned external trade in maize. Imports could have competed on a price basis with local maize in several years, for example, in 1987 (Table 12.7). However, imports were not allowed as they would aggravate the stock situation. It was impossible to export maize since minimum prices, and total costs incurred by the CFP, exceeded FOB export prices, at least at the prevailing exchange rates. The government was unwilling to subsidize maize exports.

In 1987, to reduce the uncertainty that accompanied government stockholding and maize sales, CFP introduced an automatic trigger price system. The system was designed to regulate sales from government maize stocks so as to minimize market disruptions and avoid political pressure (principally from millers) for the sale of maize. Under the trigger price system, maize is released from CFP stocks if the wholesale price in Sao Paulo exceeds a rolling average of the past 60 months plus 12 percent. CFP auctions are stopped when the price drops to the minimum price plus 5 percent. However, maize can still not be released if the trigger price does not equal, or exceed, total costs to the CFP, i.e., original purchase price plus carrying charges.

TABLE 12.5 Brazil—Quantities of Maize Purchased Under the MPP, as a Percentage of Production, 1970–1987

	Financed (EGF)	Purchased (AGF)
1970	2.6	0.0
1971	1.2	0.0
1972	1.9	0.0
1973	2.8	0.0
1974	4.8	1.0
1975	5.4	0.6
1976	7.2	0.8
1977	8.3	7.8
1978	4.3	2.3
1979	7.2	0.4
1980	7.8	0.0
1981	18.1	0.3
1982	14.5	16.2
1983	12.3	7.4
1984	8.5	2.2
1985	7.6	13.6
1986	8.3	21.1
1987	6.9	24.4

Source: World Bank, *Brazil Agricultural Sector Review,* July 26, 1990.

FIGURE 12.1 Maize — EGF and AGF as a Percentage of Production

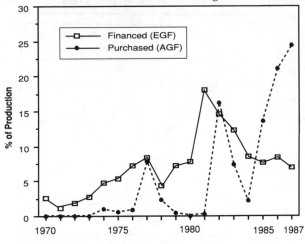

Source: World Bank, *Brazil — Agricultural Sector Review,* June 29, 1989

TABLE 12.6 Brazil—Yearly Share of Maize Under the MPP, as a Percentage of Total
Allocations, 1970–1987

	Stock Financing (EGF)	Purchasing (AGF)
1970	12.37	0.42
1971	5.83	6.90
1972	6.13	8.54
1973	13.30	2.86
1974	15.48	74.32
1975	7.49	6.10
1976	9.03	7.11
1977	9.92	38.82
1978	3.83	24.53
1979	6.68	4.94
1980	6.84	0.04
1981	14.64	2.29
1982	11.86	32.73
1983	10.06	46.21
1984	14.41	24.60
1985	12.67	23.93
1986	8.14	40.13
1987	9.93	49.82

Source: World Bank, *Brazil Agricultural Sector Review,* July 26, 1990.

FIGURE 12.2 Maize — Allocations Under the MPP Program

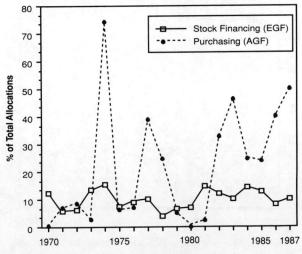

Source: World Bank, *Brazil — Agricultural Sector Review,* June 29, 1989

Since early 1989, CFP has had to pay market interest rates on the funds used to finance its maize stocks. Increases in the market price of maize must therefore exceed the average interest rate paid by CFP on its funds (by a margin sufficient to cover all other carrying costs). In the short run, only an unusual occurrence such as an agricultural calamity would raise the market price that much. Even over the period of a year, market prices have not risen enough to cover CFP's costs and interest payments. CFP has calculated that its minimum release price for maize, i.e., its average purchasing price plus all carrying costs, increased 39 percent between August 1988 and August 1989, while the benchmark wholesale price for maize in Sao Paulo increased by only 14 percent in the same period.

TABLE 12.7 Brazil—Maize Price Ratios, 1972–1988

	Farmgate to Minimum Prices (Paraná)	Minimum to Border Prices (FOB)	Farmgate to Border Prices
1972	1.10	1.09	1.20
1973	1.03	0.71	0.73
1974	0.96	0.82	0.79
1975	1.05	0.82	0.86
1976	1.01	0.85	0.86
1977	0.88	1.12	0.99
1978	1.37	0.98	1.34
1979	1.33	0.96	1.28
1980	1.43	0.84	1.20
1981	1.24	0.93	1.15
1982	0.92	1.59	1.46
1983	1.01	0.69	0.70
1984	1.30	0.58	0.75
1985	0.90	1.09	0.98
1986	1.02	1.21	1.23
1987	0.68	4.36	2.96
1988		1.03	

Note: All prices are averages for the peak harvest period, i.e., March/April/May. Border prices are as calculated by the World Bank without adjustments for taxes and at the official exchange rate.
Sources: CFP for minimum prices; Getulio Vargas Foundation (FGV) for farmgate prices and the World Bank for border prices.

Rice

As for maize, the government minimum price largely determines the price for rice. However, prices for rice have dropped as a result of large government imports, and the large quantities of rice held in government storage.

For several years in the mid-1980s, the guaranteed minimum price for rice was above the market price. In 1986, a crop failure was forecast in Rio Grande do Sul. In reaction to the predicted shortfall, Brazil purchased between 1.7 million and 2.1 million metric tons of rice (between 12 percent and 17 percent of average amounts traded yearly)—private companies imported 1.3 million metric tons, and the government imported the remainder. To stimulate rice production in anticipation of an increase in the market price, the minimum price was set well above the average market price of the previous year. With the stimulus of high prices and fortuitous good weather, the 1986/87 crop was a record one. The government ended up with about 70 percent of the total harvest, purchased at the minimum price. Following additional purchases in succeeding years, the government now holds about 5 million metric tons of rice in stock. Over 80 percent of this is inferior-quality dryland rice for which the local and international markets are limited. As for maize, under the trigger price system, the government will sell rice stocks only when the market price exceeds the average price of the prior 60 months plus 12 percent, and only as long as that trigger price exceeds total carrying costs. Early sales are therefore unlikely.

Table 12.8 shows production, trade, and stock levels for rice since 1980. The trend of steadily rising year-end stocks is quite apparent: they have grown rapidly since 1980.

Soybeans

Prices for soybeans at the export port or in the central merchandising offices in Sao Paulo are generally expressed in US dollars, and are quoted in terms of cents over or under the Chicago Board of Trade. Market information is widespread. Table 12.9 gives an example of monthly prices in 1986/87. Farmgate prices are derived by deducting from the FOB price, freight, estimated shrinkage, brokerage and taxes, and are quoted to the farmer in local currency (Table 12.10). Prices paid by processors are determined in a similar manner, with the additional subtraction of processing margin from world prices for meal. Thus, the local price of soybeans is determined by export prices minus processor demand, processing costs, marketing costs, transportation costs, the influence of government minimum prices for raw beans and oil at the retail level, and the price of meal in domestic and export markets.

TABLE 12.8 Brazil—Production, Trade and Stock Levels for Rice, 1980–1989 ('000 metric tons)

Year	Opening Stock	Production	Imports	Supply	Domestic Consumption	Surplus	Exports	Closing Stocks
1980/81	2,040	8,228	209	10,477	9,000	1,477	73	1,404
1981/82	1,404	9,155	203	10,762	9,100	1,662	18	1,644
1982/83	1,644	8,224	465	10,333	9,150	1,183	12	1,171
1983/84	1,171	8,991	91	10,253	9,200	1,053	2	1,051
1984/85	1,051	8,760	500	10,311	9,660	651	5	646
1985/86	646	9,813	2,074	12,533	10,240	2,293	6	2,287
1986/87	2,287	10,578	235	13,100	10,000	3,100	5	3,095
1987/88	3,095	11,757	190	15,042	10,500	4,542	10	4,532
1988/89	4,532	10,999	190	15,721	10,750	4,971	50	4,921
average annual growthrate	9.8%	4.2%	2.0%	5.7%	2.4%	19.3%	-6.3%	19.7%

Source: ETAC, Weekly Newsletter on Brazilian Agricultural Trade, vol. 9, no. 405, Curitiba—PR; Aug. 7, 1989.

TABLE 12.9 Brazil—Monthly Prices of Soybeans and Soybean Products in 1986/87

	Oct.	Nov.	Dec.	Jan.87	Feb.	March	April	May	June	July	Aug.	Sept.	Oct.
Beans (US$/metric ton)													
Chicago a	177.2	183.7	180.6	182.6	179.1	180.7	190.0	204.8	205.1	196.6	186.5	193.1	197.8
Premium	19.8	16.2	11.4	10.1	5.7	0.7	(1.4)	2.4	8.1	16.3	17.3	15.1	14.7
Brazil-FOB Porto Alegre	197.0	199.8	192.0	192.7	184.8	181.4	188.6	207.2	213.2	212.9	203.8	208.2	212.5
Cz$/bag equiv. b	131.9	135.2	134.8	147.0	164.7	187.9	220.8	313.7	421.7	473.3	459.3	500.8	553.3
Physical Market	143.0	147.8	143.9	141.4	150.4	164.8	201.5	314.2	426.7	473.8	541.1	631.2	650.0
Meal (US$/metric ton)													
Chicago	164.1	165.3	158.4	158.2	158.9	154.8	167.0	183.8	192.9	183.0	173.8	186.4	192.9
Premium	19.1	14.5	17.4	21.2	19.1	9.9	6.4	4.3	5.2	8.4	12.8	20.9	18.2
Brazil-FOB Porto Alegre	183.2	179.7	175.8	179.4	178.0	164.7	173.4	188.2	198.1	191.4	186.0	207.3	211.1
Campinas-Equivalent	2,801.8	2,776.3	2,905.6	3,284.3	3,758.5	3,990.1	4,810.6	6,871.2	9,035.3	9,496.2	9,897.4	11,523.5	12,612.5
Physical Market	2,508.0	2,716.4	2,902.0	2,929.0	3,227.1	3,639.2	3,662.2	5,967.7	7,995.2	8,410.2	10,740.6	12,849.5	13,019.6
Crude Oil (US$/metric ton)													
Chicago	327.6	333.8	337.3	357.8	349.0	346.3	349.4	371.9	367.7	355.1	349.4	356.7	388.7
Premium	-26.2	-14.3	-26.9	-47.8	-56.7	-64.6	-44.0	-37.7	-37.9	-48.3	-45.5	-34.6	-52.5
Brazil-FOB Porto Alegre	301.4	319.6	310.4	310.0	292.3	281.8	302.5	334.2	329.8	306.8	303.9	322.1	336.2
Sao Paulo-Equivalent	5,047.9	5,731.6	5,641.9	6,616.6	7,472.1	8,259.0	10,288.6	15,743.9	19,353.3	18,179.5	19,206.8	21,398.3	23,811.3
Physical Market	4,954.0	5,534.0	5,355.7	5,823.8	6825.0	7,630.5	8,742.5	14,587.5	17,257.1	16,243.5	18,181.0	20,394.1	23,173.8
Refined Oil in Sao Paulo Market (Cz$/case of 20 x 900 ml.)													
Import Parity Sao Paulo	149.2	155.5	170.4	192.4	242.8	271.5	332.7	487.6	562.5	570.6	605.6	699.1	783.8
Retail	147.8	152.4	153.0	154.4	183.0	209.2	222.8	304.2	467.2	482.8	483.6	519.8	523.6
Wholesale	125.6	133.2	133.2	133.2	154.5	167.2	216.3	329.5	n.a.	n.a.	n.a.	500.0	500.0

a   The Brazilian market quotes a premium, which is in essence the market determined difference between the Chicago price, at that time, and the price in the particular market concerned, in Brazil. Premiums can be positive or negative. Premiums vary between market places.

b   Canoas: Cz$/bag.

c   Cz$/metric ton, payment in 8 days.

d   Cz$/metric ton, pellets, cash payments Campinas.

e   Cz$/metric ton, payment in 30 days, inclusive of 12 percent ICM.

f   Cz$/metric ton, payment in 30 days, inclusive of 12 percent ICM.

Source: Agropec Comercial e Exportadora S.A.

TABLE 12.10 Brazil—Derivation of Farmgate Prices for Soybeans

	Bean Exports	Meal and Oil Exports	Meal	Domestic Market Oil
1	Chicago price of month (US$/bushel)	Chicago price of month	FOB mill price	CIF crude–refinery gate
2	Premium of export port (US$/bushel)	Premium of export port		
3	Sum of (1) + (2) (Cz$/metric ton)	Sum of (1) + (2) (Cz$/metric ton)		
4	Expenses:	Expenses:	Expenses:	Expenses:
	freight to port	freight to port	ICM (17% within state;12%	ICM (12%)
	port charges (about 20 diff. items)	port charges (about 20 diff. items)		financing cost 30 days
	ICM (typically 13% of (3) [a]	ICM (typically meal, 11%, oil 8% of (3) [a] outside state)		financing cost 30 days
	physical losses (.3%) [b]	physical losses (.3%) [b]		
	social security tax (.65% of (3)	social security tax (.65% of (3)	s.s. tax (.65% of (3) s.s. tax (same)	
	for. exch. brokerage (.1875% of (3) [c]	for. exch. brokerage (.1875% of (3) [c]	sales commission (.25%)	sales commission (1.00%)
	sales comm. (US$.50/m.ton) [d]	sales comm. (US$.50/mt; US$1.0/t oil) [f]		
5	Farmgate total net p. ton (3)—(4) [e]	Farmgate total net per mt beans equiv.Farmgate tot. net per mt beans eq. (3)—(4) [g] (3)—(4)		

a State tax on the Circulation of Merchandise, varies by state.
b Standard retention by port; part of port revenue.
c Varies with payment terms
d US$.25/metric ton for reselling traders
e In case of payment terms "net 30 days" at then prevailing exchange rate, a premium is given of 1.0 to 1.8% + interest at LIBOR over 30 days
f In case of payment terms "net 30 days" at then prevailing exchange rate, a premium is given of 1.0 to 1.8% + interest at LIBOR over 30 days. Standard yields are 77% of meal and 19% of oil. Standard crushing costs are the equivalent of US$12/metric ton.
g Standard yields are 77% of meal, 19% of crude oil and 96.5% of refined oil. Standard crushing costs are the equivalent of US$30/metric ton. standard refining costs are the equivalent of US$12/metric ton,

Source: Pacheco, Luiz Carlos, *A Comercialização da Soja no Brasil*, ETAC—Curitiba—1987.

Brazil tends to export its soybeans immediately after harvest, whether or not soybean prices are highest in those months. Coincidentally, that is often the case because of its counter-seasonal production, as compared with the northern hemisphere. Brazil imports for its own needs between harvests, when prices tend to be lowest, resulting in average annual export prices that are generally above average annual import prices (Table 12.11).

The soybean processing sector is highly competitive. Local firms predominate. There are many multinationals also, but they control only 25 percent of domestic crushing capacity. Due in part to tax advantages for investment, the industry has extensive overcapacity. Crush capacity is about 25 million metric tons compared to an annual harvest of about 17 million metric tons in recent years (only 14 million metric tons are processed).

Almost all export soybeans are financed and hedged offshore, often through foreign exchange accounts legally established with seed capital authorized by the Central Bank. Most additional funding is borrowed, apparently with the hedged product as collateral. There is no local futures market of importance for soybeans or any other crop. There is no futures contract in Cruzados and the individual farmer is too small to mount an offshore exporting operation.

The export trade is ostensibly controlled through CACEX's export registry. However, price variations within any 24-hour period, and variations in product quality, make this system less than watertight. "Paraguayan" exports via the port of Paranaguá in the state of Paraná go unreported. Paraguay has a free port in Paranaguá, where goods transported over land in bond can be delivered for export. Brazilian soybeans occasionally leak into this transportation system. Because exporters can benefit from the large difference between the official exchange rate and the parallel market rate and the opportunity to avoid taxes, circumventing CACEX's controls is very lucrative.

TABLE 12.11 Brazil—Average Export and Import Prices for Soybeans, 1984–1988 (US$/metric ton)

	Average Export Price (FOB)	Average Import Price (CIF)	Export/ Import (%)
1984	702	686	102
1985	631	595	106
1986	358	347	103
1987	307	334	92
1988	432	382	113

Source: Association of Brazilian Vegetable Oil Industries (ABIOVE)

Quality of soybeans is becoming increasingly important in international markets. While soybean yields in Brazil have increased, oil and protein content have declined. Average oil content 20 years ago was 20.5 percent, but it is now closer to 18.5 percent. The Variety Release Program administered with major input from EMBRAPA includes information on chemical composition but assumes "that all commercial varieties submitted meet the minimum requirement for chemical composition." The primary criterion in seed selection is yield, although several other agronomic characteristics are included in the list of criteria evaluated prior to the release of any new variety. Soybeans harvested at high moisture content are subject to rapid deterioration. Immediate drying to 13 percent moisture is required to assure safe storage under the high temperatures existing in much of the soybean producing areas. Inadequate drying capacity in the newly developed soybean production areas can create problems of quality for soybeans stored in these regions.

Notes

1. Down the Amazon and its tributaries for production from Mato Grosso and Rondonia; down the Sao Francisco River for production from Minas Gerais and Goias; on the river Tieté in Sao Paulo and down the Paraná and Paraguay rivers.

13

Future Directions

Brazilian grain producers operate in a business climate that is dominated by high risk and uncertainty compared with that of their counterparts in Australia, Canada and the United States. The root cause of much of this risk and uncertainty is the political crisis that has existed in Brazil for decades, arising principally from a failure to reach a consensus in the body politic on the essential principles that should inform and guide the people and their institutions regarding the governance of Brazil. From the political crisis has come economic chaos as evidenced by 4-digit annual inflation rates, huge foreign debts, and government policies that frequently change in quite extraordinary and unpredictable ways.

It is against this background that policies affecting the Brazilian grain sub-sector must be seen. As measured by producer subsidy equivalents (PSEs), the level of support to Brazilian grain producers compares favorably with that received by their counterparts in Australia, Canada, and the United States, and is substantially higher than that received by their neighbors in Argentina (Table 13.1).

As measured by PSEs, Brazilian wheat farmers receive more support than their counterparts in the countries shown in Table 13.1, while support to Brazilian maize, rice and soybean producers is comparable to that received by producers in the United States. See the discussion in Chapter 10, particularly Table 10.1, however, which supports the view that agriculture was actually discriminated against in Brazil.

Interestingly, despite all the apparent distortions and uncertainties, the performance of the Brazilian grain sub-sector compares quite well with that of Argentina, Australia, Canada, and the United States in terms of yield levels and growth in yields (see Tables 9.2, 9.3, and 9.5). Major gains have been made in Brazil, especially in soybeans and wheat. The level of fertilizer use is also not out of line with that in Australia, Canada and the

TABLE 13.1 Producer Subsidy Equivalents for the Major Grains in Selected Countries, Average for 1982–1986 (in percentages)

	Argentina	Australia	Brazil	Canada	U.S.A.
Wheat	-54	7.7	56	30.4	36.5
Maize	-65	n.a.	14	10.0	27.1
Rice	n.a.	13.0	49	n.a.	45.2
Soybeans	-76	n.a.	7	13.5	8.5

Source: USDA-ERS-Agriculture and Trade Analysis Division, "Estimates of Producer and Consumer Subsidy Equivalents," *Government Intervention in Agriculture, 1982-86*, Washington, D.C., April, 1988.

United States (Table 13.2). In addition, as noted earlier, agricultural output in Brazil increased significantly because of an expansion in the area seeded to grains.

Policies and Institutional Framework

A fundamental problem in Brazil is that the traditional policy and institutional framework affecting the grain sub-sector is not fiscally sustainable and is in need of radical change. At the root of the problem is the role played by government. The government has been performing poorly functions that are genuinely public sector responsibilities, while assuming functions that could better be performed by the private sector, at least in a reformed economy. The Brazilian government clearly has an essential role to play in supporting the grain sub-sector, but one that differs in important ways from the role it has traditionally assumed. Given the Brazilian government's limited capacity to perform the functions that cannot be left to the private sector, it would be well advised to support private sector initiatives in all areas where the public sector has no clear comparative advantage. The most important policy changes suggested revolve around the appropriate public and private sector roles.

TABLE 13.2 Consumption of Fertilizer per Hectare of Arable Land and Permanent Crops for Selected Countries, 1987, kg N, P_2O_5, K_2O

	Argentina	Brazil	Australia	Canada	U.S.A.
Nitrogen	2.6	12.4	7.9	25.8	50.0
Phosphate	1.4	21.0	17.6	13.8	19.6
Potash	0.4	19.9	3.1	8.6	23.6
Total	4.4	53.3	28.6	48.2	93.2

Source: FAO Fertilizer Yearbook, 1989.

Macroeconomic Policies

A high priority for the Brazilian grain sub-sector is the implementation of monetary and fiscal policies that will bring inflation to a level where it no longer plays a role in investment decisions. Putting in place a unified market-determined exchange rate system should be an integral part of these macroeconomic reforms. It is essential that stability be introduced into the economic and political environment in which farmers operate. Risk and uncertainty must be considered as farmers make plans for investment and production. Uncertainty adds to production cost as surely as if the government placed an even higher tax on farmers' use of inputs. Macro policies must be altered to create a favorable climate for private production and marketing decisions. In such a changed macroeconomic environment, land would be held more and more for its agricultural production possibilities and less and less as a hedge against inflation.

Stability in the macroeconomic environment, including control of inflation, is a necessary, though by no means a sufficient, precondition for the development of efficient financial markets, including those in rural areas, with the capacity and willingness to finance the short-, medium-, and long-term needs of the grain sub-sector. The availability of such financing will be essential if Brazil is to realize its full potential in grain production. Furthermore, macroeconomic stability, including control of inflation, would greatly facilitate the development of an organized private market for financing grain stocks (see below).

Trade Policy

A neutral and liberal trade regime that removes quantity restrictions and export quotas and has a low and uniform level of tariffs, together with a liberalized exchange rate regime, would greatly enhance the prospects for Brazil's agricultural sector, including the grain sub-sector.

Tax Policy

A neutral tax system that relies on income taxes at the Federal level and land taxes at the local level is suggested. Neutrality in the tax regime, both within and between sectors, is viewed as important for the same reasons as neutrality in the trade regime is seen as desirable, namely, for the effect it would have on resource allocation and on the public perception of fairness. Maintenance of the neutrality principle in tax policy has implications for the grain sub-sector that reach far beyond on-farm production into the private financing of grain stocks (see below).

Transport Policy

High transportation costs are affecting the competitiveness of Brazilian grains in both the domestic and export markets. A major product of improved macroeconomic policies would be an enhanced ability on the part of the public sector to finance needed public sector investments in water, rail, and road transport systems, and an increase in the willingness and capacity of the private sector to operate such systems. Better integration of the various modes of transport is also critical to lowering transport costs in the marketing of grains.

Targeted Food Subsidies

Replacing generalized food subsidies with subsidies directly targeted to the poor, both in urban and rural areas, would better serve both the poor and the grain producers. If properly implemented, this policy change could generate budgetary savings, while simultaneously enhancing the access of the poor to food. Food stamps and feeding programs for eligible groups are economically more efficient than generalized subsidies, which inflate costs by providing cheap food for many who are not in economic need.

Evaluation of Marketing Systems

At a time when Brazil may be considering reforms in the marketing of grains, it is important to evaluate the four alternatives of: (1) free-market or competitive model—public regulation (e.g., through a grain commission), but otherwise maximum reliance is placed on the private sector; (2) marketing board model (the Canadian and Australian Wheat Boards); (3) grain board model (the Argentine case, where the Grain Board combines regulation, marketing functions and consumer protection under one agency); and (4) the Brazilian model, where the public sector role is pervasive, but the responsibilities are distributed among many, many different government agencies.

The performance of a marketing system can only be evaluated if performance norms are specified in qualitative, if not quantitative, terms. Performance norms in theory are generally broad and vague and intended to be comprehensive. When applied to individual industries, they must be made more explicit. The performance norms selected for this comparison include: (1) price level; (2) price stability; (3) bargaining power and pricing strategies; (4) marketing costs and efficiency; (5) entry and exit; and (6) resource allocation.

The level of prices in international markets is set by interaction of the buyers and sellers in the market. Although there are policies and restrictions that influence price in almost every country, these tend to counteract one another such that the overall world price level is set by forces of supply and demand. Therefore, any of the four models of market organization (free market model, marketing board, grain board, Brazilian model) would face the same international prices and presumably could have little influence on world price levels.

However, the different forms of market organization can influence the extent to which world prices are transmitted back to individual producers through the marketing chain. Unless the public sector subsidizes an income transfer to producers, there is no a priori reason to believe it would give a larger share of the world market price to the farmer than would a free market system. In fact, most studies show that the marketing costs (including handling, distribution and risk shifting) are higher for public-sector managed marketing systems than for a competitive industry consisting of many firms, each seeking its own best interest. Higher marketing costs generally mean lower returns to producers. One must, therefore, conclude that the free market model would be equal to or better than the marketing board, grain board, or Brazilian models with respect to world prices as well as farmgate prices, in the absence of public-sector subsidies.

Price Stability. Price stability (defined as long-run prices received) is similar to the discussion of price level in the preceding section. Supply and demand in world markets will move prices month by month and year by year. The exporting or importing country has no alternative but to follow these prices if it is to continue to trade in the world market. Only if it has sufficient volume and storage capacity to influence world price levels could it have any impact upon stability. A study by Hill and Mustard comparing the United States and Canada indicated that the Canadian Wheat Board had a larger share of the world market when prices declined and world supplies increased than when supply was low and prices high.[1]

This suggests that the Canadian Wheat Board, by accentuating the swings, tended to destabilize rather than stabilize.

On the question of short-run price stability, the results are different. Fluctuations in a free market such as the Chicago Board of Trade occur day by day and even minute by minute. This type of instability is reduced or even eliminated by a system where prices are fixed by government decree in the domestic market. Prices at the local farm level under a free market system may also vary from firm to firm, region to region, and day to day. These fluctuations can be controlled with the authority of the government, although at a cost of efficiency and responsiveness. Therefore, on the criterion of short-run price stability, the competitive model would rank below the performance of a controlled market.

Bargaining Power and Pricing Strategies. It has been argued that many centrally planned economies purchasing grain would prefer to purchase from single entities in the exporting country. However, there is no evidence to indicate that the U.S.S.R. or China have purchased preferentially on the basis of ideology or of the number of sellers in the exporting country. The U.S.S.R., in fact, has repeatedly demonstrated its ability to use the Chicago Board of Trade to its advantage. It has also been argued that a single committee will make better decisions as to when and where to sell than does the aggregated decisions of thousands of individual firms. The study by Hill and Mustard shows that Canada was a more aggressive seller in the market than the United States in periods of low prices than in periods of high prices.[2]

Its market share was highest in years when prices were lowest, and lowest in years when prices were highest. This may be less the result of Canadian strategy than of the U.S. actions to store grain across crop years during periods of excess supply in the world markets. Although causality is difficult to establish, the data do not support the hypothesis that the Canadian Wheat Board has exhibited superior marketing strategies or has generated consistently higher prices for its producers.

It has been suggested on occasion that public-sector importing agencies such as the Japanese Food Agency may exert monopsony power in buying imports and, thus, lower the food import bill relative to what the private sector would pay. CTRIN, however, seems more intent on achieving self-sufficiency in Brazil in wheat than on lowering the import price for wheat, even if this policy fosters inefficient growth in the agricultural sector.

Marketing Costs and Efficiency. Efficiency in this context should be defined to mean performance of all of the basic marketing functions at the lowest possible cost of resources. Costs of marketing can be minimized by not providing marketing services, so efficiency comparisons are preferable to comparison of marketing costs. Marketing services include: information on prices and quantities, transportation, risk shifting, and regulatory functions. A 1980 study by Hill and Dollinger showed that marketing costs for wheat (including transportation) were US$25.60 per metric ton in Canada and US$14.55 per metric ton in the United States.[3]

When quality, geographical area, and export destination were matched for Canada and the United States, farm prices and export prices were equal for the two countries. This relationship was maintained in Canada only by subsidizing transportation to compensate for the extra US$11.02 per metric ton incurred in the marketing activity. McCalla and Schmitz also concluded that technological development in Canada was lagging behind that of the U.S. in terms of transportation, export facilities, and grain handling because the Canadian Wheat Board failed to provide economic incentives for increasing efficiency.[4]

Comparison with the Argentine system shows the same relationship. Technology used in transportation and handling is less under the Argentine Grain Board than is found in the free market sector of the grain industry. Cost of handling and marketing and the inefficiency of labor utilization has put the Board's export elevators at a sufficient disadvantage to the free market to require heavy subsidies even to keep them operating.

The conclusion from the available studies is that the free market system generates a more efficient marketing system than any achieved by the Brazilian model, grain boards, or marketing boards in the various countries. The severity of competition often results in capital losses to the inefficient firms. Many are forced out of the industry. However, the ones that remain are the most efficient and the process of "natural selection" results in a lower cost of marketing for the entire industry than has been achieved under any centrally planned marketing operation. The many daily decisions made by thousands of individual firms, each seeking its own individual gain, provides far more information and expertise than can be gained by a centralized committee of a few experts regardless of how well trained or well paid these commission members are.

Entry and Exit. By definition, a grain board, a marketing board, and the Brazilian model restrict entry. Restriction is essential to exclude the possibility of lower-cost firms capturing market share. Exit of the less efficient firms is also precluded in the design of most of these systems. Costs are often averaged or prices set to cover the highest-cost firm. There is little penalty for poor management or wrong decisions, because agencies such as CTRIN do not have a binding budget constraint. The entry and exit of firms into the grain trade under a free market economy will often result in capital losses to individuals but a gain to the system in total. Under the proper regulatory environment, the ease of entry guarantees the absence of excess profits while encouraging maximization of efficiency. A grain board can eliminate all profits if that is the policy which it chooses to follow. However, the politicization of most government operations encourages the transfer of income from one sector to another for purposes deemed important in the overall economy. Agencies such as CTRIN and CIBRAZEM seldom operate solely for the benefit of producers. Even the marketing boards used in Canada, South Africa and Australia are not always operated in the best interest of producers. Subsets of goals such as survival of the institution and positions for individuals within the organization sometimes comprise the criterion of "service to producers" . The competitive system also provides the necessary incentive and opportunity for exit of inefficient firms, while public-sector agencies are committed to maintaining their facilities and position regardless of relative costs or efficiencies of alternatives.

Resource Allocation. A free market price is the best allocator of resources and goods among alternative competing uses. It meets the economic criteria of perfect competition, although it is often criticized for inability to meet social and political objectives. However, the efficiency of the market-oriented system will generate more total income to be used to meet these social and political objectives than if these non-economic goals are used to distort the market system and resource allocation. A responsive and flexible system in terms of facilities for handling and shipping of grain is essential to minimizing costs and maximizing returns to producers. This flexibility, lacking in the Brazilian model, marketing boards, and grain boards, is clearly evident in the ability of U.S. firms to make rapid changes in grain flows and mode of transport in response to every opportunity to arbitrage over time, form and space.

Conversely, the Brazilian model, through the actions of the CFP (which operates the Minimum Price Program), CIBRAZEM and other institutions, encourages resource misallocation, and probably long-term environmental damage, through policies that encourage grain production in some frontier areas that otherwise would not be competitive. Sustainability becomes a critical issue in some of these frontier areas because grain production is being encouraged through non-market-based incentives.

Social objectives such as equalizing producer returns often conflict with optimum resource allocation. Several studies have shown the less-than-optimal allocation of resources in Canada with grain being produced in areas in conflict with the principle of comparative advantage as a result of equalized prices (i.e., cross subsidization among crops and regions) throughout the growing area. Similar distortions have occurred in the U.S. production system as a result of price supports. To the extent that prices, or price supports, are established by government decisions, the United States has instituted a limited form of grain board. It is generally agreed that those price distortions created by government decisions have resulted in increased costs of production and decreased efficiency in production and marketing (Part Five).

If the arguments favoring the free market or competitive model in grain production and marketing are found to be compelling for Brazil, then the issues and preconditions necessary for the efficient functioning of a primary and secondary market for private financing of privately-held grain stocks need to be explicitly addressed.

Establishment of an Environment Supportive of the Private Holding of Grain Stocks

Macroeconomic instability reduces the desire of investors to hold instruments like bankers' acceptances and hinders the development of most formal markets for financing stocks of commodities or for lending and borrowing commodities at low transaction costs (i.e., futures markets).

Thus, some degree of *macroeconomic stability* is a necessary, but not a sufficient condition for the successful development of any formal market for financing stocks of commodities. This is because tax policies as well as government credit policies can hinder the development of a formal market, notwithstanding macroeconomic stability.

In the area of *tax treatment*, the experience within Brazil indicates that analysis of the taxation of participants in the agricultural marketing and financial system needs to be undertaken and actions implemented simultaneously with the issuance of regulations permitting various forms of commodity-backed or linked securities. In general, a good principle to follow in considering reforms in this area is to assure that tax treatment is as neutral as possible in regard to its effects upon the forms of commodity-linked financing vehicles available, and in regard to the fiscal deficit. Also, participants should be assured of the stability of the new tax regime.

Comprehensive examination of all forms of government intervention in the process of pricing, marketing, or financing commodities is necessary. This would include restrictions on the balance sheets of private economic agents imposed by governments in the context of various credit allocation programs, special government subsidized credit programs for producers, price support programs, and subsidized credit given to exporters by the Central Bank in the context of promoting pre-export financing. Reforms to limit government intervention of these types should be undertaken soon after or in conjunction with a macroeconomic stabilization program. Moreover, within this area–where possible–price support programs should be dismantled first, while simultaneously government credit resources allocated to the agricultural sector should be priced as close to "market" prices as possible.[5]

Reforms to the structure, integrity, and security of the *warehousing system*, as well as associated legal reforms (e.g., changes in the warehousing law, or laws governing bonding/insurance) can be undertaken prior to efforts to liberalize. In fact, reforms in this area can be justified even in periods of macroeconomic instability because the lead-time for institutional reforms to the warehousing system is often longer than that for policy reforms. A useful analogy is that of reforms to the financial system where reforms to the system of market supervision should generally be undertaken prior to widespread liberalization. Brazil's experience indicates that actions to ensure: (1) issuance of uniform documents by warehouses; (2) adequate systems of inspection of public and private warehouses; (3) development of bonding and insurance systems; and (4) adequate central registration procedures for warehouses, etc., are some of the essential preconditions. If these institutional arrangements are not in existence, they will greatly reduce the desire of financial institutions or other institutions (e.g., corporations in the case of commercial paper backed by or linked to commodities) to undertake agricultural stock financing.

The timing of reforms in the area of *custody, clearing and settlement* depends critically on the extent of development of the financial markets. In this regard, systems in Brazil are greatly advanced in relation to those in many smaller countries in Latin America, or even vis-a-vis some of the larger Latin American countries (e.g., Argentina). Hence, modifications of existing Brazilian clearing and settlement systems to permit trading of commodity-linked or backed securities do not present additional technical problems. Thus, reforms in this area can be implemented quickly, after warehousing system reforms have been implemented.

In the area of *links to other markets*, the experience of Brazil indicates that careful thought should be given to what links are most important to establish between stock financing on the one hand, and cash, futures or international markets on the other hand. A guiding principle would be to permit maximum flexibility to participants in the agricultural marketing system and in the financial system to engage in a sufficiently diverse set of contracts—thereby permitting the more efficient management of various forms of risk and commodity inventories. Thus, for example, links between the financing of stocks and hedging operations should not be discouraged in the context of drafting regulations or laws regarding the financing of stocks and documents issued by warehouses (i.e., warehouse receipts). The use of warehouse receipts as collateral in posting margin for futures contracts should be permitted as long as the warehouse custody system is secure, and does not permit double-pledging of the agricultural commodities.

Finally, as in the area of institutional arrangements, *legal and regulatory reforms* necessary to permit maximum flexibility to economic agents to finance the storage of goods under well-defined property rights will often require a longer lead time to both investigate and implement than other types of economic policy reforms. The following kinds of legal and regulatory reforms may be required:

- clearly defining the rights of the parties to commodity-linked securities and making certain that, in the event of default, the rights to the financed agricultural stocks are assured;
- adopting uniform documents acceptable to most domestic and international market participants so that the capacity to trade the instruments internationally could exist;
- implementing legal preconditions to enhance the origination and secondary market trading of warehouse documents and instruments involving the pledging of such documents; and
- establishing legal preconditions to properly define the rights of parties in cases of bonding and insurance of warehouses, and issuing documents used to create financial instruments traded in the secondary market.

Several major recommendations follow from the above findings. First, reforms to enhance the private financing of grain stocks should gradually phase in legal and institutional reforms. Such reforms could be undertaken prior to, or at the same time as, adoption of economic liberalization measures aimed at a reduction of all forms of government intervention in the financing or pricing of commodities, since their implementation will further reduce the transaction costs to parties engaging in the financing of agricultural stocks or the management of commodity inventories.

Second, the particular sequence of institutional, legal, and regulatory reforms recommended is as follows:

- after study of the existing warehouses, revise the warehouse statutes and regulations;
- inspect existing warehouses and prepare a list of those approved under the new legal framework;
- provide for a uniform system of warehouse documents acceptable to most producers and traders; and
- provide for an adequate bonding or insurance framework for the warehouses.

Third, at the stage of adopting economic policy reforms to the agricultural market, finance and tax systems, it is recommended that (due to the intragovernmental coordination needed) a special inter-agency group be established that will develop a reform package that will permit the natural evolution of a *private* market for financing stocks of agricultural commodities. (Private stock financing is discussed in further detail in Appendices 1 to 3).

Regulatory Environment—Establishment of a Brazilian Grain Commission

All markets require regulatory controls in order to ensure efficient operation. Regulations to control and enforce contracts, to guarantee performance bonds, to set grades and standards, and to control predatory practices by firms are all examples of regulations that are essential to efficient market operation. These regulations can best be handled by an agency, or agencies, completely segregated and insulated from daily buying and selling activities.

In Canada, the Canadian Grain Commission, which is independent and separate from the Canadian Wheat Board, is responsible for control and supervision of grain quality and for all aspects of grain handling. It licenses grain handling facilities, sets grain standards, provides inspection, including of weighing equipment, handles complaints and ensures the quality of Canadian grains. It is supported by fees levied for its services.

The United States relies on several agencies to regulate the industry and to protect producers and consumers from unfair or non-competitive pricing practices. The Federal Grain Inspection Service of the USDA has responsibility for grading standards and inspection procedures for all official grades in domestic and export markets. Federal agencies also supervise weighing, especially in the export market, but also wherever official weights are requested on a grading certificate. The Food and Drug Administration has responsibilities with respect to phyto-sanitary regulations, food safety and purity. Several other agencies are responsible for export licensing and reporting, and for the delivery of grain to LDCs. The Foreign Agricultural Service provides supervision of trading practices, records export sales, and supports staff in overseas markets. Although many of these agencies lack the coordination that might increase their efficiency, their role and their responsibility to a particular sub-sector is generally clear and explicit.

The federal government of the United States is responsible for quality control standards, for universal access to all navigable waterways, and for the enforcement of antitrust regulations. The Export Trading Act of 1982, however, limits the antitrust liability for grain export companies. Cooperatives have always had less stringent requirements with respect to antitrust regulation than other forms of corporation. The federal government is also responsible for maintaining waterways and the federal highway system. In both waterways and highway systems, the federal government provides the basic roadbed and charges user fees to recover a portion of the fixed cost. Operation on those roadbeds is by private companies competing without restrictions in terms of size and rates.

In addition, many of the regulations pertaining to grain are developed and implemented at the state level. Some state governments also have regulations related to delayed pricing, and supervise the operation of country elevators with respect to speculation and to weights and measures.

The Brazilian system has delegated the responsibilities described above among so many different agencies that coordination is impossible and consistent policies unachievable. Selecting those responsibilities that can best be handled by government and giving them to a Federal-level grain commission may provide the necessary coordination and, at least, the opportunity for internal consistency of the policies. Those functions that can best be handled by the market system should then be left to the market with an appropriate regulatory framework established by laws. It is, therefore, recommended that the regulatory responsibilities be given to a Grain Commission and that the marketing responsibilities be left to the free market. Regulations which ensure freedom of entry and a "level playing field" for competitors whenever profit-making opportunities arise are the guarantees required to avoid monopolies and the associated evils of excess profits and exploitation.

The proposed Brazilian regulatory agency, the Brazilian Grain Commission, would have at least the following regulatory responsibilities:

- licensing and inspection of warehouses;
- issuance of uniform documents by warehouses;
- central registration procedures for warehouses;
- registration of all participants in the trade;
- standardization of all trading procedures through the design and administration of all trading, transport, and handling documentation;
- organization of forms and terms of contracts, as well as their control and regulation;
- issuance of Brazilian quality certificates;
- maintenance of grain movement statistics;
- seed production; and
- maintenance of an export registry.

The poor performance of the Argentine system (see Part 1) was blamed, in part, on a single agency trying to achieve conflicting goals because a single agency was trying to meet the needs of both producers and consumers. The internal conflicts and necessary compromises often resulted in ineffective or contradictory policies. The performance of the public agencies in Brazil represents the opposite extreme. Brazil has a plethora of agencies all trying to resolve the same problem in a different way, resulting in conflicts, overlapping responsibilities, and inefficiencies. The lack of clear directives and distinct assignment of responsibilities has led to numerous rules and regulations that are subject to change by any of the many agencies operating in the agricultural marketing system.

Even the export market is controlled by many agencies such as CONCEX, CACEX and SEAP. Inspection and grading is everybody's responsibility and nobody's responsibility. While there are federal grades and standards, the trading contracts of the exporters supersede these. In addition, many states have their own set of grading standards, which are also generally ignored. Grading may be done by the Domestic Inspection Service (CLASPAR) within a state or at the port, in conjunction with private inspection agencies. The private agencies (e.g., Societe Generale de Surveillance, SGS) appear to have final say in issuing certificates.

These are only a few of the illustrations of the confusion and inefficiencies in the marketing system that result directly from failure of the government to organize a rational program of market rules and regulations where they are needed. The creation of a Grain Commission could be a step in the right direction if it is given the authority now shared by numerous other agencies. If it becomes simply another layer in an already over-bureaucratized system, nothing will be gained.

Pricing and marketing decisions should be left to the market participants. However, more information is needed about these prices at the country level. Encouragement of grain exchanges, operated by private industry and regulated by the Grain Commission, could provide a better environment in which to determine and disseminate prices and price information. A department within the Grain Commission could be given responsibility for continual reporting of prices to marketing firms and producers. Price information at the present time is distributed sporadically by some of the state agencies and by trade associations. There is no systematic reporting of prices in a manner similar to that of the U.S. Market News Service.

Implicit within these recommendations is the removal of any implicit or explicit limitations on the use of the futures markets by farmers, country elevators, processors, and exporters. Responses from operators in the Brazil soybean market demonstrated a range of perceptions about the legality of hedging. These responses varied all the way from "there are no limits" to "no one but exporters are allowed to hedge." Whether or not these are legally enforced limitations, the perception should be removed and the opportunity for use of the futures markets be publicized in order to shift risk from the operating firms. Risk and uncertainty is one of the costs of doing business. To the extent that risk can be shifted through hedging, marketing costs can be reduced and returns to producers increased.

GATT Negotiations

Because the Brazilian grain sub-sector is closely linked to the world grain trading system, the outcome of the Uruguay Round of GATT negotiations can have a significant impact if trade liberalization is achieved. The potential effects of removing trade barriers external to Brazil may be almost as significant as improving the workings of Brazilian government policies.

Development of Improved Production Technology

The Brazilian grain sub-sector compares quite well with that of Argentina, Australia, Canada and the United States in terms of yield levels and growth in yields. As indicated earlier (Chapter 10), EMBRAPA made a major contribution to the Brazilian grain sub-sector, particularly in the development of soil treatments for the cerrado soils of the Center/West, improved yields for wheat and improved soybean varieties. In order to maintain Brazil's future competitiveness in grains, support for research, both public and private, should be given high priority by the government.

Notes

1. Lowell D. Hill and A. Mustard, "Economic Considerations in Industrial Utilization of Cereals," in Y. Pomeranz and Lars Munck, ed., Cereals: A Renewable Resource, St. Paul, Minn: American Association of Cereal Chemists, 1981, p.25.

2. Hill and Mustard, "Economic Considerations."

3. Noreen L. Dollinger, "Comparative Analysis of Transport and Handling of Export Wheat in Canada and the U.S." , Unpublished M.S. Thesis, University of Illinois, 134 pp., November 1980.

4. Alex McCalla and Andrew Schmitz, "Grain Marketing Systems: The Case of the United States vs. Canada," American Journal of Agricultural Economics, 61:2 (May 1979) 199–212.

5. Problems can arise if interest rates are not being liberalized generally in the economy, because borrowers will often obtain "subsidized credit" from other direct or indirect government-owned financial institutions.

Appendix 1

Developing Private Markets for Agricultural Stock Financing[1]

Introduction

This chapter reviews and analyzes the issues and preconditions necessary for the efficient functioning of a primary and secondary market for private financing of the stocks of agricultural commodities. To make the discussion more concrete, reference is made to Brazil's experience and to the conditions necessary to permit the financing of stocks of commodities through the use of bankers' acceptances that are backed by storage receipts issued by warehouses. In spite of the focus on acceptances, the issues raised should be considered in the context of legislative or regulatory changes permitting any form of capital market instrument (e.g., commodity-backed commercial paper or commodity-indexed liabilities, such as certificates of deposit or export notes indexed to commodity prices) designed to finance the stocks of agricultural commodities. Some of the more important implications for the sequencing of reforms in developing countries characterized by differences in the economic policy environment or in the stage of development of the institutional and legal infrastructure necessary to support the private provision of agricultural stock financing are highlighted.

Macroeconomic Policies

General

Monetary and Fiscal Policy. Monetary and fiscal policies that result in rapid and volatile inflation rates coupled with large expected fiscal deficits that require the government to pay extremely high (ex-ante and ex-post) real rates of interest on its domestic debt can greatly reduce the incentives for private financing of the stocks of agricultural commodities. This can occur for several reasons.

First, in such an environment private securities are crowded out because the yields offered on government securities make investors less likely to have interest in private fixed-income securities, such as bankers' acceptances, even if these are indexed.

Second, fear of confiscation by the government forces economic agents to undertake an increasing amount of transactions on an informal basis, inclusive of extensions of credit for financing of stocks. Such contractual arrangements preclude the organization of an orderly, formal primary or secondary market for an instrument such as bankers' acceptances.

Third, given macroeconomic policies that lead to high and volatile inflation and interest rates, the maturity of all financial contracts, inclusive of indexed contracts, tends to shorten particularly when the index used is provided by a noncredible government that does not use a widely accepted and independently verifiable measure of inflation. Under these circumstances, issuance of indexed bankers' acceptances of maturities greater than several days, as would often be needed to finance stocks of commodities, exposes the originating commercial bank to significant interest rate or valuation risks vis-a-vis government securities that are offered sometimes with a maturity of only one day.

Finally, in some extreme cases (e.g., Argentina) macroeconomic instability and the threat of future confiscations of wealth result in a situation where producers in the marketing system refuse to sell stocks of commodities unless payment is made in goods or dollars. The implication for stock financing of the noncredibility of the domestic currency as a store of value is that pre-export finance will often be used in order to provide dollar financing for the stocks of goods, where this can be arranged. However, even this form of finance can be increasingly rationed when macro-economic instability results in the expectation that the government will not permit, or will tax heavily, the exportation of the tradeable good being financed.

In sum, some level of minimum stability in the macroeconomic environment is a critical precondition for the gradual development of any organized private market of substantial size for financing the stocks of agricultural commodities. Although such stability may be a necessary condition it is by no means a sufficient condition.

Exchange Rate Regime. An additional link between macroeconomic policies and private financing of stocks of commodities relates to exchange rate policy and the existence or nonexistence of markets to manage own-currency foreign exchange risk (i.e., between the local currency and dollar). For example, in countries that maintain an official exchange rate that may not move in line with domestic inflation, and where no market exists to manage the risks of movements in this rate or in the black market exchange rate (through the use of forward contracts or futures contracts), development of a liability indexed to the prices of tradeable goods has been viewed attractively by investors as a good substitute for a dollar asset or in some instances as a good means of hedging own-currency exchange risk.[2]

Under these conditions financial institutions (that own warehouses) have been able to purchase tradeable goods for future delivery to the warehouses and simultaneously issue (often to multinational corporations) a form of security indexed to the price of the commodity. Thus, in this transaction, a form of production and/or stock financing is provided to a producer that is ultimately financed in the domestic capital markets.

The Case of Brazil

Monetary and Fiscal Policy. Instability in Brazilian macroeconomic policies affected the prospects for the development of a primary and secondary market for bankers' acceptances in several ways.

First, given the uncertainty surrounding future government actions, large expected future deficits and inflation rates that approached almost 80 percent per month by early 1990, commercial banks that might originate acceptances, larger agribusinesses that might borrow under the program, and investors were all maintaining very short-dated portfolios of assets and liabilities (often of no greater duration than between 7-15 days). Simultaneously, the average maturity of the government's internal debt had fallen from over a year toward the beginning of 1988 to less than five months by the end of October 1988 and to about one day by the end of 1989. Private banks were typically issuing CDs (often at annual compounded rates below that of the government's overnight cost of borrowing) and investing these funds in the overnight (repurchase) market. In such an environment it is unclear that acceptances having a minimum maturity of 60 days with a discount will be able to compete with government securities, or with other forms of deposits that must be earmarked for agriculture (e.g., such as so-called green savings accounts; see below).

Second, as a result of the increased perception by commercial banks and investors that the government would not maintain a *stable* tax regime in regard to financial or commercial (i.e., agricultural) transactions or would intervene in the process of engaging in private contracts, informal market transactions through intercompany loans became more popular. Hence, acceptances will have to be competitive with a variety of formal and informal alternatives, before commercial banks or investors will have an interest in using this instrument.

Exchange Rate Considerations. Finally, although the general macroeconomic instability worked against the development of a private market for financing stocks based on storage warrants, it was clear that this instability created an interest by borrowers in creating a *secure* system for trading short-dated commodity-linked securities, linked to prices of tradeable goods. During 1989, investment banks developed means of privately placing this type of paper in the market for creditworthy clients through the creation of so-called export or DEX notes whose return was linked to commodity prices. Such notes were issued in domestic capital markets and were used to finance advance purchases of agricultural commodities by the affiliated trading company subsidiary of the same financial group. In this way, producers did obtain a form of credit (i.e., cash prior to delivery of commodities). The financial institution took the underwriting risk associated with selling participations in the commodity-linked instruments of sufficient value to finance the cash purchase of goods that were stored at warehouses owned by the financial group until the time of export. Finally, the investor (often a multinational) purchased these notes in part as a means of hedging the risk of maxi-devaluations in the dollar/cruzado rate. This was possible to the extent that prices of tradeable goods were highly positively correlated with those exchange rate movements, thereby serving as a form of cross-hedge.

Alternatively, given existing tax incentives, a variety of financial groups acquired and held inventories of tradeable raw agricultural goods or industrialized products, often by creating trading companies that operated under the umbrella of a holding company (i.e., often within a financial institution such as a commercial bank). In this way, the financial institution was able to diversify its portfolio across financial assets and commodities and was not subject to credit risk as it physically controlled the commodities.

Tax Policies

General

The existence of different forms of contracts for financing the stocks of agricultural commodities and the desirability of changing other regulations or legislation to permit other forms of private financing for the stocks of agricultural commodities are not independent of the existing structure of the tax system. In this regard, numerous tax-related issues arise in regard to different forms of explicit and implicit taxes levied by government on participants in an agricultural marketing system and upon potential investors in commodity-backed or -linked securities or upon institutions in a position to originate different forms of securities that could be demanded by the investor. Although not exhaustive, experience has shown that the following types of considerations are important: (1) asymmetric corporate income tax treatment of industrial versus agricultural companies; (2) tax treatment of commercial versus financial transactions; (3) various forms of export taxes; (4) various forms of implicit taxes (e.g., reserve requirements, forced investments, lending limits, etc.).

Corporate Income Tax. In many countries, the corporate income tax treatment of agricultural companies versus financial or nonfinancial companies differs, with that for the former often being less onerous than for the latter. In these cases, commercial banks having trading company subsidiaries viewed as agricultural companies will tend not to have an incentive to originate bankers' acceptances. This is because such transactions will involve an extension of a collateralized loan, the income from which will accrue to the bank, and will be taxed at a higher corporate income tax rate. Instead it will be preferable for the trading company subsidiary to engage in different forms of contracts with producers that involve payment in advance in expectation of future delivery so that the subsidiary is viewed as undertaking a cash transaction that affects income.

Commercial Versus Financial Transaction Taxes. In some countries, tax treatment of financial transactions versus commercial transactions can create incentives for not organizing a formal market for financing agricultural stocks of commodities. For example, if there are taxes on capital gains associated with financial transactions, and commercial transactions are only subject to corporate income taxes at a lower rate, incentives will exist, ceteris paribus, to book commercial transactions in an agricultural company. Moreover such asymmetries can be created through other forms of taxation applied to financial institutions (e.g., withholding tax on interest payments on loans extended). In this context a banker's acceptance that is originated by a commercial bank involves an extension of credit against the storage receipts, permitting recourse to the goods. If interest on the loan is subject to withholding taxes that are in excess of the income tax rate for agricultural companies, further incentives are created for multinational corporations to engage in special advance payment contracts of the type described above. These contracts can be booked as commercial transactions, thereby minimizing tax liability of the corporation, despite the fact that an implicit interest rate exists under such transactions.

Export Taxes. Both explicit taxes on exports and implicit taxes through the use of multiple exchange rates can contribute to the lack of organized domestic private markets for domestic financing of stocks of agricultural commodities regardless of the incidence of this tax. For example, if the incidence of the tax falls upon the exporter his demand for the good will fall, ceteris paribus, and the extent to which the exporter will need to arrange for pre-export financing will also fall, thereby reducing one form of stock financing. Similarly, if efforts are made to shift the tax to the investor in the form of offering a lower return on his investment, it is likely (depending on the elasticity of substitution of acceptances with other financial assets) that this other source of finance for the carrying of stocks would also fall.

Implicit Taxes. A variety of implicit taxes if applied differently across institutions involved in marketing the commodity, or upon intermediaries or corporations extending financing, could affect the preferences of both borrowers and investors in the context of private financing of commodities. Of particular importance in many countries are reserve requirements, forced investments in government securities earning non-market yields, lending limits, eligibility to use the commodity-linked or backed security as collateral in countries with well-developed repurchase contract markets, etc. For example, if reserve requirements are not applied to acceptances originated by a bank while they are applied to other bank liabilities issued, "incentives" are created for issuance of this form of stock financing as a means for financial institutions to avoid the implicit tax of reserve requirements. Such an exemption can only be justified on fiscal grounds if the *loss* in revenue due to non-uniform taxes on different forms of bank liabilities is more than offset by the *reduction* in government-subsidized credit and price supports extended to finance agricultural commodities. A similar type of analysis could be applied to the other less traditional forms of implicit taxes noted above. In cases where several forms of implicit taxes are applied simultaneously to operations of a bank or other institution originating acceptances (or commodity-backed or -linked assets), an analysis would have to be performed of the overall extent of taxation relative to other funding alternatives.

In sum, a review of the tax regime within a given country is an essential precondition to the drafting of regulations or legislation that will permit issuance of different commodity-backed or -linked notes by different forms of financial institutions. Although general principles are difficult to establish in this area, it is clear that efforts should be made to make the investor's decision as well as that of the institution originating the acceptance or other instrument neutral in regard to tax treatment. In general, tax treatment should not encourage the development of informal markets that often increase transaction costs for all parties, or that can reduce the amount of information economic agents can obtain from prices, or increase the costs of acquiring such information.

The Case of Brazil

Corporate Income Taxes. Differences in the corporate income tax rates applied to corporations that are financial versus agricultural companies and differences in the tax treatment of capital gains associated with "financial" versus "commercial" transactions have both contributed to the slow development of an organized domestic private market for financing stocks of agricultural commodities. For example, income earned by agricultural companies (that can include export trading companies trading agricultural commodities) is taxed at a rate of between 3 percent to 6 percent whereas the corporate income tax rate is normally close to 40 percent for nonagricultural corporations depending upon whether they are financial or nonfinancial. The implications of this form of tax treatment have been that financial and nonfinancial corporations have found it profitable to create subsidiaries that specialize in the trading of agricultural commodities. Moreover, it is more profitable for such corporations to record transactions on the books of the agricultural commodities company and, as is made clear below, to treat the transaction as commercial as opposed to financial. Thus bankers' acceptances based on storage warrants that can only be "originated" and "booked" on the balance sheet of commercial banks would not be subject to such preferable income tax treatment.

Commercial versus Financial Transaction Taxes. Another tax incentive that works against development of "formal" domestic provisions of stock financing is the tax treatment of "financial" versus "commercial" transactions. For example, as of January 1989, capital gains on all financial transactions regardless of maturity were taxed at a rate of 25 percent applied to the difference between a government index (OTN fiscal) and the market value of the security. By contrast, "commercial transactions" are not subject to a capital gains tax, and revenue accruing from such transactions is included in income that is taxed at *year end*, an accounting treatment that confers substantial benefit in a high-inflation environment. In addition, the financial transaction tax (IOF) is still in effect under which amounts loaned by financial institutions are subject to withholding taxes, while such taxes are not applied in the context of "commercial transactions" that constitute a purchase or sale. The implication of these tax provisions for the private domestic provision of agricultural stock financing (through acceptances or other instruments) is two-fold.

First, differences in the real-after tax yields in booking "commercial" versus "financial" transactions have resulted in the arrangement of transactions where both "borrowers and lenders" have benefitted by booking so-called commercial transactions that involve an implicit extension of credit. For example, many trading companies engage in *anticipated payment contracts* in which they provide cash against future delivery of goods by the producer or warehouse person to a specified grain elevator. Often cash is advanced before the goods have been harvested, or even before the time at which planting begins. This is booked as a "cash transaction" even though *implicitly* credit has been extended to the extent that this is not a *spot* transaction. Thus, from the vantage point of *financial* institutions originating acceptances or other forms of commodity-linked securities, the yield on acceptances would have to be sufficiently high to compensate for the tax advantages inherent in extending credit implicitly through booking a commercial transaction.

Second, the fact that commercial transactions are exempt from the financial transactions tax (i.e., IOF) also creates disincentives for banks to originate bankers' acceptances or other commodity-linked or -backed instruments. This is because the origination of such instruments often entails the granting of a loan or time draft by the commercial bank to the borrower against the storage warrant, a transaction that would be subject to the IOF tax. The alternative of extending "informal" anticipated payment contracts will be preferred, which amounts to a form of implicit supplier's credit.

Export Taxes. Various forms of implicit (i.e., dual exchange rates) or explicit export taxes have existed historically in Brazil; however, in recent years this form of fiscal finance has not been used actively in Brazil for the tradeable agricultural commodities eligible to be financed under Brazil's bankers' acceptance regulations.

Implicit Taxes. At the time of the original design of regulations permitting issuance of bankers' acceptances in Brazil, reserve requirements were imposed on all time deposits, and it was envisioned that acceptances deemed eligible under the bankers' acceptances program would be exempt from required reserves. However, at present, no reserve requirements are imposed on time deposits, even though these deposits are the key marginal source of funds for commercial banks and investment banks. For these purposes, time deposits include overnight deposits, while, in contrast, the higher US reserve requirements on demand deposits apply to deposits with maturities as long as six days. Thus, a key incentive for originating acceptances based on storage warrants (i.e., an exemption from reserve requirements) is no longer present.

Also, in Brazil as opposed to a variety of other countries (e.g., USA), repurchase contracts using private securities as collateral are not subject to reserve requirements so that a reserves exemption for engaging in repurchase agreements with bankers' acceptances would not be of value for banks. However, in the Brazilian context, differing leverage requirements vis-a-vis capital are applied for the use of private versus government securities in repurchase transactions (e.g., 3 to 1 for private securities and up to 30 to 1 for government securities). Although the 3 to 1 limit is not currently binding for many large banks (the ones that might be expected to originate acceptances), or for pension funds held by banks, some banks indicated (as did smaller banks), that creation of a special sublimit for use of bankers' acceptances as collateral in repurchase agreements would create incentives for participation in the program by medium-sized banks.

Government Intervention in Marketing Systems

General

Government Credit Programs. Many countries maintain substantial directed and subsidized credit programs targeted at the agricultural sector. Although some of these programs may be justified in the context of targeting subsidies to particular individuals, provision of this form of finance (at non-market rates) to finance stocks of commodities has often reduced the incentives for private provision of such short-term credit. In subsidized agricultural credit schemes, the government tends to become the only marginal source of new credit.

Price Support Programs. Price support programs, like directed and subsidized credit programs, can undermine the extent to which private financing arrangements will flourish for two reasons.

First, in many instances, the volume of funds the government devotes to purchasing and storing stocks of agricultural commodities directly substitutes for private financing arrangements.

Second, if the minimum price rather than the world price is used in order to determine the initial value of the loan to be made by the bank against collateral when a bankers' acceptance is originated, and the minimum price is substantially below the world price, the demand for this form of financing can fall. The demand falls because the borrower will receive far less financing than would be available under a free market regime with no minimum domestic price for the commodity. Even if this factor is compensated for through adjustment of other terms of the loan (e.g., lower interest rates), the policy will still be problematic because distortions will be introduced into the pricing of credit purely as a result of real sector intervention (i.e., the minimum price policy).

Real Trade Intervention. In addition to intervention taking the form of explicit export taxes, many countries have either closed export registers, not permitting the exportation of goods, or have engaged in outright liquidation of private contracts involving the financing of tradeable goods. Often these actions are taken as part of macroeconomic stabilization plans or in the context of husbanding scarce hard currency reserves. Under these circumstances, investors as well as banks or corporations that might originate acceptances or commercial paper will tend to fear that their security in the transaction — a tradeable good that can be sold on world markets — will no longer be sufficient security. This leads to an increase in the spreads required to extend this form of financing to holders of commodities. This problem will be particularly acute when government interventions are repeated frequently and the government has little credibility. In addition, such interventions adversely affect trade finance extended in support of pre-export financing that often serves as an important source of short-term financing for the stock of tradeable commodities.

Intervention in Complementary or Substitute Markets. In addition to intervention in the market for the commodity to be financed, or in the credit markets, interventions in complementary or near substitute markets can have an impact on the emergence of private markets for the private financing of agricultural commodities. A good example of this phenomenon relates to the existence of interventions in explicit or implicit (i.e., futures) markets for borrowing and lending commodities. In this case, government liquidation of these contracts in the context of a stabilization plan increases the transaction costs to firms of hedging, or alternatively assuring access to the commodity at minimum costs, because such government actions result in the non-existence of an organized market for borrowing and lending commodities at low transaction costs. In many cases, processors will want to finance stocks of commodities that are too "large" because they will fear that the costs of obtaining greater access to the good will be very high. Alternatively, the industrial structure of firms may be altered, through increased vertical integration among institutions in the agricultural marketing system.

The nonexistence of well-organized markets for borrowing and lending commodities will not permit participants in the agricultural marketing system to economize on commodity inventories, implying that they may wish to use the credit markets to a greater degree to finance stocks. However, this should not be viewed as an indication of the successful development of the private financing of stocks of commodities, but instead as a problem in that the full set of contracts does not exist to permit the most efficient management of commodity inventories. In addition, development of well-organized and regulated commodity futures markets or formal markets for trading and marking to market warehouse receipts also permits better management of commodity price risk as well as permitting greater accessibility to a given good at low transactions costs.[3]

Under these conditions greater amounts of finance may be provided to holders of stocks of commodities if they can hedge their positions.

In sum, efforts to eliminate or drastically reduce government credit and price support programs and interventions in the trade of agricultural goods, or in related financial markets, are important preconditions for the success of private financial arrangements.

The Case of Brazil

Government Credit Programs. Direct or indirect government provision of credit to the agricultural sector in general, regardless of the specific use of funds (i.e., for commercialization, investment, etc.), has been declining.

During 1987, some official banks were authorized to issue so-called green savings accounts (cadernetas verdes) which were special *tax free* savings accounts paying 6 percent over the government-calculated index of inflation (i.e., now BTN). To raise funds in this way (through what amounts to a government subsidy), these institutions must direct these funds to the agricultural sector. In addition, mandatory applications of sight deposits (net of reserve requirements and taxes) to agriculture under government regulations were increased to a range of 60 percent to 90 percent, depending on the size of the bank.[4]

Finally, the Government's program of subsidized loans to finance stock-holding (EGF) was increased so that the Government could reduce the funds allocated to crop purchases under a separate crop purchase program (i.e., AGF).

Price Support Programs. The AGF program, under which the Government stands ready to purchase goods offered at minimum prices, will continue to affect the development of a bankers' acceptance market if the value of the credit made available to the producer is linked to the minimum prices set by the Government. The policy of linking the size of financing made available to the value of goods based on administered prices limits the attractiveness of instruments like bankers' acceptances, as so-called minimum prices often lie substantially below the world market prices, thereby further limiting the value of the loan that can be made against the commodities posted as collateral.

Real Trade Intervention. The Government has often intervened in the past to either close the export registries for traded goods or to severely limit export licensing. Such actions have had a negative effect on the development of private arrangements to finance storage of commodities because investors or banks that originate commodity-backed or -linked instruments would want assurance that the goods could be sold offshore freely.

Intervention in Complementary or Substitute Markets. In Brazil, the most extensive government interventions have been in complementary markets such as the Brazilian domestic commodity futures markets. These interventions have taken the form of outright liquidation of futures contracts in the course of macroeconomic stabilization plans that contain incomes policies involving the freezing of all prices and wages (e.g., the Cruzado Plan of 1986). The result has been that producers, processors, or other agents in the marketing system for different commodities have not used the domestic futures markets extensively. Thus, the liquidity of these markets (i.e., the ability of participants to offset positions without large changes occurring in the price of the contract) has been very limited even for nearby contracts one month ahead. This has resulted in a greater need for smaller companies in processing industries such as soybeans, orange juice, etc., to either integrate into production or in some cases into storage. In spite of the fact that agents in the agricultural marketing system may have to carry greater stocks on average with, ceteris paribus, a greater associated demand for financing, this effect has been outweighed by the negative effect upon financing extended by banks, due to the inability of borrowers (e.g., producers) to manage commodity price or output risks, particularly if the transaction costs of accessing offshore futures exchanges are high.[5]

Institutional Issues

Although close attention is often paid by governments of developing countries to many of the economic preconditions that are necessary for creating an environment conducive to *private* storage financing, less attention is devoted to the equally important area of institutional issues and preconditions. Under this heading the following issues need to be examined: (1) the structure of the warehousing system and particular institutional characteristics that should be in place if private financing of stored agricultural goods is to flourish; (2) institutional characteristics of custody, settlement and clearing arrangements that are especially important in the context of commodity-backed or -linked financings; and (3) monetary policy and Central Bank discount window operations.

Warehousing System

General

The initial hurdle to be surmounted is the existence of a system of warehouses capable of storing stocks of commodities and issuing credible, legally binding receipts for those commodities. Some regions might lack adequate storage facilities, while other regions might have a surplus of such facilities.

Uniformity of Documents. Even in countries in which adequate warehouse systems exist, all warehouses might not issue comparable documents as receipts. Privately owned and controlled warehouses might issue receipts in a form little changed for decades; cooperative warehouses might issue different kinds of receipts altogether; and government-operated warehouses might issue yet a different kind of receipt. Federal or State authorities should establish basic legal requirements so that all receipts issued by warehouses are uniform.

Inspection Procedures. Depending on the kind of warehouse system, inspection procedures might vary widely. Private or cooperative warehouses might not be audited by chartered accountants or their equivalents, and government inspectors might visit only government-operated warehouses. Uniform inspection procedures should be adopted by governmental authorities to ensure the legitimacy and integrity of all warehouse receipts issued.

Ownership Issues. Inspection and insurance procedures might be adequate for the establishment of privately operated remote or field warehouses in regions lacking sufficient independent storage capacity. Cooperative warehouses already exist in some developing countries, and municipal, state or provincial government-operated warehouses also might exist. Federal or central government-operated warehouses exist in many developing countries. If none of these types of warehouses exists, or if the existing warehouses are inadequate, then commercial banks or other financial intermediaries might create, purchase, or operate the warehouses in conjunction with governmental authorities or establish a quasi-public corporation to own and operate warehouses issuing receipts.

Registration Procedures. The warehouses certified by whichever authority coordinates the stock financing program (e.g., the Bankers' Association, Central Bank, or government warehouse inspection unit) should be compiled in a list of qualified warehouses. Existing warehouses should be encouraged to apply for such certification.

Bonding and Insurance Requirements. In order to ensure the viability and financial capacity of the warehouse person to honor receipts he or she issues, credible surety bonds or insurance provided by third persons who inspect the warehouses should be required. Banks operating in countries lacking such bonding and insurance companies could resort to direct ownership or control of warehouses to satisfy these concerns. All warehouses issuing receipts used to finance agricultural stocks should be certified by governmental authorities or an efficient quasi-public entity. Existing warehouses desiring to issue valid receipts also should be subject to such certification.

Technological Developments. Advances of computer access and technology in developing countries might create an environment in which it would be feasible to prescribe uniform standards for issuance, transfer, and redemption of warehouse receipts supporting the financing of agricultural stocks. If networks of computers exist, then a centralized clearing and reconcilement mechanism for warehouse receipts could be created.[6]

Increased computerization of warehouse operations also might facilitate centralized monitoring of the inventories of goods in storage by the responsible authorities, regardless of whether those stocks are financed.

The Case of Brazil

Uniformity of Warehouse Documents. Three different groups of warehouses might be interested in issuing storage documents for use in bankers' acceptance or other forms of commodity-linked financings. However, each of these groups operates under a different law and ordinarily issues a different set of storage documents, as follows:

Warehouse	Documents
Armazens Gerais (privately owned warehouses)	Conhecimento de Deposito plus warrant (technically both documents are negotiable)
Cibrazem (publicly owned warehouses)	Certificado de Deposito (non-negotiable) Conhecimento de Deposito plus warrant (these negotiable documents can be requested but are rarely asked for)
Cooperatives	Certificado de Deposito or mere warehouse receipt (non-negotiable)

Each of these warehouse groups may issue non-negotiable storage documents. Cibrazem warehouses may issue negotiable conhecimentos de deposito (i.e., certificates of deposit) and warrants but usually do so only for large producers. There is no system of storage documentation that is consistent across all classes of warehouses and for all classes of producers that would participate in a uniform bankers' acceptance or other commodity-linked financing program. The implication of this situation is that the set of possible participants may be limited purely because of differences in the documents issued by different types of warehouses.

Of particular importance are the documents issued by cooperative warehouses. It is not possible, under the applicable law for Armazens Gerais (Decreto No. 1102, November 21, 1903), for cooperative warehouses that buy and sell goods of the types that they store also to issue negotiable certificates of deposit and warrants. This problem could be overcome, according to government officials, if the law for cooperatives were changed so as to permit these institutions to issue these documents. However, such changes would need to be coupled with special arrangements for the bonding and insuring of so-called "field" warehouse persons that would guarantee to banks originating acceptances that cooperatives owning warehouses were not pledging the same goods twice as collateral.

Ownership Issues. Participation by cooperatives in any private storage financing market is further complicated by the fact that many cooperatives that wish to participate also own their own warehouses. This would make them ineligible borrowers under the existing Brazilian regulations governing bankers' acceptances. Cooperatives could borrow against goods stored in their own warehouses if Cibrazem's inspection procedures at warehouses were amended to authorize field warehousing at cooperatives. The field warehouse would have to be registered as a general warehouse and would have to be properly bonded and insured.

Registration Procedures. Cibrazem is engaged in preparation of a cadastro (register) of warehouses qualified to participate in Brazil's officially recognized bankers' acceptances program. It appears that an encouraging beginning has been made in this effort, but much remains to be done in registering qualified warehouses, especially if field warehouses are created to enable cooperatives to participate in these markets. Moreover, current registration procedures probably should be modified because many private sector warehouses and agribusinesses suggested that very efficiently run new warehouses have sometimes not been granted a registration number because they only had bulk storage facilities and could not provide for bag storage.

Inspection Procedures. Inspections of warehouses in Brazil are weak, but reforms have been undertaken in the Banco do Brasil's program of inspection under which it acts as agent for Companhia de Financiamento da Producao (CFP).[7]

The inspections of Armazens Gerais are supposed to be performed by local Juntas Comerciais (a stronger version of U.S. Chambers of Commerce), but those inspections have been weak in recent years. There is no significant confidence in the current inspections of armazens gerais, but there is a general consensus that private warehouse conditions are best in the states from Minas Gerais southward. In spite of this situation, a number of private warehouse companies indicated that they were examining the technical feasibility of setting up container storage facilities in the interior and shipping agricultural products to their own warehouses. Such procedures could overcome some of the security problems associated with storing goods in public warehouses in the interior.

Supervision of warehouses and warehouse inspections also needs further improvement. For example, the role of Banco do Brasil as inspector of cooperatives and on-the-farm warehouses might create a conflict of interest if Banco do Brasil were a major lender to customers or owners of those warehouses and if reported defects in those warehouses weakened the credit or legal quality of Banco do Brasil's own assets. Thus far, the Banco Central has not engaged in on-site inspections of warehouses in the banking supervision department (DEFIS). Instead, the rural credit department of the Central Bank (DERUR), which might not have the required degree of incentives to discover defalcations in warehouses, has performed the on-site supervisory inspections. The banking supervision department does not receive detailed reports on the operations of warehouses owned by or affiliated with banks but focuses instead on the net income and net worth positions of those warehouses, which are the factors affecting most directly the safety and soundness of banking operations. Also, the banking supervision department does not inspect the warehouse inspection units of Banco do Brasil on the understandable ground that such inspections are conducted only in an agency capacity for CFP and do not put Banco do Brasil's own funds at risk. Unfortunately, the degree of warehouse inspection and supervision necessary to inspire and sustain public confidence in the existing warehouses did not occur under existing arrangements as of 1989.

Bonding and Insurance Procedures. Bonding and insurance arrangements for warehouses have not been common in Brazil and costs of obtaining this service have been high by comparison to many other countries. Questions arise as to the size of the actuarial risk involved in providing bonding and insurance services to warehouses in Brazil as well as the reasons why market participants have not found it profitable to develop systems for pooling risks associated with providing this service.

Technological Developments. Although computer technology is advanced and is applied well in the financial market in the context of custody, settlement and clearing of financial transactions, little attention has been given in Brazil to the computerization of warehouse receipts so as to permit computerized trading of these receipts and the marking to market (using commodity prices) of the positions in commodities carried.

Custody, Clearing, and Settlement

General

The ability of a market for the financing of stocks of agricultural commodities to grow in volume and to permit agents in the marketing system, or investors, the flexibility to manage the risks of such financing, requires that positions in bankers' acceptances or commercial paper linked or backed by commodities be offset at low transaction costs and with minimum credit risk. A key precondition for this is the existence of adequate clearing, settlement, and custody services that result in a minimization of credit risk and transaction costs in trading these securities. Several issues arise that must be considered in ascertaining whether important institutional pre-conditions can be met that are somewhat unique to commodity-backed or -linked instruments.

Custody. In the context of transactions backed by or linked to commodities, several issues related to custody arise. First, in the context of an acceptance or a commodity-linked or -backed commercial paper transaction, custody arrangements may involve the tracking of as many as three documents, although typically only two documents are essential.[8]

What is needed in this situation is an independent, centralized depository (or linked depositories) that can track ownership of these documents, transfer ownership if the documents are negotiable (see below), and assure the immobilization of these documents.[9]

In fact, in countries that already have designated depository banks or centralized depositories for all forms of private securities (i.e., bills of exchange, equity, debentures, commercial paper, etc.) it will be possible also to include acceptances or commercial paper linked to or backed by commodities as other securities within the centralized depository. Similarly, it will often be possible to permit inclusion of other documents and their separate trading if they are not attached to the security being traded and such trading is legally permitted.[10]

Second, for the custody system to be linked properly to the system for settlement and clearing of *cash* and *physical* securities, it is important that each document emitted in the context of originating an acceptance or (some other instrument financing commodities) have a unique identification number. In addition, a unique identification number should also appear on each document that indicates the warehouse issuing the storage documents (i.e., warrants, etc.) or the accepting bank (i.e., bankers' acceptances) or issuing corporation (e.g., financial corporation). Such a system would permit the clearing and settlement corporation to coordinate with the custody system (e.g., centralized depository or custodian banks) to track the locations and ultimate owners of each document, a capacity that is important in reconciling trades that for some reason do not clear due, for example, to a default by a counterparty, etc.[11]

Finally, the system of custody to be set up can also provide services other than safekeeping. These can include (1) settlement (see below); (2) collection and distribution of dividends, distribution of monies, etc., on behalf of members; (3) pledging of securities; (4) money handling; and (5) checking to see that delivery of securities has been made in the context of matching payment of cash with delivery of securities. Of the above functions, the role of the custody system in permitting the safe pledging of securities or documents (i.e., storage warrants) as collateral to secure loans, or as margin in futures and options trades, is particularly important in the context of developing private markets for financing stocks of commodities, or local commodity futures exchanges.

For example, in the context of the origination of acceptances or other commodity-linked or backed securities, storage warrants have to be pledged as collateral by the borrower, or as collateral in the context of posting margin in a futures trade if such markets exist. Without a centralized depository (or linked set of depository banks) borrowers would have to physically deliver the warrants or other documents to the lender's custodian who must verify the documents with the warehouse as well as the market value of the goods, etc. Warrants pledged must be segregated from other warrants, and securities often need to be re-registered. This process is labor intensive and increases transaction costs as well as the possibility for abuses in trading (such as double leveraging using the same collateral).

Clearing. This concept refers to the process of determining the accountability for the exchange of money and securities of the counterparties to a securities trade. In the context of commodity-backed or linked securities, as in the case of securities generally, this will require that three issues be investigated.

First, what system for trade comparisons exists and is there a private clearing corporation or a corporation associated with an exchange that can undertake this function? Depending on the level of development of financial markets it may not be difficult to use a system for comparing trades that is used on the exchanges or if relevant by a clearing corporation if such institutions exist.

Second, is a clearance system (e.g., trade for trade or various forms of netting) already present, and is it the correct system in light of the expected volume of transactions? Systems for clearing transactions run the gamut from trade for trade systems, where following execution of the trade the buyer and seller directly exchange securities and funds so that securities are processed one by one, to netting procedures (bilateral or multilateral) that reduce processing costs of clearing securities trades but are often only worth implementing when the volume of trading grows above certain minimum levels.[12]

Third, would it be wise to permit the clearing organization to offer counterparties to trades in these securities trade guarantees whereby the clearing organization stands between the buyer and seller of the securities? In many countries provision of these guarantees creates incentives for the clearing organization to police the actions of all members of the clearing organization or of indirect correspondents (warehouses emitting documents, banks issuing securities, etc.).

Settlement. Efficiency of the market for financing stocks of agricultural commodities, as well as the secure secondary market trading of commodity backed or linked securities, requires that a reasonable system for settling securities trades be in place. In this context settlement refers to the completion of a transaction where securities and corresponding funds are delivered to the appropriate accounts. To meet these objectives several issues need to be investigated.

First, how linked is the payment system to the system for settling the securities trade, and is a double-entry bookkeeping system in place that will permit settlement of securities trades without physical movement of securities? In cases where these systems are not linked, or where settlement of the securities trade does not occur simultaneously or close to the payment date, credit risks and transaction costs are higher. Problems in this area can greatly affect the willingness of agents in the agricultural marketing system, or in financial institutions, to participate in the origination and trading of securities linked or backed to commodities.

Second, is it legally possible to borrow and lend securities and the documents associated with the creation of different commodity-backed or -linked securities, and is such lending and borrowing adequately regulated by requiring marking to market? In many countries such transactions are undertaken illegally in the context of informal transactions. Such transactions should be allowed and properly regulated because they permit increases in the efficiency of settling securities trades and, more importantly, in the efficiency of commodity inventory management by agents in the agricultural marketing system. For example, efficiency of commodity inventory management will be improved by permitting economic agents to borrow and lend commodities in a secure environment through the use of warehouse receipts. A warehouse person or commodity processor in a marketing system could obtain some of the same benefits that would be realized from an organized commodity futures exchange.

The Case of Brazil

Clearing and Settlement. Acceptances and negotiable storage documents could and probably should be entered into a book-entry settlement and clearing system. This kind of system already exists in Brazil for securities not issued by the Government. CETIP has been designated as the book-entry settlement and clearing agency for bankers' acceptances in Brazil. However, it does not actually maintain physical custody of documents.

Custody. An independent third-party custodian of the physical documents should be agreed upon by Banco Central and CETIP (or Andima). For example, a set of private banks could be designated as the custodian of the physical documents, while CETIP would account for all book-entry registration and transfers of the acceptance documents. In order to make possible the accurate record-keeping and retrieval of documents pledged or traded separately through CETIP, each document (the acceptance, certificate of deposit, and warrant) should have a unique identification number. Identification numbers for accepting banks and warehouses issuing the storage documents also should appear on those documents. Then the clearing and settlement organization (i.e., CETIP in Brazil for privately originated securities) could carefully track the location (ownership) of each of the bankers' acceptance documents. In the secondary market, the registered owner of the acceptance and the certificate of deposit should be the same, while the accepting bank should be the registered owner of the warrant.

Monetary Policy and Discount Window Operations

General

The financing of storage of agricultural stocks by bankers' acceptances creates monetary policy issues that the Central Bank must consider because bankers' acceptances are bank liabilities having approximately the same monetary policy consequences as the issuance of a bank certificate of deposit of the same maturity.[13]

Thus, if 60-day bankers' acceptances secured by warehouse receipts representing the storage of grain were issued, the Central Bank should treat them as though 60-day certificates of deposit were issued because the issuing bank must pay the face amount of the instrument on the 60th day, without fail.

Open-Market Operations. If the Central Bank conducts open-market operations in private-sector financial instruments, then it might wish to give the approved form of bankers' acceptances at least the same favorable degree of treatment in regard to collateral for repurchase agreements that is given to other classes of banks' assets. The circumstances might vary from country to country, but in principle there is no logical or economic requirement that Central Banks' open-market operations discriminate either in favor of or against bankers' acceptances or any other kind of private-sector financial asset.

Rediscounts. The Central Bank might wish to make available a rediscounting facility for the approved type of bankers' acceptances or other forms of the stock financing mechanism. In countries with well-developed private capital markets, such a Central Bank rediscount facility is not strictly necessary, but such a facility might be judged convenient from the perspective of the Central Bank, accepting banks, or investors holding the paper eligible for rediscounting. If financing transactions for the storage of agricultural stocks, other than bankers' acceptances, are contemplated, such as commercial paper secured by warehouse receipts, or other capital market instruments, then it is less clear that a Central Bank rediscount facility is required because it is not necessary to raise funds to cover a bank's liability with respect to the other possible forms of financing. The Central Bank would have to issue regulations defining the conditions and limits of such access.

Pricing Central Bank Liquidity Assistance. Whatever Central Bank liquidity assistance is provided, whether to banks or to non-banks, in regard to storage-related paper, that assistance has to be priced correctly if it is not to have the perverse effect of tending to attract all the outstanding eligible paper into the hands of the Central Bank. A "penalty rate" is necessary to accomplish this objective at the rediscount facility (e.g., overnight or bank certificate of deposit rate, plus an appropriately determined spread). In open-market operations, in theory, the Central Bank would attract fewer offers of eligible, storage-related paper with a penalty rate (i.e., an above-market rate of discount offered to sellers) and no more than a random distribution of its fair share of such paper if the applicable rate of discount were exactly the market rate. The Central Bank theoretically would attract all the eligible paper in the market to itself, which would defeat the purpose of creating a privately funded, secondary capital market for such paper, if it offered a subsidy rate (i.e., if it demanded a rate of discount lower than the market rate from offerors of eligible paper).

The Case of Brazil

Open-Market Operations. The Brazilian Central Bank is legally prohibited from conducting open-market operations in private securities. Hence there is no direct role for the open market desk to play through purchases and sales of acceptance paper. However, the open market desk would have to coordinate with the discount window in the following areas: (1) to provide information on the positions of dealers/non-bank dealers on acceptances; (2) to consult regularly with discount window personnel about information obtained in dealer surveillance reports that relate to the acceptance market; and (3) establish computer links between discount window operations in Brasilia and open market operations to permit a better sharing of information and consistency in operations done through the window versus open market operations.

Rediscounts. The discount window function that is managed within the Central Bank in Brasilia, like the discount window operations in most developed country banking systems, operates primarily as a liquidity facility. At present both the setting of the rate for discount window borrowings and non-price rationing guidelines (often related to the frequency of borrowings by a particular bank) are employed in managing the facility.

The discount window of Banco Central do Brasil exists primarily as an overnight facility for liquidity advances only (no rediscounts to final maturity). An advance can be readily renewed for a second time, but additional renewal requires specific approval by Directors of the Central Bank and the borrowers' motives are viewed with suspicion. Acceptable collateral at the window includes government securities, commercial loans, and trade finance obligations backed by letters of credit (CDs are excluded). Base lending rates are determined by the National Monetary Council and are based on averages of market rates (open market repurchase agreements, interbank lending, and demand and time deposit rates), plus a substantial coefficient. Collateral is held by the borrower (against a trust-type receipt) and is not evaluated by the discount window. On-site reviews of collateral or documentation are sometimes performed but not systematically.

By contrast to Brazilian discount window procedures, US discount window procedures require that collateral used against advances ordinarily be held at the Central Bank. Frequent borrowers from the Federal Reserve's discount window often provide a pool of proposed collateral for early evaluations (see Appendix 2). This issue is mainly critical in regard to pledges of *private* collateral for discount window advances because in Brazil (as in the U.S.) government securities are entirely in book-entry form, and registration, clearing, and settlement take place on the Central Bank computer system, SELIC.

The practice of not holding collateral at the Central Bank when making advances under the financing facility to support warrants associated with a bankers' acceptances transaction probably should be reviewed by the Central Bank, particularly if private securities such as acceptances are being pledged.[14]

Pricing of the Central Bank Facility. It should be observed that until trading in bankers' acceptances begins, the daily base rate to use in recalculating the daily rate charged should be either the CDB rate quoted by Andima based on the CETIP clearing system for private securities or the LFT rate. In addition, a substantial premium should be charged (perhaps as much as 500 to 700 basis points above the base rate) in order to discourage use of this facility. The size of this premium should be based on a number of considerations, including the extent of dispersion of the individual rates around the mean base rate, over time. Moreover, officials running the discount window might also use non-price rationing procedures to assure that eligible banks did not continually use this facility.

Legal and Regulatory Issues

General

Rights of the Parties. The legal issues raised in the financing of agricultural stocks are discussed in detail in Appendix 3. Briefly, however, it appears to be a necessary legal precondition that, in any country offering bank advances secured by commodity-linked paper, federal or uniform state or provincial laws be adopted to ensure the integrity of all warehouse receipts issued, and define the rights and liabilities of each party to the warehouse receipt. Generally, applicable laws should provide mechanisms for the continued security of stocks under storage, the legitimacy of warehouse receipts representing such stocks, and the legal protection of holders of acceptances or commercial paper secured by such stored stocks or warehouse receipts. Issues that need to be addressed in any country in connection with such secured transactions generally include:

- Insuring the validity of a warehouse receipt under local law. Some countries might not allow enforceable claims against warehouse persons on the basis of certain types of warehouse receipts. Other countries might have effective legal protection against overissuance of warehouse receipts, while offering no effective legal remedy for failure to honor an original warehouse receipt in the context of competing claims to a defined quantity of stored commodities. Some countries might have effective legal provisions for guaranteeing the integrity of warehouse receipts, on paper, but still might lack a sufficient enforcement mechanism and infrastructure of qualified, certified warehouses to permit an effective bankers' acceptances or commercial paper program from being implemented. Local law must also provide stiff penalties or criminal sanctions for fraudulent issuances of receipts or similar practices.

- Certainty of validation of legal title to stored agricultural stocks upon bankruptcy of the issuer of warehouse receipts. Warehouse receipts issued by owners or producers of agricultural stocks might not be beyond creditors' challenge in bankruptcy proceedings if the warehouse person became insolvent. That is why surety bonds or insurance for warehouse person might become an important issue, but such bonds or insurance might not be available in particular countries.

- Protecting the rights of a holder of a receipt to reclaim stored agricultural stocks on presentation of the warehouse receipt. Depending upon the types of warehouse receipts issued, the rights of the holder of a receipt to reclaim stored agricultural stocks from the warehouseman might vary from country to country. For example, if the Anglo-Saxon practice of issuing unitary warehouse receipts were followed, then the holder ordinarily could expect to be able to reclaim stored stocks merely by presenting the receipt. But if the Civil Law practice of issuing two sets of warehouse documents (typically, a "certificate of acknowledgment of deposit" that actually conveys title or risk of loss for stored goods, plus a "warrant" issued to the initiator of storage as a record of the initial monetary value of the stored goods), it becomes necessary to unify the two documents (present both the "certificate of deposit" and the "warrant") to insure that the warehouse person releases stored goods.

- Providing the holder of the warehouse receipt (the issuer of the banker's acceptance or commercial paper) with a first priority lien on the stored agricultural stocks. The holder of a banker's acceptance or commercial paper might be deemed to have a lien on the stored agricultural stocks, in some countries, because of the written evidence of the debt of the drawer or issuer constituted by the acceptance or promissory note. In other countries, no lien on the stored agricultural stocks would be presumed to exist in the absence of the required set of warehouse documents. The "warrant" alone might constitute a lien in favor of a banker or holder of commercial paper, but usually it would cover only the initial value of the stored goods, and that value might fluctuate. Accepting banks might not have a lien on the stored goods that would be effective against third-party creditors of the drawer or of the warehouse person unless they held the original warehouse receipt (Anglo-Saxon practice), or the required set of warehouse documents (Civil Law practice). If the optional centralized clearing or settlement mechanism were used and if that mechanism became a book-entry system, then the accepting bank and any subsequent holder of bankers' acceptances, and any holder of commercial paper, ought to have the warehouse documents registered in its name in order to achieve a first priority lien or similar degree of legal protection that the holder of an original set of warehouse documents could have.

Separation of Warehouse Documents. Warehouse documents are *not* required to be joined physically to bankers' acceptances under United States or United Kingdom law, except at the time of initial acceptance. Thereafter, they may be separated. In developing countries, however, with less fully developed sets of property rights under an effective rule of law, it would appear to be the wiser choice to *require* the physical attachment of warehouse documents to bankers' acceptances (or to commercial paper) at all times or, alternatively, to have original documents immobilized in a central clearing or settlement mechanism and to have book-entry receipts only (which would have to be re-registered with each transfer to be fully effective) traded in a secondary market.

Legal Preconditions for Secondary Market Trading. In addition to the foregoing legal issues, the following points should be resolved in connection with creating a suitable legal framework for secondary market trading of commodity-backed or commodity-linked bankers' acceptances or commercial paper:

- Whether local law allows the trading of securities, commercial paper, etc., whose value is linked to commodity prices or to a commodities index. For example, some countries prohibit the private ownership or trading of gold-linked instruments. Inquiry should be made to determine whether comparable restrictions would affect ownership or trading of debt instruments secured by or whose nominal value was linked to the prices of agricultural stocks.

- Local law must ensure that the bank or other entity issuing the banker's acceptance or commercial paper is primarily liable to the holder for the face value of the acceptance or promissory note. On issuance of the banker's acceptance or commercial paper, the producer or owner of the warehouse receipt should remain liable for payment on account of the instrument at maturity, but the holder of the banker's acceptance or warehouse receipt should look first to that issuer for payment. This has the effect of making the banker's-sacceptance or commercial paper more tradeable and acceptable in the market because investors would look to the issuer's credit worthiness rather than the producer's. Moreover, if the issuer is a bank or a large trading company, its credit standing and reputation should be of greater consequence than the producer's or account party's.

- A determination must be made as to whether nonfinancial firms may issue commercial paper or accept drafts or bills of exchange (create bankers' acceptances), or whether such activities are to be restricted to banks or other financial intermediaries. The Anglo-Saxon practice is rather expansive on this point: one does not have to be a bank to be able to issue a banker's acceptance. Additionally, do restrictions exist under local law regarding investments in such instruments?

- Under local law, bankers' acceptances and commercial paper must be negotiable. Negotiability is explained in more detail in Appendix 3, but essentially it ensures easy transferability of bankers' acceptances and commercial paper by proper endorsement.

- The legal status (and the practical enforceability) of surety bonds or warehouse person's insurance for the persons who would issue warehouse receipts must be ascertained. If such bonds or insurance are prohibited, do not exist, or are unenforceable, then is it feasible to consider creating a system of government-supervised warehouses, or of bank-owned warehouses, that could issue credible warehouse receipts?

- Additional issues addressed more fully in Appendix 3 include: limitations on the maturity of bankers' acceptances, ready marketability of particular agricultural stocks sought to be financed, and the independence of warehouses issuing valid warehouse receipts that may be used as collateral for bankers' acceptances and commercial paper.

The Case of Brazil

Rights of Parties. It is contemplated under the regulations governing transactions in bankers' acceptances that the accepting banks retain the certificates of deposit and warrants and that only the acceptances be sold in the secondary market. However, it is not clear that the investors would acquire any right to attach or liquidate the underlying stored goods if, under Brazilian law, the receiver or liquidator of the failed bank would have the prior claim on stored goods covered by negotiable storage documents still in the failed bank's possession. In other words, the investor in (holder of) an acceptance would have no right to receive the underlying stored goods unless the Banco Central, in extrajudicial liquidation procedures, granted such preferential rights to investors in acceptances. The existing bankers' acceptance regulations in Brazil create a situation in which the normal legal rights of the parties under the law and custom of merchants and applicable international conventions and protocols would be as follows:

- accepting banks should retain both the certificates of deposit and warrants issued by warehouses; and
- investors should receive acceptances that provide only a monetary claim on the accepting bank and no direct legal claim on the underlying stored goods.

The above definition of "rights of parties" affects investor incentives. For example, investors in the private sector proved less than eager to invest in acceptances versus competing instruments unless they could obtain a clear legal claim on, and the capacity to liquidate, the underlying stored goods quickly. The investors' interest would be increased by their capacity to take delivery of stored goods in lieu of demanding payment of acceptances at maturity or more so if they could link the return on the security to movements in the price of the tradeable good. For the banks' protection, they should retain the warrants until the acceptances mature. However, to improve investor interest, the certificates of deposit could be attached to or, in a book-entry custody and clearing system, traded together with the acceptance in the secondary market.

Legal Rights to Issue Commodity-backed Instruments. The class of entities capable of creating acceptances probably should be expanded to include non-bank entities because, in fact, some grain trading companies might have greater credit standing in the financial markets than some of the banks participating in the acceptances market. Such possibilities have been limited in Brazil due to regulations related to Brazil's commercial code, which does not permit nonfinancial corporations to issue commercial paper. These restrictions are evaded to the extent that firms issue debentures with repurchase clauses that can be linked to commodity prices, or engage in intra-corporate borrowing illegally (the mutto).

Collateral Liquidation Procedures. It appears that, while Brazilian law is clear as to the respective rights of holders of bankers' acceptance documents to obtain access to and to liquidate the underlying stored goods, actual collateral liquidation procedures may be too slow for normal commercial transactions in agricultural commodities. Thus, for example, if a producer defaulted on his loan (reimbursement agreement with the accepting banks), the accepting bank might be unable to foreclose upon and liquidate the collateral in a timely manner, even if it held the conhecimento de deposito and warrant, due to delays in obtaining any necessary enforcement assistance from Brazilian courts. Most lawyers and bankers indicated that delays in collateral liquidations were (and probably still are) a significant problem in Brazil.

Links to Other Markets

General

A variety of issues needs to be examined in considering the links to other explicit or implicit markets for either financing stocks of commodities, borrowing or lending commodities (e.g., through use of futures contracts) or the cash or international capital markets. These links are often important to establish because they increase the efficiency with which agents in a marketing system can manage the different risks inherent in processing goods and financing the storage of such goods or the possible sources of financing.

Cash Markets. Two institutional links to cash markets merit attention. First, in some countries, negotiating the cash sale of goods by the borrower is permitted at any time within the period during which the goods have been pledged as collateral in obtaining financing. In this case, documents are issued by the warehouse that permit such sales at any time (i.e., certificate of deposit); however, precautions would have to be taken by the bank or corporation, in the case of commercial paper linked to or backed by commodities, so that the value of its security interest (i.e., the claim on the assets, e.g., commodities, of the borrower) would not be diluted. There is evidence that in many countries financing will not be provided to producers or other agricultural borrowers unless the lender (i.e., bank or corporation) obtains both a lien on the stored goods and the right to sell the goods in the event that the borrower defaults. Under these conditions the financing would require the borrower to relinquish the right to sell the goods in the cash market.

Second, as has been noted above, cash market contracts for the commodities to be financed need to be carefully understood and examined prior to the formulation of policy reforms that would permit the introduction of different forms of financing. Examples include payment in advance contracts, dollar- or domestic currency-denominated price-to-be-fixed contracts under which the seller is given the option of picking a price at any time up until final delivery, and offshore or onshore forward contracts. In all these cases the contracts negotiated between buyer and seller are not spot contracts but, rather, provide intertemporal options to the seller and buyer over some finite period of time. As such, many cash contracts can be viewed as a combination of a loan or credit extension and a final sale or purchase at some future date. Although the existence of the specific form of these contracts may be driven by tax or other regulatory, legal, or technological considerations, implementation of policy reforms without an understanding of the forms of these contracts can result in omission of key changes in regulations that would permit greater private-sector stock financing, even through greater use of special forms of cash contracts.

Futures Markets. The link between the system for stock financing and offshore and onshore futures markets is also important. Two links of special importance need to be investigated.

First, in some countries the extension of agricultural stock financing is linked to "development of futures markets" through regulations that indicate that more credit can be extended if the producer or holder of commodities is hedged on the onshore or offshore futures exchanges. However, in general, no "incentives" need be embodied in regulations so as to encourage lenders (e.g., banks or investors in the case of commercial paper) to insist that borrowers (e.g., producers or trading corporations, etc.) hedge commodity price or output risk. Such arrangements should constitute part of prudent risk management on the part of credit departments of the banks, or by investors themselves.

Second, links will often need to be established between the documents issued by the warehousing system, the banks or corporations trading securities linked or backed by these warehouse documents, and the futures exchanges. The warehouse receipt could conceivably be used as an eligible form of collateral in posting margin in order to take a position in commodity futures contracts or could be directly borrowed and lent in countries that legally permit such transactions. However, care needs to be exercised not to permit double leverage, where the same receipt would be pledged both to obtain stock financing and as initial margin in a commodities futures contract. Avoidance of this type of practice will require good supervision of exchanges as well as the existence of a good custody, clearing, and settlement system.

Pre-export Finance and International Markets for Bankers' Acceptances. In many developing countries, links between trade finance and the domestic financing of stocks of agricultural commodities will be important to investigate in designing reforms.

First, an analysis of how short-term external debt facilities operate will be important to understand in some of the highly indebted developing countries. This is because the operation of such facilities, or their nonexistence, can have differing implications for the development of domestic markets for agricultural stock financing. For example, if the developing country is subject to constraints imposed by the existence of short-term external debt facilities, then whether there is a facility that supports trade finance, including pre-export financing, might affect determinations of the eligibility of pre-export finance to satisfy foreign lenders' commitments to maintain their trade finance lines. In many developing countries, pre-export finance is extended on a revolving basis, ranging from 30 days to as long as 360 days (e.g., Argentina), and competes with domestic sources of stock financing.

Second, an issue that must be investigated is whether acceptances originated in Brazil, based on storage warrants, would be eligible for rediscount with either the Federal Reserve Bank of New York or with the Bank of England. If such acceptances were eligible, which is either not likely or might be likely only subject to stringent conditions, then Brazil or any other developing country would increase the amount of trade finance made available to it on terms that might be more preferable than through general borrowings. In general, foreign storage-based bankers' acceptances may be technically eligible for rediscount with the Federal Reserve but in face *are not* purchased in the open market by the Federal Reserve (ineligible for purchase).

The Case of Brazil

Links to the Cash Market. Design of the bankers' acceptances regulations may also have implications for the future operation of the cash Bolsas (designed to hold auctions for agricultural commodities). At present producers have two choices: they can either negotiate cash sales directly on the Bolsa, in which case both the warrant and conhecimento would be required at the time of the transaction, or the producer can obtain EGF credit from the Banco do Brasil. In the latter case, if the producer cannot repay the EGF loan, the Government (through Banco do Brasil, acting as agent) purchases the goods at the minimum price determined by CFP under the AGF program.

The Government is now proposing that producers who cannot repay an EGF credit line be given a further choice of either having the Bolsas act as agent of the Banco do Brasil to liquidate the collateral by auctioning off the goods or selling the goods back to the Government at the minimum price under AGF, whereupon CFP would auction the goods through the Bolsa. Any changes considered in the legal structuring of acceptance transactions, or redefinition of documentation required of banks, should reflect careful consideration of how such changes would impact on the Government's ability to employ the Bolsas to minimize the costs to the Government of the EGF/AGF program.

Links to Futures Markets. Final regulations issued by the Central Bank did not explicitly incorporate incentives into the regulations to promote hedging on both domestic or offshore futures exchanges.[15]

This is because it was felt that such arrangements more appropriately would be left to the parties within a bankers' acceptance or commodity-backed or linked financing transaction.

There are other ways in which the design of the bankers' acceptance market regulations can impact upon the desire of participants in a bankers' acceptance transaction to use the futures markets. For example, both the Bolsa de Meradorias (BMSP) and the Bolsa de Mercantil y Futuro (BM&F) in Sao Paulo trade commodity futures contracts. On both these exchanges warrants issued by warehouses to producers and/or banks technically can be used as eligible collateral in posting initial margin. Thus, under the existing bankers' acceptances regulations, both the warrant and the conhecimento de deposito would have to be kept at the commercial bank, and it would not be possible for the producer or the warehousing company (as agent for the producer) to pledge the warrant as initial margin on domestic futures exchanges.[16]

Due to these considerations it will be important to ascertain how to structure bankers' acceptance transactions legally so as not to *preclude* parties to the transactions from hedging the "price" risk on the domestic commodity futures exchanges by separating the documents issued by warehouses.[17]

Pre-Export Finance and Domestic Stock Financing. Originally, it was intended that a contract for the export of goods be exhibited to a lending institution before an exporter could obtain pre-export financing under existing dollar trade credit lines for Brazil. Now, an exporter only needs to be licensed by CACEX and to have a contrato de cambio (foreign exchange contract, an asset in the hands of a Brazilian exporter) to be able to receive pre-export financing. The export contract does not have to exist when the foreign exchange is released from the Central Bank. For example, up to 50 percent of all soybean exporters currently are financed under so-called adiamantes contrato de cambio (ACC) credit lines. Similarly, soybean crushers that export processed soybean products have obtained substantial amounts of ACC financing for stocks of soybeans that they store before processing.

In order to retain foreign exchange under the ACC program, the exporter actually must close an export contract and ship the goods within 180 days after receiving foreign exchange, with an additional 10 days grace allowed for shipment (export) of the goods. This comparatively generous provision for pre-export financing under the ACC program provides strong competition for any program that would permit the domestic financing of exportable goods. This has been even more the case in the Brazilian context where domestic interest rates have been higher than the exchange rate-adjusted foreign interest rates, so that this form of financing permits the exporter to invest funds in the financial markets before purchasing commodities. That factor, together with fewer documentation requirements under ACC, may make it difficult for any bankers' acceptances program to take root as long as ACC credit is available in large quantities on a pre-export basis.

In addition to the advantages of ACC financing from the vantage point of documentation requirements, some banks felt that ACC financing provided a more secure form of financing because it was *related* to exports and because nonperformance by the exporter carries severe penalties; often CACEX will revoke the exporter's license. Thus, it was argued by some Brazilian bankers and exporters that this arrangement provided more security than possession of the title documents issued by a warehouse in the name of the producer, as further distinguished from the case of a bankers' acceptance originated on the basis of storage warrants issued by warehouses.

Findings and Recommendations

Findings

The case of Brazil vividly illustrates that macroeconomic instability and lack of credibility associated with large expected future fiscal deficits reduce the desire of investors to hold instruments like bankers' acceptances and hinders the development of most formal markets for financing stocks of commodities or for lending and borrowing commodities at low transaction costs (i.e., futures markets). Thus, some degree of macroeconomic stability is a *necessary*, but not a sufficient condition for the successful development of any *formal* market for financing stocks of commodities. This is because tax policies as well as government credit policies can hinder the development of a formal market, notwithstanding macroeconomic stability. Brazil's experience until the end of 1989 suggests that actions should be taken to stabilize the economy, either prior to or in conjunction with other economic policy actions (reductions in real trade intervention other forms of intervention, or tax reform) that can affect incentives for financing stocks of agricultural commodities.

In the area of *tax treatment*, the experience within Brazil indicates that analysis of the taxation of participants in the agricultural marketing and financial system needs to be undertaken and actions implemented simultaneously with the issuance of regulations permitting various forms of commodity-backed or linked securities. In general, a good principle to follow in considering reforms in this area is to assure that tax treatment is as neutral as possible in regard to its effects upon the forms of commodity-linked financing vehicles available, and in regard to the fiscal deficit. Also, participants should be assured of the stability of the new tax regime.

Comprehensive examination of all forms of government intervention in the process of pricing, marketing, or financing commodities is necessary. This would include restrictions on the balance sheets of private economic agents imposed by governments in the context of various credit allocation programs, special government subsidized credit programs for producers, price support programs, and subsidized credit given to exporters by the Central Bank in the context of promoting pre-export financing. Reforms to limit government intervention of these types should be undertaken soon after or in conjunction with a macroeconomic stabilization program. Moreover, within this area—where possible—price support programs should be dismantled first, while simultaneously government credit resources allocated to the agricultural sector should be priced as close to "market" prices as possible.[18]

Reforms to the structure, integrity, and security of the *warehousing system*, as well as associated legal reforms (e.g., changes in the warehousing law, or laws governing bonding/insurance) can be undertaken prior to efforts to "liberalize." In fact, reforms in this area can be justified even in periods of macroeconomic instability because the lead-time for institutional reforms to the warehousing system is often longer than that for policy reforms. A useful analogy is that of reforms to the financial system where reforms to the system of market supervision should generally be undertaken prior to widespread liberalization. Brazil's experience indicates that actions to ensure: (1) issuance of uniform documents by warehouses; (2) adequate systems of inspection of public and private warehouses; (3) development of bonding and insurance systems; and (4) adequate central registration procedures for warehouses, etc., are some of the essential preconditions. If these institutional arrangements are not in existence, they will greatly reduce the desire of financial institutions or other institutions (e.g., corporations in the case of commercial paper backed by or linked to commodities) to undertake agricultural stock financing.

The timing of reforms in the area of *custody, clearing and settlement* depends critically on the extent of development of the financial markets. In this regard, systems in Brazil are greatly advanced in relation to those in many smaller countries in Latin America, or even vis-a-vis some of the larger Latin American countries (e.g., Argentina). Hence, modifications of existing Brazilian clearing and settlement systems to permit trading of commodity-linked or -backed securities do not present additional technical problems. Thus, reforms in this area can be implemented quickly, *after* warehousing system reforms have been implemented. Even in smaller countries where stock exchanges are present, incorporating clearing and settlement functions (with custody handled by specialized trust departments of commercial banks) to permit the trading of acceptances or other forms of instruments linked to commodities can be arranged rapidly. However, the key constraints would involve the size of capital investment (in computers, etc.) that the Exchange would be willing to undertake to expand its clearing and settlement systems given the volume of commodity-backed securities to be traded.

In the area of *links to other markets*, the experience of Brazil indicates that careful thought should be given to what links are most important to establish between stock financing on the one hand, and cash, futures or international markets on the other hand. A guiding principle would be to permit maximum flexibility to participants in the agricultural marketing system and in the financial system to engage in a sufficiently diverse set of contracts—thereby permitting the more efficient management of various forms of risk and commodity inventories. Thus, for example, links between the financing of stocks and hedging operations should not be discouraged in the context of drafting regulations or laws regarding the financing of stocks and documents issued by warehouses (i.e., warehouse receipts). The use of warehouse receipts as collateral in posting margin for futures contracts should be permitted as long as the warehouse custody system is secure, and does not permit double-pledging of the agricultural commodities.

The effects of introducing commodity-backed financial instruments *for monetary policy* should not be substantial, as long as the Central Bank does not engage in the provision of credit at non-market rates to banks originating acceptances. Permitting the issuance and trading of commodity-linked securities in a tax-neutral environment may be sufficient to enable a market to develop without any form of explicit or implicit Central Bank support. In most cases, if the Central Bank does rediscount commodity-backed instruments, it should price the rediscounts at a penalty rate above the "market" rate for financial instruments of comparable maturity until trading in such instruments is sufficient to establish a privately determined market rate. Also, the Central Bank's presence through use of a rediscount facility should be phased out quickly after the market is operational.

Finally, as in the area of institutional arrangements, *legal and regulatory reforms* necessary to permit maximum flexibility to economic agents to finance the storage of goods under well-defined property rights will often require a longer lead time to both investigate and implement than other types of economic policy reforms. The following kinds of legal and regulatory reforms may be required:

- clearly defining the rights of the parties to commodity-linked securities and making certain that, in the event of default, the rights to the financed agricultural stocks are assured;
- adopting uniform documents acceptable to most domestic and international market participants so that the capacity to trade the instruments internationally could exist;
- implementing legal preconditions to enhance the origination and secondary market trading of warehouse documents and instruments involving the pledging of such documents (see Appendix 3); and
- establishing legal preconditions to properly define the rights of parties in cases of bonding and insurance of warehouses, and issuing documents used to create financial instruments traded in the secondary market.

The Brazilian experience indicates that, as in the case of institutional reforms, legal and regulatory reforms could be implemented before imposition of a macroeconomic stabilization plan. The Brazilian experience highlights the importance of reforms in all these areas because:

- the rights of parties to a banker's acceptance or commodity-linked instrument under default are not clear under Brazil's law for extrajudicial liquidation of financial institutions because the investor in a bank-issued commodity-backed instrument would stand in line with all other general creditors of the failed bank;
- collateral liquidation procedures are extremely inefficient, in part because of problems in the enforcement of the bankruptcy laws and disorganization of the court system;
- the warehousing law (of 1903) does not establish a standard set of documents to be issued by all types of warehouses; and
- existing laws and regulations permitting the origination and trading of commodity-backed securities do not clearly delineate the kinds of parties that can originate or trade these securities as well as the regulatory agencies with authority to regulate such securities.

Recommendations

Several major recommendations follow from the findings above.

First, countries contemplating reforms to enhance the private financing of stocks of agricultural commodities should gradually phase in legal and institutional reforms. Such reforms could be undertaken prior to, or at the same time as, adoption of economic liberalization measures aimed at a reduction of all forms of government intervention in the financing or pricing of commodities, since their implementation will further reduce the transaction costs to economic agents in engaging in the financing of agricultural stocks or the management of commodity inventories.

Second, the particular sequence of institutional, legal, and regulatory reforms recommended is as follows:

- after study of the existing warehouses, revise the warehouse statutes and regulations;
- inspect existing warehouses and prepare a list of those approved under the new legal framework;
- provide for a uniform system of warehouse documents acceptable to most producers and traders; and
- provide for an adequate bonding or insurance framework for the warehouses.

Third, at the stage of adopting economic policy reforms to the agricultural market, finance and tax systems, it is recommended that (due to the intragovernmental coordination needed) a special inter-agency group be established that will develop a reform package that will permit the natural evolution of a *private* market for financing stocks of agricultural commodities.

Fourth, it is recommended that more research be undertaken to examine the links between institutional/legal reforms and economic behavior of agents in the agricultural marketing system, with special reference to how current arrangements affect transaction costs of both managing commodity inventories; links between institutional/legal reforms in these areas and the economic policy reforms noted could be examined in a variety of countries.

Notes

1. This appendix was authored by Thomas Glaessner (World Bank), Joseph Reid (University of Baltimore Law School) and Walker Todd (Cleveland Federal Reserve). The views expressed are solely those of the authors and do not necessarily represent those of the institutions with which they are affiliated.

2. In this context own-currency risk refers to movements in the domestic currency against the dollar. In many developing countries organized or informal forward contracts for managing exchange rate risk do not exist although the demand for this contract is often present particularly among multinational companies that may be searching for means of managing these risks.

3. For a much more detailed treatment of the economic function of futures markets and the importance of such contracts to processors in managing commodity inventory (as opposed to a contract with which to hedge), see Williams (1989).

4.   This increase in mandatory allocations of sight deposits to agriculture *may not* reflect an increase in the overall volume of forced lending to agriculture to the extent that sight deposits at commercial banks have been reduced significantly with the increase in inflation rates and the increased availability of interest-earning demand deposits (i.e., remunerada accounts).

5.   The costs of offshore hedging are likely to be high where the country has exchange controls or other forms of capital account restrictions.

6.   Such a clearing system could be operated by the government or a private entity cooperatively owned by either commercial or other providers of agricultural stock financing.

7.   The CFP provides minimum-price non-recourse loans for commodities stocks and has administered a minimum price purchase program for stocks in the past.

8.   These documents include the acceptance issued by the bank and the warehouse receipt issued the borrower and used as collateral by the bank issuing the acceptance (conhecimento de deposito). In some countries (e.g., Brazil, France) these documents must be followed closely and an additional warrant or certificate is needed. This warrant or certificate is sometimes used by the borrower (i.e., producer, warehouse person, etc.) to negotiate the sale of the goods.

9.   The concept of immobilization refers here to the use of a depository as a place to safekeep the documents associated with an acceptance transaction. The securities are immobilized in that the settlement system will permit double entry bookkeeping or accounting that will not necessarily require the physical movement of documents between custody accounts. In fact some depositories are now characterized by what is referred to as de-materialization (Denmark, Norway, France) where holders of securities cannot go to a depository and obtain physical documents proving ownership. Instead, the depository transfers ownership between counterparties through use of a book-entry system such as those used in settling securities transactions. Note that in many countries current laws would not permit such dematerialization. Moreover, in the case of bankers acceptances, computerization of documents issued by warehouses that constitute part of the acceptance transactions can be costly (if many warehouses are present). However, the goal in most countries should be a centralized depository where all securities issued were dematerialized.

10.   An example is the case of a banker's acceptance for which the warrant issued by the warehouse serves as the lien of the bank in the case of default by the borrower (producer), but the warehouse may issue other documents (e.g., certificate of deposit) that can be used by the borrower to negotiate sale of the goods and remit funds to the lender during the life of the acceptance.

11.   Under some custody systems fungibility is possible under which all securities of the same issue (e.g., all acceptances originated on the same date) are filed or warehoused together and are completely interchangeable. This can lead to significant economies of scale in markets where a lot of trading takes place in a given security. In this environment specific securities are not identified with a specific beneficial owner. In the context of commodity related transactions fungibility can raise special issues if the commodities stored in different warehouses as represented by the warrant are not homogeneous or if the warehouses are not judged to maintain the same standards for storing the goods. In this case, banks originating

acceptances and, more importantly, investors, where the security is designed so that they have title to goods, will not want a claim on any bank or associated warehouse. Thus for participants to realize the savings in transactions costs afforded by fungible custody systems may be more difficult.

12. Without some form of netting the broker dealer will typically require cash (or bank loans) to cover the total value of all his trades due for settlement on a given day. Any form of netting system reduces cash requirements and the associated need to assemble additional collateral. Under net-settlement systems, the obligations between two or multiple counterparties are batch processed to obtain one net-settlement figure indicating the amount of securities and cash to be transferred between counterparties to a trade. In the case of netting systems legal issues will often need to be investigated because the counterparties to a trade can change and the clearing organization can itself become the counterparty or guarantor of finality of settlement.

13. A review of the history of the New York bankers' acceptances market and discount window and open-market operations, prepared by Larry Aiken, appears as Appendix 2.

14. The problems inherent in not taking physical custody of collateral when private securities are pledged may be partially mitigated by the Central Bank's role as receiver in liquidations of insolvent banks, as this role is extra-judicial (not subject to court review) in Brazil.

15. Hedging of "commodity prices" can be undertaken on offshore exchanges, however, the Central Bank limits the positions that can be taken. Specifically, futures contracts can only be sold. No purchases of these contracts are permitted on offshore exchanges.

16. A separate question related to this discussion is whether the "warrant" or "conhecimento" would provide more security to the futures exchanges from a legal perspective because the conhecimento is the title document. By contrast, the warrant represents a lien on the value of the goods on a fixed date.

17. Permitting both the certificate of deposit and the warrant to be traded separately might not preclude various parties from hedging.

18. Problems can arise if interest rates are not being liberalized generally in the economy, because borrowers will often obtain "subsidized credit" from other direct or indirect government-owned financial institutions.

Appendix 2

History of the Bankers' Acceptance Market in New York

Besides the special exclusion from reserve requirements granted, the Federal Reserve system, in the earlier years of the market, sought to encourage the use of bankers' acceptances (BAs) by posting preferential rates of discount for the instrument and by standing ready to purchase all acceptances offered at those rates. Until the late 1920s, the Federal Reserve (Fed) held up to 60 percent of all outstanding acceptances. Then the Fed stopped buying BAs at preferential rates and the market foundered through the depression and war years. After the second world war, acceptance finance picked up somewhat without overt Federal Reserve encouragement but did not expand substantially until after 1955 when the New York Fed began using acceptances as part of open market operations. These activities were at market rates of discount and holdings were limited to a modest portion of the total available. U.S. trade with other countries, however, was flourishing. The Federal Reserve system signified a special interest in this private market instrument by providing its own guarantee of payment for all BAs purchased on behalf of foreign official accounts maintained at the New York Fed. The Central Bank assured a ready supply of the instrument for its own use and for customer purchases by paying dealers in the open market an additional amount above that obtained for sales to other investors, in return for their endorsements on BAs delivered. By the mid-1970s the Fed's guarantee and the payment for dealer endorsements had been discontinued as unnecessary and inappropriate. By the mid-1980s, the Fed ceased using the instrument in open market operations because such support and encouragement was viewed as no longer necessary. Currently, the Fed of New York purchases and sells acceptances only as agent for its customer accounts, and retains only a residual regulatory role in the market. Physical delivery of the instruments, involving inspections to verify the eligibility and negotiability of each security received (no borrower's or seller's custody of the instruments was allowed), have always been a requirement of participation in this market.

In the mid-1930s the Fed, as a neutral participant, took over an ongoing survey initiated by the issuing banks to track, on a monthly basis, the size and nature of the financings done on an acceptance basis. The New York Fed continues to solicit this data from the banking community. The Board of Governors is among the parties interested in the data; the Board uses it in the preparation of business statistics and monetary aggregates, and verification of related data submitted by banks in Quarterly Call Reports. The dealer market is able to make judgments as to the overall availability of BAs and their individual roles in that market.

The dealers in acceptances with whom the New York Fed is authorized to transact trades report daily to the open market desk the details of their trading activity by types of customers, positions retained and their offering rates at the close of the day.

* * *

[Two sections from this appendix have been omitted here because of permissions difficulty. The next numbered page is p.241.]

Appendix 3

Legal Preconditions for Agricultural Storage Financing[1]

Introduction

The purpose of this appendix is to elucidate in cogent terms the legal requirements necessary for financing stored agricultural products. It is not the author's intent to discuss all forms of agricultural finance, such as long-term real estate or mortgage lending or direct loans financing crop production; however, the emphasis is on the legal preconditions endemic to primary and secondary market financing for stored agricultural products. Generally, financings for stored agricultural products are of a short-term nature. In virtually all instances, the agricultural products are stored in a warehouse which issues warehouse receipts as evidence of such storage. Because the warehouse receipt is functionally equivalent to the stored agricultural products, it is a secure instrument which can be used as collateral for bank financing.

In the primary market, warehouse receipts are used as collateral for typical bank loans. Since a warehouse receipt is clear evidence of ownership of the stored agricultural products and is functionally equivalent to the products under storage, the warehouse receipt is willingly held by a bank lender as security for a loan. The bank's interest is in insuring it has clear title to the agricultural products on default of the bank loan. Also, in the event the warehouse which issued the corresponding warehouse receipt is subject to insolvency, the bank must be certain it has a right to the stored products as against any creditor of the warehouse or any other party. The principal legal precondition for this primary market is to ensure that the warehouse receipt is functionally equivalent to ownership of the stored agricultural product, and that the bank creditor holding such warehouse receipt is able to liquidate its interest in such agricultural products as against any other party if its borrower defaults on the loan or other extension of credit secured by the warehouse receipt.

If a producer is of sufficient size and has a substantial volume of stored agricultural products, banks extend a less costly form of credit called bankers' acceptances. A banker's acceptance, in simple terms, is a short-term time draft drawn by the producer and accepted by the bank. The bank's liability or credit is placed before the producer's (drawer's) which allows the draft or banker's acceptance to be sold in the money markets based on the bank's credit rather than the drawer's. The second portion of this paper focuses on the legal preconditions essential for the creation of a secondary market involving the use of warehouse receipts for bankers' acceptance financing.

Legal Preconditions for Primary Market Agricultural Storage Finance

In the primary market, warehouse receipts are used as collateral for bank loans. There are two main legal concerns and requirements. First, it is necessary to develop a regulatory apparatus to protect the depositor of the stored agricultural product, or one to whom a warehouse receipt is transferred. This ensures that the warehouse issuing the warehouse receipt will have the stored agricultural product or its fungible equivalent when the warehouse receipt is presented for delivery of the stored agricultural product. Second, it is essential that a statutory scheme exist which clearly defines the rights and liabilities of each party having an interest in the warehouse receipt and the stored agricultural products. In the United States, the regulatory apparatus for licensing, examining and regulating warehouses is the responsibility of the Federal government. State laws establish, define and regulate the rights and liabilities of each party to the warehouse receipt.

Types of Warehouses. There are four different kinds of warehouses storing agricultural products in the United States: (1) independent; (2) field; (3) subsidiary; and (4) owner-operated.[2]

- An *independent* warehouse is a public warehouse where the owner acts only as custodian and has no ownership interest in any of the stored agricultural products. This arrangement is preferable to bank creditors, and in most instances banks will not extend credit unless the warehouse receipt is issued by an independent warehouse.
- A *field* warehousing arrangement is where a public warehouse stores agricultural products principally of the depositor (producer) on the depositor's premises by leasing property for storage from the depositor. The public warehouse must maintain complete control of the property at all times, only use its own employees, segregate the leased storage property by the erection of proper partitions, place

prominent signs of sufficient size throughout the property to give clear notice that the property is being operated by the warehousing company, and essentially, insure total dissociation from the depositor (producer). Banks provide financing based on warehouse receipts issued by field warehouses, although independent warehouses are more common and preferable.

- *Subsidiary* warehouses involve the operation of a warehouse by a company wholly-owned or controlled by a parent organization through stock ownership or interrelated management. The goods stored are primarily those of the parent organization.
- An *owner-operated* warehouse is where a merchant, dealer, manufacturer, producer, or operator runs a public warehouse auxiliary to his/her principal business and stores his/her own agricultural products as well as those of the public.

In terms of legal structure, it is necessary for warehouses to be either *independent* or *field* warehouses. The owner of the warehouse should have no interest in any of the stored agricultural products, although under the field warehousing arrangement, storage may take place on a depositor's property provided the warehouse is on property leased to and fully controlled by a warehouse company with no relationship to the depositor. If these basic requirements are met in terms of legal structure of the warehouse, banks will be more willing to provide loans secured by warehouse receipts issued by such warehouses.

Regulation of Warehouses. In the United States, the Federal government regulates and licenses all warehouses storing agricultural products and issuing warehouse receipts. For the most part, only warehouse receipts issued by Federally-licensed warehouses can be used as collateral for a bank loan or bankers' acceptance financing. The purpose of Federal regulation is to foster confidence in warehouse receipts for agricultural products stored in order to facilitate financing and also to establish standards for safe storage of agricultural products. A Federally-licensed warehouse must submit an application showing that (1) the warehouse is suitable for storing the particular agricultural commodity; (2) the warehouse is equipped with necessary facilities for properly caring for the licensed commodity at all times; (3) personnel connected with the warehouse are competent and trustworthy; (4) personnel are sufficiently capable of grading, weighing, inspecting and classifying agricultural products received for storage; and (5) the warehouse possesses sufficient net assets. The warehouse must also satisfy certain bonding and surety requirements established by the Department of Agriculture, the Federal agency responsible for licensing, examining and inspecting Federally-licensed warehouses. The bonding requirements are normally at a level which ensures coverage of all warehouse receipts outstanding. Federally-licensed warehouses are examined and inspected at frequent invervals.

These examinations assess the financial condition of the warehouse and make careful inventories of the warehouse, checking outstanding warehouse receipts against the amount of stored agricultural products. Licensed warehouses pay the Department of Agriculture reasonable fees for each examination and inspection, annual licenses to operate, and licenses for personnel to classify, inspect, grade, weigh and sample agricultural products.[3]

Under Federal law, licensed warehouses may not discriminate between persons desiring to store agricultural products in their facilities provided the product is the kind and grade customarily stored by the warehouse and is in suitable condition for warehousing. It is a criminal felony offense with a fine of up to US$10,000, or double the value of the products involved if such double value exceeds US$10,000, or imprisonment up to ten years for any person who forges, alters, or issues counterfeit warehouse receipts. These penalties also apply to anyone fraudulently issuing warehouse receipts where agricultural products are not in storage, removing stored products without a receipt, and using a fraudulent receipt for purposes of securing a loan. Additionally, it is a misdemeanor fine of US$500 or six months imprisonment to falsely or fraudulently classify or grade agricultural products for storage.

Minimally, the primary market for financing agricultural products requires an effective regulatory apparatus to ensure the integrity of the warehouse receipt. Any financial institution extending credit based on the security of a warehouse receipt must have a high degree of confidence and certainty that in the event of borrower default the stored agricultural products will be available on liquidation. Such confidence can only be sustained if all warehouses storing agricultural products and issuing warehouse receipts are licensed, examined, regularly inspected and otherwise subjected to regulatory oversight so lenders can be certain proper practices are followed. Criminal sanctions are also necessary for anyone issuing false or fraudulent warehouse receipts and other practices which do not ensure the integrity of all warehouses issuing warehouse receipts. The prime importance of an aggresive regulatory apparatus is that banking instituitions are more willing to make loans secured by warehouse receipts issued by warehouses which are properly licensed and regulated.

Warehouse Receipts. A valid warehouse receipt must have the following essential terms: (1) location of the warehouse; (2) date of issuance of the warehouse receipt; (3) consecutive number of the receipt; (4) statement whether the agricultural products received will be delivered to the bearer or to a specified person or his order (negotiable) or to a specified person (nonnegotiable); (5) rate of storage charges; (6) description of the agricultural products received in terms of quantity, grade and classification; (7) warehouse person's lien, if any; (8) statement whether receipt is issued for agricultural products of which warehouse person is owner; and (9) signa-

ture of the warehouse person or agent. In addition to the foregoing required terms, warehouse receipts generally include an exculpatory clause
disclaiming the warehouse person's liability for loss, damage or delay
caused by Acts of God, war, civil or military authority, insurrection, riot,
strikes, labor disputes, fire, sprinkler leakage, flood, wind, storm, vermin,
change of temperature and similar events. The warehouse person remains
liable for any loss or damage caused by his/her failure to exercise ordinary
care or diligence. Whether or not an exculpatory clause is included on the
warehouse receipt, it is imperative that all warehouse receipts used as collateral for bank financing contain all of the essential terms.

Rights and Liabilities of the Parties to a Warehouse Receipt. In addition to a regulatory apparatus ensuring the integrity of warehouse receipts
and the proper operation of warehouses storing agricultural products, a
statutory scheme which clearly defines the rights, duties and liabilities of
each party to a warehouse receipt is also a necessary legal precondition for
the use of warehouse receipts in the primary market. Generally, a licensed
warehouse person is required by Federal law to deliver without unnecessary delay stored agricultural products on presentation of a valid, unal-
tered warehouse receipt. Whereas the Federal government regulates the
issuance of warehouse receipts and the operation of warehouse storing agricultural products, state law, as embodied in the Uniform Commercial
Code (UCC), governs the rights and liabilities of each party to the warehouse receipt or any transferee thereof.[4]

- *Issuer Liability.* The warehouse person issuing a warehouse receipt
  is liable for damages caused by nonreceipt or misdescription of the
  stored agricultural products. The issuer of the warehouse receipt
  may limit his/her liability in this regard by expressly and conspicu-
  ously so indicating any uncertainty in description or receipt of
  stored agricultural products on the face of the warehouse receipt.
  The issuer must also exercise ordinary care in management and control of the stored agricultural product. As stated above, the issuer
  usually disclaims certain liability by including an exculpatory
  clause on the reverse side of the warehouse receipt.
- *Transferability.* Besides insuring the integrity of warehouse receipts,
  another important legal precondition is the necessity for easy transferability of the warehouse receipt. In this regard, there are two
  kinds of warehouse receipts: negotiable or nonnegotiable. Under
  common law, all warehouse receipts were nonnegotiable. This
  means that the warehouse receipt was payable to a named or specified person. In this situation, there is no direct obligation or duty of
  the warehouse person to hold possession or deliver stored agricultural products until notice of transfer of the warehouse receipt has
  been communicated to the warehouse person from the transferor or
  transferee of the warehouse receipt. Until such notice, the purchaser

of the warehouse receipt acquires no rights against the warehouse person. Additionally, a transferee of a nonnegotiable warehouse receipt may be subject to claims or defenses to the warehouse receipt, such as liens of the warehouse for unpaid storage charges. These limitations on transferability do not apply to negotiable warehouse receipts. Thus, it is a necessary legal precondition that warehouse receipts be negotiable.

- *Negotiability.* A warehouse receipt is negotiable if the stored agricultural products are deliverable to bearer (person in possession of the instrument) or to the order of a named person. The warehouse receipt should clearly indicate on its face whether it is negotiable or nonnegotiable. Banks prefer negotiable warehouse receipts as security for extensions of credit because such warehouse receipts, if duly negotiated, give the Bank or its subsequent transferee rights superior to those acquired by the original depositor or producer storing agricultural products and receiving the warehouse receipt as evidence thereof. A warehouse receipt is duly negotiated if it is properly endorsed (no endorsement is required if goods are deliverable to bearer), purchased in good faith (honestly) without notice of any defense against or claim to it on the party of any person, and for value. If these requirements are met, the warehouse receipt is said to be duly negotiated to a transferee or holder and, as a result, such holder takes the warehouse receipt free from any claims or defenses except "real" defenses. Real defenses include legal challenges to a negotiable instrument such as lack of capacity, duress or illegality, fraud and certain misrepresentations, discharge in insolvency or bankruptcy proceedings, or such other discharges or defenses of which the holder of the negotiable instrument has notice. Extensions of credit by banks on the security of warehouse receipts rarely are affected by these kinds of extraordinary events if the lending bank exercises due diligence in assessing the creditworthiness and salient aspects of the storage transaction giving rise to the warehouse receipt.

A holder of a negotiable warehouse receipt which has been duly negotiated acquires rights in the stored agricultural products superior to any other party except: buyers of fungible agricultural products in the ordinary course of business which are sold and delivered by the warehouse person who is also in the business of buying and selling such agricultural products; one with a perfected security interest in the stored agricultural products; and, the warehouse person to the extent he/she has any liens or charges, provided such liens or charges are clearly set forth on the face of the warehouse receipt. Besides these three limited instances, a holder of a negotiable warehouse receipt duly negotiated acquires title to the stored agricultural products and has a right to receive delivery of such products on presentation of the warehouse receipt to the warehouse person. Regulations or statutory provisions allowing for the negotiability and free transferability of warehouse receipts are essential to protect the right of holders of such receipts to obtain possession to the stored agricultural products on presentation of the warehouse receipt to the warehouse person. Without negotiability and free transferability, banks are less willing to lend on the security of warehouse receipts.

- *Priorities on Liquidation.* In the event of default by a producer receiving extensions of credit secured by warehouse receipts or a warehouse person, a lender must be certain it will have a priority interest as against all other parties to the stored agricultural products represented by warehouse receipts. Under the Uniform Commercial Code, one normally has a priority interest to stored agricultural products to the extent their security interest in the stored agricultural products or the warehouse receipt is perfected. After completing a security agreement, a security interest is perfected by taking possession of the negotiable warehouse receipt or by filing a financing statement which contains a general description of the collateral covered by the security agreement. This financing statement must be filed in the appropriate local county recording office or with the particular state's Secretary of State. Under the first to file rule, one who perfects its security interest by taking possession or by filing first, takes priority over other creditors who file subsequently even though these creditors obtained a security interest in the underlying collateral before the party first perfecting obtained their security interest. However, the UCC provides for automatic perfection of a negotiable warehouse receipt for a period of 21 days after the security interest attaches if one has possession of the negotiable warehouse receipt. The security interest attaches when the collateral (warehouse receipt) has been pledged with the secured party under an oral agreement or a security agreement has been signed by the debtor. In almost all circumstances, banks will take possession of the negotiable warehouse receipt, require endorsement by the borrower (producer) and require a signed security agreement before making loans secured by negotiable warehouse receipts.[5]

To promote bank financing of warehouse receipts in the primary market, it is necessary that the warehouse receipt be negotiable and that a mechanism exist for the bank to acquire a priority security interest in the warehouse receipt and the stored agricultural products. It is necessary that each locality, state or political subdivision have a uniform recording and filing system so that lenders or other secured parties can determine the extent to which the warehouse receipt or stored agricultural products are subject to superior liens. The following hypothetical illustrates the clear benefits of such a system:

A bank (Bank A) loans a producer US$50,000 on the security of its grow-
ing corn crop. The bank executes a security agreement and perfects its security interest by filing a financing statement in the local recording office.
After harvesting the corn crop, the producer stores it in a local warehouse
and receives a negotiable warehouse receipt as evidence thereof, intending
to store the corn only for such time as necessary for the price of corn to rise.
Without Bank A's knowledge, the producer then pledges the negotiable
warehouse receipt to another bank (Bank B) which extends credit on the
security of the warehouse receipt. If the producer defaults on its loans to
Bank A and Bank B, which bank has priority to the stored corn? Under the
UCC, Bank A would have priority because it perfected its security interest
before the negotiable warehouse receipt was issued and Bank B took possession. Bank B could have protected itself by reviewing the local recording and filing system to ascertain whether a prior perfected security
interest existed with respect to the producer's corn crop before it lent on
the security of the warehouse receipt.[6]

This hypothetical problem illustrates that without a local recording and
filing system and a method for perfecting security interests, banks and
other lenders will not be able to prevent double-financing by producers.

Legal Preconditions for Secondary Market Financing of Agricultural Stocks[7]

The principal mechanisms and methods for financing the storage of agricultural products in the secondary market are bankers' or trade acceptances. A banker's acceptance is a negotiable time draft drawn by a bank
customer and accepted by a bank. A trade acceptance is a draft drawn by
a customer and accepted by a nonbanking institution. The effect of acceptance is to place the bank's or nonbank acceptor's liability on the draft in
place of the drawer's. A holder of a banker's acceptance looks first to the
acceptor for payment at maturity and only secondarily to the drawer (customer). Bankers' and trade acceptances are useful financing mechanisms
for agricultural producers because they allow less creditworthy borrowers
to use the bank's or other acceptor's credit to finance the storage of agricultural products. Although very creditworthy nonbank companies do issue trade acceptances, for secondary market purposes bankers'
acceptances are preferable. Because a banker's acceptance is a negotiable
instrument which trades in the money market like commercial paper and
similar short-term securities, it is easily transferable and a preferred instrument for investors.

In a typical banker's acceptance transaction, a producer in possession
of a validly issued warehouse receipt pledges the warehouse receipt to a
bank which has agreed to allow the producer to draw drafts which are ac-

cepted by the bank. This agreement is usually reflected in an acceptance credit agreement executed by the bank and the drawer (producer). The acceptance credit agreement usually requires the drawer to pay the bank the face amount of the banker's acceptance or allows the bank to deduct a commensurate amount from the drawer's depoit account at the bank either several days before maturity of the acceptance or on sale and termination of storage of the agricultural products evidenced by the warehouse receipt. The accepted draft is discounted to the drawer (producer) and then either held in the accepting bank's portfolio until maturity or rediscounted in the money markets to an investor, usually another commercial bank, an investment bank, the central bank or any other money market investor. If the banker's acceptance is held in the accepting bank's portfolio until maturity, it is similar to other extensions of credit subject to the accepting bank's lending limits. However, if the banker's acceptance is rediscounted in the money market to other investors or the central bank, the bank has effectively funded its original discount to the drawer in the secondary market. The distinct and apparent advantage to the accepting bank of rediscounting the banker's acceptance in the secondary market is that the bank has financed the drawer's (producer's) storage transaction through the money markets rather than using its own funds. Thus, the amount of bankers' acceptances rediscounted in the secondary market by the accepting bank are generally not subject to lending limits provided certain requirements discussed below are met.

The first requirement is to ensure that the warehouse receipt has been validly issued and represents agricultural products under storage. The foregoing discussion of primary market financing of warehouse receipts delineates the regulatory apparatus and statutory scheme required to make certain warehouse receipts are validly issued and the issuing warehouse's operations are proper.

Assuming the warehouse receipt is validly issued, it is then necessary that, on acceptance, the accepting bank's liability become primary as against the drawer. A holder of a banker's acceptance, notwithstanding the financial viability of the drawer, should be able to receive payment on maturity from the accepting bank. Only on default of the accepting bank should a holder have to look to the drawer for payment. Under the UCC, an acceptor engages that he will pay the instrument according to its tenor at the time of his engagement and that the drawer will pay the amount of the draft upon dishonor by the acceptor. Thus, the acceptor is primarily liable on the draft, while the drawer remains secondarily liable. This requirement is essential for a banker's acceptance to appeal to investors because such investors only need to assess the creditworthiness of the accepting bank rather than that of the drawer's.

Additionally, it is a necessary legal precondition that the banker's acceptance be a negotiable instrument. The foregoing discussion of negotiability would apply to bankers' acceptances in the same manner as warehouse receipts. However, the significant difference is that, provided there are no "real defenses" to the banker's acceptance, payment should be required at maturity, notwithstanding a secured party's prior perfected security interest in the stored agricultural products. To promote bankers' acceptances as a secondary market financing mechanism, it is essential that a holder receive payment on presentation of the banker's acceptance to the accepting bank at maturity. Whether the drawer (producer) has satisfied its obligations under the acceptance credit agreement and repaid the face value of the draft banker's acceptance prior to its maturity should not affect a holder's ability to receive payment at maturity from the accepting bank.

In the United States, the main impetus for creating a secondary market in bankers' acceptances is that the central bank (the Federal Reserve System) will rediscount bankers' acceptances so long as they are "eligible". The concept of eligibility makes certain that funds will be available at the accepting bank to pay the holder of the acceptance the face value thereof at maturity. In designing or promoting an active secondary market in bankers' acceptances, it is thus necessary for the central bank to have authority to rediscount "eligible" acceptances. The ability to rediscount bankers' acceptances at the central bank enhances a commercial bank's desire to use bankers' acceptance financing as a funding mechanism by adding sufficient liquidity to the secondary market. Also, an effective and efficient bank examination process must be in place to monitor the accepting bank's use of acceptances in "eligible" transactions. The clear benefits of acceptance financing to an accepting bank is that the bank need not use its own funds to lend to the borrower (producer), but simply uses its credit and thereby raises funds in the market which are downstreamed to the borrower by rediscounting the banker's acceptance either at the central bank or in the market. Additionally, eligible bankers' acceptances which are rediscounted in the money market or at the central bank should not be subject to applicable lending limits and reserve requirements since the accepting bank's own funds are not used and the banker's acceptance is secured until maturity by a warehouse receipt which can be expeditiously liquidated in the event of default by the drawer (producer) under the acceptance credit agreement. Other than these minimal legal preconditions, adopting and strictly enforcing requirements of eligibility will insure that bankers' acceptances created to finance the storage of agricultural products actually finance such storage, and are not used for working capital purposes by the drawer (producer). The following minimal requirements for eligibility are essential.

Marketability of Stored Agricultural Product. A fundamental requirement for eligibility should be that the particular agricultural products financed by a banker's acceptance are "readily marketable staples".

Generally, readily marketable staples are agricultural products (or raw materials) which are nonperishable and have a wide, ready market. Such agricultural products have a wide, ready market if daily price quotations are available. This assures that in the event of default by the drawer or, if there are significant price declines in the stored agricultural product while the banker's acceptance is outstanding, the accepting bank will be able to require more collateral from the drawer or quickly sell the stored agricultural products represented by a warehouse receipt securing the banker's acceptance. The agricultural product should be nonperishable for the obvious reason that there is protection to the accepting bank because perishable products could potentially lose their entire market value at any time the banker's acceptance is outstanding.

Maturity of Banker's Acceptance. It is equally important that the maturity of the banker's acceptance correspond to the duration of the storage of the agricultural product so that the storage transaction is self-liquidating. The basis of this self-liquidating requirement is twofold: First, the banker's acceptance should finance only the storage of agricultural products pending a reasonable immediate sale, shipment or distribution of those products by the producer in the market rather than for speculative or working capital purposes; and secondly, on sale of such agricultural products by the drawer (producer), proceeds of such sale should be immediately remitted to the accepting bank to cover payment of the banker's acceptance at maturity. As long as the banker's acceptance is self-liquidating and secured the entire period by the accepting bank's possession of the warehouse receipt while the banker's acceptance is outstanding, it is sufficiently certain that the accepting bank will not have to use its own funds but only those resulting from the storage pending the sale transaction to pay the face value of ther bankers' acceptance to the holder thereof at maturity. Thus, the maturity of the banker's acceptance should correspond to the duration of the storage of agricultural products which it is secured by.

An effective method of ensuring that the banker's acceptance is self-liquidating is to limit "eligible" acceptances to a maturity of six months. Although most bankers' acceptances will be of shorter duration because the underlying storage pending sale is less than six months, the short-term nature and self-liquidating aspects of the bankers' acceptances are more readily assured if their maturity is limited to six months from the date of acceptance.

Independent Warehouse Person. As discussed above in connection with primary market financing of warehouse receipts, it is preferable that only warehouse receipts issued by licensed, independent warehouses be acceptable for bankers' acceptance financing. This ensures the integrity of the warehouse receipt and more effectively protects the accepting bank's security interest in the stored agricultural products. In some situations, however, warehouse receipts issued by a field warehouse should be eligible as security for bankers' acceptance financing provided applicable regulations prevent the drawer (producer) from having any access to or control over the stored agricultural products. In both the independent and field warehousing arrangements, the warehouse person should be sufficiently bonded to ensure payment on the warehouse receipt should stored agricultural products be destroyed or damaged.

In addition to the foregoing requirements, the face of the banker's acceptance should clearly delineate the specific nature and duration of the transaction which it is secured by. This is accomplished by having an eligibility stamp which places this information on the banker's acceptance. It is also necessary to ensuring that the accepting bank has a filing system which can efficiently determine the particular transaction and commensurate warehouse receipt securing the banker's acceptance. Additionally, any banker's acceptance rediscounted in the money market should be endorsed by both the drawer (producer) and the accepting bank. This requirement protects a holder of the banker's acceptance in the event of insolvency and liquidation of the accepting bank. As indicated above, so long as the drawer (producer) remains secondarily liable, the holder of the banker's acceptance is more certain he/she can receive any payment due to maturity.

Summary

In sum, the principal legal preconditions for primary and secondary market financing of the storage of agricultural products are twofold: First, it is necessary that a regulatory and statutory apparatus exists which ensures the integrity of the issuance of warehouse receipts. A holder of a warehouse receipt or a bank lending on the security of warehouse receipts must be assured that the corresponding stored agricultural products will be available for resale in the market in the event a borrower (producer) defaults on an extension of credit. This assurance is readily provided if an effective system exists to license, regulate and inspect at frequent intervals all warehouses issuing warehouse receipts. Secondly, to promote and sustain secondary market financing of stored agricultural products, banks or other sufficiently creditworthy commercial lenders must have authority to create acceptances. In addition to such authority, a statutory and regulatory scheme is necessary to exempt "eligible" acceptances from a bank's lending limits to the extent the acceptance is rediscounted in the market. Assuming bankers' acceptances are "eligible" by being secured through their existence by validly issued warehouse receipts, by financing only readily marketable staples, by being self-liquidating, and by limiting their maturity to the time period of the underlying transaction or to a maximum of six months, the central bank should have authority to rediscount such "eligible" acceptances both to enhance their use and sale in the private credit markets, and more significantly, to monitor the quality of such "eligible" acceptances to ensure all established guidelines are satisfied. The foregoing legal preconditions are minimal requirements for establishing and sustaining both primary and secondary market financing for the financing of agricultural stocks. In many particular cases, more strict requirements may be necessary depending upon the particular economic and regulatory environment affecting the storage of agricultural products.

Notes

1. This appendix was prepared by Mr. Joseph Reid of the University of Baltimore Law School.

2. *See* W. Schneider, *Field Warehousing As A Facility For Lending Commodities* (1941).

3. United States Warehouse Act, 7 U.S.C. Sections 241 *et seq.* (1980).

4. *See* U.C.C. Article 7 (10th ed. 1987); 78 AM. JUR. 2d *Warehouses* Sections 52 *et seq.* (1975); R. Henson, *Documents of Title Under the U.C.C.* (1990); Clark, B., *The Law of Secured Transactions*, Chapter 8 - "Agricultural Financing" (1988); Jensen, L., *Agricultural Lending in the 1980's: An Insurance Company's Perspective*, 18 MEM. ST. U.L. REV. 353 (1988); Hawkland, *The Proposed Amendments to Article 9 of the U.C.C. - Part 1: Financing the Farmer*, 76 COMM. L.J. 416 (1971); Clark, *Some Problems in Agricultural Financing Under the U.C.C.*, 39 U. COLO. L. REV. 352 (1967); Note, *Agricultural Financing Under the U.C.C.*, 12 ARIZ.L.REV. 391 (1970).

5. *See* Henson, *supra*; Clarke, *supra*; Hamilton, N., "Article 9 and Agricultural Finance," Chapter 8, in BENDER'S UNIFORM COMMERCIAL CODE SERVICE, Vol. 1D, *Secured Transactions Under the U.C.C.* (1990).

6. Clarke, *supra* at 8-4.

7. *See* Ryan, R. "Letters of Credit and Bankers' Acceptances," *Practising Law Institute*, Course Handbook Series, No. 450 (1988); Todd, W.F., "An Introduction to Bankers' Acceptances in the 1980's," in *International Banking: United States Laws and Regulations*, Chapter 1 (American Bankers Association, International Division) (1984); Reid, J., "The Federal Reserve System's Regulation of Bankers' Acceptances," *The Letter of Credit Law Institute* (1986).

14

Statistical Annex

TABLE 14.1 Brazil—Geographic Distribution of Wheat Production, 1980—1988 (metric tons)

Regions	1980	1981	1982	1983	1984	1985	1986	1987	1988	Production Breakdown 1988	Average Annual Growthrate
North											
Rondonia	0	0	0	0	0	0	0	0	0	0.00%	
Acre	0	0	0	0	0	0	0	0	0	0.00%	
Amazonas	0	0	0	0	0	0	0	0	0	0.00%	
Roraima	0	0	0	0	0	0	0	0	0	0.00%	
Para	0	0	0	0	0	0	0	0	0	0.00%	
Amapa	0	0	0	0	0	0	0	0	0	0.00%	
Subtotal:	0	0	0	0	0	0	0	0	0	0.00%	
Northeast											
Maranhao	0	0	0	0	0	0	0	0	0	0.00%	
Piaui	0	0	0	0	0	0	0	0	0	0.00%	
Ceara	0	0	0	0	0	0	0	0	0	0.00%	
Rio Gr. do Norte	0	0	0	0	0	0	0	0	0	0.00%	
Paraiba	0	0	0	0	0	0	0	0	0	0.00%	
Pernambuco	0	0	0	0	0	0	0	0	0	0.00%	
Alagoas	0	0	0	0	0	0	0	0	0	0.00%	
Sergipe	0	0	0	0	0	0	0	0	0	0.00%	
Bahia	0	0	0	0	0	0	0	0	0	0.00%	
Subtotal:	0	0	0	0	0	0	0	0	0	0.00%	

Regions	1980	1981	1982	1983	1984	1985	1986	1987	1988	Production Breakdown 1988	Average Annual Growthrate
Southeast											
Minas Gerais	20,862	15,912	22,999	27,890	23,724	12,929	23,199	16,535	26,923	0.49%	0.64%
Espirito Santo	0	0	0	0	0	0	0	0	0	0.00%	
Rio de Janeiro	0	0	0	0	0	0	0	0	0	0.00%	
Sao Paulo	195,314	131,556	134,000	174,347	113,060	295,995	311,454	319,800	309,600	5.60%	11.86%
Subtotal:	216,176	147,468	156,999	202,237	136,784	308,924	334,653	336,335	336,523	6.09%	10.85%
South											
Parana	1,350,006	915,000	1,025,000	1,066,000	1,113,009	2,642,153	2,950,000	3,318,200	3,040,000	55.02%	18.40%
Santa Catarina	9,033	8,620	13,656	9,881	11,854	44,000	152,009	160,120	138,404	2.50%	54.23%
Rio Gr. do Sul	1,016,243	1,072,914	516,790	797,422	611,632	933,510	1,739,340	1,783,449	1,586,921	28.72%	10.32%
Subtotal:	2,375,282	1,996,534	1,555,446	1,873,303	1,736,495	3,619,663	4,841,349	5,261,769	4,765,325	86.25%	15.45%
Center West											
Mato Gr. do Sul	110,000	65,395	112,641	159,365	108,775	317,664	460,000	498,656	420,000	7.60%	28.31%
Mato Grosso	59	100	107	3	298	162	542	843	783	0.01%	49.11%
Goias	0	2	1,270	1,168	633	557	1,539	745	2,190	0.04%	
Distr. Fed.	96	132	482	624	172	227	87	763	444	0.01%	18.02%
Subtotal:	110,155	65,629	114,500	161,160	109,878	318,610	462,468	501,007	423,417	7.66%	28.31%
Total Brazil:	2,701,613	2,209,631	1,826,945	2,236,700	1,983,157	4,247,197	5,638,470	6,099,111	5,525,265	100.00%	15.80%

Note: A value "0" denotes "not available"
Source: IBGE for basic data

TABLE 14.2 Brazil—Geographic Distribution of Maize Production, 1980—1988 (metric tons)

Regions	1980	1981	1982	1983	1984	1985	1986	1987	1988	Production Breakdown 1988	Average Annual Growthrate
North											
Rondonia	106,976	114,065	136,434	97,282	158,912	147,664	189,134	173,112	240,925	0.96%	9.73%
Acre	21,726	24,130	25,770	19,597	41,724	25,770	26,778	33,445	40,669	0.16%	6.61%
Amazonas	10,203	7,907	5,635	3,460	1,730	2,738	3,457	4,010	4,264	0.02%	-10.62%
Roraima	5,762	14,479	2,513	591	6,107	7,183	3,902	4,026	1,984	0.01%	-7.58%
Para	76,742	79,983	141,152	67,385	159,174	134,587	178,852	219,154	312,557	1.25%	17.76%
Amapa	522	1,161	1,108	864	701	799	574	461	597	0.00%	-5.87%
Subtotal:	221,931	241,725	312,612	189,179	368,348	318,741	402,697	434,208	600,996	2.40%	11.94%
Northeast											
Maranhao	270,583	144,470	303,592	90,041	268,656	125,141	302,231	110,478	345,881	1.38%	0.83%
Piaui	73,548	43,365	123,316	25,507	157,429	259,033	351,288	116,268	381,188	1.52%	26.18%
Ceara	96,000	21,600	153,349	17,531	257,573	165,070	274,503	74,711	435,545	1.74%	24.56%
Rio Gr. do Norte	2,669	2,210	5,743	1,980	86,656	50,307	76,050	8,293	84,038	0.34%	54.67%
Paraiba	33,981	26,208	26,065	24,954	199,186	157,501	181,977	63,547	191,300	0.76%	29.04%
Pernambuco	59,042	52,200	94,738	11,895	301,945	196,199	253,645	59,101	241,368	0.96%	19.68%
Alagoas	8,832	8,257	35,478	3,981	40,754	49,018	48,419	10,114	53,214	0.21%	19.97%
Sergipe	3,310	16,345	78,260	2,458	75,059	94,451	104,174	27,466	67,879	0.27%	34.67%
Bahia	282,495	191,075	312,251	133,959	84,177	430,073	312,690	150,744	405,640	1.62%	3.23%
Subtotal:	830,460	505,730	1,132,792	312,306	1,471,435	1,526,793	1,904,977	620,722	2,206,053	8.82%	12.65%

Regions	1980	1981	1982	1983	1984	1985	1986	1987	1988	Production Breakdown 1988	Average Annual Growthrate
Southeast											
Minas Gerais	3,008,788	2,912,874	3,030,924	2,674,869	2,556,393	3,015,115	3,266,247	3,336,890	3,282,269	13.12%	1.72%
Espirito Santo	205,293	221,520	222,540	154,236	213,852	230,512	249,300	255,724	218,113	0.87%	2.19%
Rio de Janeiro	45,684	55,044	73,281	73,268	69,500	67,955	70,498	54,072	56,463	0.23%	1.07%
Sao Paulo	2,335,800	2,752,800	3,164,000	3,164,000	2,866,742	2,900,881	3,093,600	3,732,500	3,691,140	14.76%	4.45%
Subtotal:	5,595,565	5,942,238	6,490,745	6,066,373	5,706,487	6,214,463	6,679,645	7,379,186	7,247,985	28.98%	2.99%
South											
Parana	5,466,967	5,363,109	5,430,000	5,018,870	5,400,000	5,803,713	4,331,546	7,641,800	5,560,000	22.23%	1.38%
Santa Catarina	3,009,995	3,162,590	2,628,756	1,687,355	2,345,209	2,159,049	1,951,299	2,419,200	2,371,200	9.48%	-3.45%
Rio Gr. do Sul	3,162,033	3,808,793	3,147,246	3,174,771	3,567,360	3,558,591	1,937,656	3,873,498	2,537,036	10.14%	-2.77%
Subtotal:	11,638,995	12,334,492	11,206,002	9,880,996	11,312,569	11,521,353	8,220,501	13,934,498	10,468,236	41.85%	-0.87%
Center West											
Mato Gr. do Sul	188,396	232,636	257,902	236,443	262,220	327,334	320,743	649,515	647,635	2.59%	15.76%
Mato Grosso	142,572	190,765	288,324	319,238	318,477	410,500	529,072	683,334	789,786	3.16%	22.43%
Goias	1,751,507	1,666,946	1,921,842	1,722,912	1,719,918	1,690,770	2,646,400	3,034,700	2,990,000	11.95%	7.89%
Distr. Fed.	2,646	2,376	3,858	3,769	4,684	7,200	19,192	50,484	61,545	0.25%	53.24%
Subtotal:	2,085,121	2,092,723	2,471,926	2,282,362	2,305,299	2,435,804	3,515,407	4,418,033	4,488,966	17.95%	10.66%
Total Brazil:	20,372,072	21,116,908	21,614,077	18,731,216	21,164,138	22,017,154	20,723,227	26,786,647	25,012,236	100.00%	2.72%

Source: IBGE for basic data

260

TABLE 14.3 Brazil—Geographic Distribution of Rice Production, 1980—1988 (metric tons)

Regions	1980	1981	1982	1983	1984	1985	1986	1987	1988	Production Breakdown 1988	Average Annual Growthrate
North											
Rondonia	178,394	217,030	188,714	100,576	181,847	220,548	279,058	213,322	245,025	2.06%	4.75%
Acre	21,711	24,884	27,761	19,085	44,813	27,792	38,218	41,009	42,801	0.36%	9.11%
Amazonas	7,706	7,234	4,840	1,277	2,178	3,218	4,374	3,310	1,869	0.02%	-11.44%
Roraima	25,718	44,830	18,524	4,235	15,409	15,689	11,359	10,644	8,823	0.07%	-12.86%
Para	154,663	138,434	175,481	102,865	148,991	133,530	180,540	148,915	184,627	1.55%	2.10%
Amapa	597	2,071	1,844	1,884	1,431	1,408	550	440	649	0.01%	-11.05%
Subtotal:	388,789	434,483	417,164	229,922	394,669	402,185	514,099	417,640	483,794	4.07%	2.93%
Northeast											
Maranhao	1,281,316	690,951	1,575,030	431,195	1,145,503	622,877	1,291,982	595,829	1,325,671	11.15%	-0.56%
Piaui	76,807	87,585	213,103	54,480	200,057	266,807	387,936	162,496	409,115	3.44%	20.79%
Ceara	18,000	30,600	70,525	30,077	82,597	89,420	153,446	113,967	160,563	1.35%	29.14%
Rio Gr. do Norte	878	2,039	1,782	1,336	8,742	8,592	12,736	2,082	7,960	0.07%	27.71%
Paraiba	7,221	7,912	7,666	3,645	13,261	14,871	21,597	17,642	28,124	0.24%	20.77%
Pernambuco	5,406	10,207	14,910	10,709	15,688	20,041	36,087	30,148	25,940	0.22%	21.97%
Alagoas	14,680	13,105	15,537	11,379	13,634	18,096	26,261	17,318	27,100	0.23%	8.33%
Sergipe	19,030	16,577	24,758	22,739	26,912	29,087	29,420	27,730	39,411	0.33%	8.78%
Bahia	60,200	40,250	57,441	54,053	33,080	66,513	108,489	52,174	97,828	0.82%	7.25%
Subtotal:	1,483,538	899,226	1,980,752	619,613	1,539,474	1,136,304	2,067,954	1,019,386	2,121,712	17.84%	4.25%

Regions	1980	1981	1982	1983	1984	1985	1986	1987	1988	Production Breakdown 1988	Average Annual Growthrate
Southeast											
Minas Gerais	833,829	688,847	731,721	778,656	594,307	850,974	950,908	909,359	885,264	7.44%	2.85%
Espirito Santo	57,942	57,034	71,790	74,795	85,244	97,970	122,057	120,430	105,110	0.88%	10.44%
Rio de Janeiro	84,085	87,534	92,419	95,735	95,978	104,709	123,745	95,139	97,461	0.82%	2.56%
São Paulo	420,000	379,890	463,500	617,400	398,790	508,111	543,130	552,160	511,665	4.30%	3.45%
Subtotal:	1,395,856	1,213,305	1,359,430	1,566,586	1,174,319	1,561,764	1,739,840	1,677,088	1,599,500	13.45%	3.40%
South											
Parana	638,000	493,632	257,229	368,313	242,570	296,000	206,000	342,844	327,138	2.75%	-7.11%
Santa Catarina	428,868	404,068	374,0783	95,613	453,057	446,366	450,899	504,756	550,521	4.63%	3.67%
Rio Gr. do Sul	2,293,386	2,455,360	2,589,885	2,220,497	3,119,013	3,207,046	2,987,503	3,561,498	3,881,290	32.63%	6.67%
Subtotal:	3,360,254	3,353,060	3,221,192	2,984,423	3,814,640	3,949,412	3,644,402	4,409,098	4,758,949	40.01%	4.67%
Center West											
Mato Gr. do Sul	504,212	452,233	339,315	450,796	381,660	323,993	276,013	465,987	355,698	2.99%	-3.36%
Mato Grosso	1,175,041	941,577	1,002,243	790,469	672,671	521,776	794,182	922,384	1,018,560	8.56%	-2.49%
Goias	1,455,406	920,593	1,396,899	1,081,295	1,037,760	1,115,240	1,358,400	1,501,040	1,546,240	13.00%	2.85%
Distr. Fed.	12,624	13,849	17,558	18,649	12,170	8,482	9,786	12,477	9,532	0.08%	-5.50%
Subtotal:	3,147,283	2,328,252	2,756,015	2,341,209	2,104,261	1,969,491	2,438,381	2,901,888	2,930,030	24.63%	-0.07%
Total Brazil:	9,775,720	8,228,326	9,734,553	7,741,753	9,027,363	9,019,156	10,404,676	10,425,100	11,893,985	100.00%	3.01%

Source: IBGE for basic data

TABLE 14.4 Brazil—Geographic Distribution of Soybean Production, 1980—1988 (metric tons)

Regions	1980	1981	1982	1983	1984	1985	1986	1987	1988	Production Breakdown 1988	Average Annual Growthrate
North											
Rondonia	0	0	0	0	0	0	0	0	0	0.00%	
Acre	0	0	0	0	0	0	0	0	0	0.00%	
Amazonas	0	0	0	0	0	0	0	0	0	0.00%	
Roraima	0	0	0	0	0	0	0	0	0	0.00%	
Para	0	0	0	0	0	0	0	0	0	0.00%	
Amapa	0	0	0	0	0	0	0	0	0	0.00%	
Subtotal:	0	0	0	0	0	0	0	0	0	0.00%	
Northeast											
Maranhao	96	112	430	487	7,604	9,012	13,881	8,864	25,916	0.14%	113.02%
Piaui	0	0	20	0	781	0	1,333	1,681	9,095	0.05%	
Ceara	0	0	0	0	0	0	0	0	0	0.00%	
Rio Gr. do Norte	0	0	0	0	0	0	0	0	0	0.00%	
Paraiba	0	0	0	0	0	0	0	0	0	0.00%	
Pernambuco	0	0	0	0	0	0	0	0	0	0.00%	
Alagoas	0	0	0	0	0	0	0	0	0	0.00%	
Sergipe	0	0	0	0	0	0	0	0	0	0.00%	
Bahia	2,224	1,019	354	4,200	35,929	75,600	140,418	148,313	375,313	2.06%	131.30%
Subtotal:	2,320	1,131	804	4,687	44,314	84,612	155,632	158,858	410,324	2.26%	126.15%

Regions	1980	1981	1982	1983	1984	1985	1986	1987	1988	Production Breakdown 1988	Average Annual Growthrate
Southeast											
Minas Gerais	289,542	273,874	390,390	477,222	554,082	882,607	796,530	809,040	936,24	5.15%	18.11%
Espirito Santo	0	0	0	0	0	0	0	0	0	0.00%	
Rio de Janeiro	0	0	0	0	0	0	0	0	0	0.00%	
Sao Paulo	1,099,058	1,032,000	993,300	966,000	870,703	960,386	918,036	923,400	1,001,900	5.51%	-1.43%
Subtotal:	1,388,600	1,305,874	1,383,690	1,443,222	1,424,785	1,842,993	1,714,566	1,732,440	1,938,142	10.65%	4.87%
South											
Parana	5,400,192	4,983,210	4,200,120	4,315,000	4,121,000	4,413,000	2,600,000	3,810,000	4,800,000	26.39%	-3.62%
Santa Catarina	718,764	648,196	534,652	405,397	578,769	563,882	498,034	455,339	535,600	2.94%	-3.36%
Rio Gr. do Sul	5,737,170	6,088,344	4,220,579	5,268,869	5,415,494	5,711,149	3,269,024	4,995,218	3,631,281	19.96%	-4.64%
Subtotal:	11,856,126	11,719,750	8,955,351	9,989,266	10,115,263	10,688,031	6,367,058	9,260,557	8,966,881	49.30%	-3.98%
Center West											
Mato Gr. do Sul	1,322,082	1,347,447	1,537,341	1,801,000	2,006,835	2,558,720	1,965,013	2,283,898	2,463,979	13.55%	8.54%
Mato Grosso	117,173	224,901	365,501	611,258	1,050,095	1,656,039	1,921,053	2,389,033	2,828,745	15.55%	49.53%
Goias	455,794	382,713	560,916	693,106	847,510	1,356,240	1,127,560	1,063,990	1,500,000	8.25%	17.94%
Distr. Fed.	13,709	25,551	32,444	39,808	51,990	91,787	83,809	90,056	81,920	0.45%	25.57%
Subtotal:	1,908,758	1,980,612	2,496,202	3,145,172	3,956,430	5,662,786	5,097,435	5,826,977	6,874,644	37.79%	18.88%
Total Brazil:	15,155,804	15,007,367	12,836,047	14,582,347	15,540,792	18,278,422	13,334,691	16,978,832	18,189,991	100.00%	2.36%

Note: Data for 1987 and 1988 estimated. A value "0" denotes "not available"
Source: IBGE for basic data

TABLE 14.5 Brazil—Geographic Distribution of the Production of Major Grains, 1980—1988 [a] (metric tons)

Regions	1980	1981	1982	1983	1984	1985	1986	1987	1988	Production Breakdown 1988	Average Annual Growthrate
North											
Rondonia	285,370	331,095	325,148	197,858	340,759	368,212	468,192	386,434	485,950	0.80%	6.79%
Acre	43,437	49,014	53,531	38,682	86,537	53,562	64,996	74,454	83,470	0.14%	7.93%
Amazonas	17,909	15,141	10,475	4,737	3,908	5,956	7,831	7,320	6,133	0.01%	-10.74%
Roraima	31,480	59,309	21,037	4,826	21,516	22,872	15,261	14,670	10,807	0.02%	-11.83%
Para	231,405	218,417	316,633	170,250	308,165	268,117	359,392	368,069	497,184	0.82%	9.29%
Amapa	1,119	3,232	2,952	2,748	2,132	2,207	1,124	901	1,246	0.00%	-8.84%
Subtotal:	610,720	676,208	729,776	419,101	763,017	720,926	916,796	851,848	1,084,790	1.79%	6.88%
Northeast											
Maranhao	1,551,995	835,533	1,879,052	521,723	1,421,763	757,030	1,608,094	715,171	1,697,468	2.80%	-0.08%
Piaui	150,355	130,950	336,439	79,987	358,267	525,840	740,557	280,445	799,398	1.32%	23.02%
Ceara	114,000	52,200	223,874	47,608	340,170	254,490	427,949	188,678	596,108	0.98%	25.12%
Rio Gr. do Norte	3,547	4,249	7,525	3,316	95,398	58,899	88,786	10,375	91,998	0.15%	47.98%
Paraiba	41,202	34,120	33,731	28,599	212,447	172,372	203,574	81,189	219,424	0.36%	27.72%
Pernambuco	64,448	62,407	109,648	22,604	317,633	216,240	289,732	89,249	267,308	0.44%	20.05%
Alagoas	23,512	21,362	51,015	15,360	54,388	67,114	74,680	27,432	80,314	0.13%	14.07%
Sergipe	22,340	32,922	103,018	25,197	101,971	123,538	133,594	55,196	107,290	0.18%	18.01%
Bahia	344,919	232,344	370,046	192,212	153,186	572,186	561,597	351,231	878,781	1.45%	12.20%
Subtotal:	2,316,318	1,406,087	3,114,348	936,606	3,055,223	2,747,709	4,128,563	1,798,966	4,738,089	7.82%	9.13%

Regions	1980	1981	1982	1983	1984	1985	1986	1987	1988	Production Breakdown 1988	Average Annual Growth rate
Southeast											
Minas Gerais	4,153,021	3,891,507	4,176,034	3,958,637	3,728,506	4,761,625	5,036,884	5,071,824	5,130,698	8.46%	3.73%
Espirito Santo	263,235	278,554	294,330	229,031	299,096	328,482	371,357	376,154	323,223	0.53%	4.34%
Rio de Janeiro	129,769	142,578	165,700	169,003	165,478	172,664	194,243	149,211	153,924	0.25%	1.95%
Sao Paulo	4,050,172	4,296,246	4,754,800	4,921,747	4,249,295	4,665,373	4,866,220	5,527,860	5,514,305	9.10%	3.36%
Subtotal:	8,596,197	8,608,885	9,390,864	9,278,418	8,442,375	9,928,144	10,468,704	11,125,049	11,122,150	18.35%	3.54%
South											
Parana	12,855,165	11,754,951	10,912,349	10,768,183	10,876,579	13,154,866	10,087,546	15,112,844	13,727,138	22.64%	1.78%
Santa Catarina	4,166,660	4,223,474	3,551,142	2,498,246	3,388,889	3,213,297	3,052,241	3,539,415	3,595,725	5.93%	-1.93%
Rio Gr. do Sul	12,208,832	13,425,411	10,474,500	11,461,559	12,713,499	13,410,296	9,933,523	14,213,663	11,636,528	19.20%	0.05%
Subtotal:	29,230,657	29,403,836	24,937,991	24,727,988	26,978,967	29,778,459	23,073,310	32,865,922	28,959,391	47.77%	0.55%
Center West											
Mato Gr. do Sul	2,124,690	2,097,711	2,247,199	2,647,604	2,759,490	3,527,711	3,021,769	3,898,056	3,887,312	6.41%	8.97%
Mato Grosso	1,434,845	1,357,343	1,656,175	1,720,968	2,041,541	2,588,477	3,244,849	3,995,594	4,637,874	7.65%	17.52%
Goias	3,662,707	2,970,254	3,880,927	3,498,481	3,605,821	4,162,807	5,133,899	5,600,475	6,038,430	9.96%	8.03%
Distr. Fed.	29,075	41,908	54,342	62,850	69,016	107,696	113,174	153,780	153,441	0.25%	23.28%
Subtotal:	7,251,317	6,467,216	7,838,643	7,929,903	8,475,868	10,386,691	11,513,691	13,647,905	14,717,057	24.28%	10.72%
Total Brazil:	48,005,209	46,562,232	46,011,622	43,292,016	47,715,450	53,561,929	50,101,064	60,289,690	60,621,477	100.00%	3.55%

a. includes: wheat, maize, rice and soybeans.
Source: IBGE for basic data.

TABLE 14.6 Brazil—Geographic Distribution of the Yield of Major Grains, 1980—1988 [a] (metric tons)

Regions	1980	1981	1982	1983	1984	1985	1986	1987	1988	Average Annual Growthrate
North										
soybeans	0.000	0.000	0.000	0.000	0.000	0.000	0.000	0.000	0.000	
maize	1.268	1.219	1.363	1.185	1.271	1.285	1.375	1.329	1.409	1.31%
rice	1.438	1.346	1.462	1.295	1.436	1.421	1.566	1.415	1.391	0.41%
wheat	0.000	0.000	0.000	0.000	0.000	0.000	0.000	0.000	0.000	
Northeast										
soybeans	1.168	0.360	0.572	0.631	1.366	1.190	1.424	0.880	1.523	10.89%
maize	0.379	0.249	0.408	0.218	0.613	0.589	0.627	0.249	0.684	7.28%
rice	1.163	0.682	1.258	0.620	1.389	1.165	1.493	0.755	1.465	3.74%
wheat	0.000	0.000	0.000	0.000	0.000	0.000	0.000	0.000	0.000	
Southeast										
soybeans	1.920	1.791	1.856	1.984	1.747	1.949	1.874	1.974	1.945	0.58%
maize	1.903	1.945	2.114	2.172	1.936	2.197	2.217	2.346	2.455	2.85%
rice	1.459	1.199	1.456	1.700	1.234	1.711	1.779	1.682	1.743	3.62%
wheat	1.144	1.042	1.000	1.292	0.903	1.902	1.465	1.795	1.780	7.88%

Regions	1980	1981	1982	1983	1984	1985	1986	1987	1988	Average Annual Growthrate
South										
soybeans	1.714	1.785	1.472	1.727	1.620	1.709	1.185	1.768	1.502	-1.65%
maize	2.262	2.404	2.140	1.899	2.147	2.300	1.733	2.399	2.154	-0.72%
rice	2.940	3.243	3.316	2.997	3.597	3.708	3.618	3.800	4.105	3.73%
wheat	0.845	1.176	0.621	1.166	1.175	1.578	1.480	1.852	1.634	10.59%
Center West										
soybeans	1.689	1.647	1.820	1.944	1.697	1.965	1.827	2.054	2.066	2.51%
maize	2.091	1.894	2.065	2.047	2.072	2.167	2.435	2.564	2.621	3.73%
rice	1.211	0.965	1.220	1.162	1.076	1.299	1.288	1.321	1.394	2.92%
wheat	0.901	0.814	0.695	1.383	0.981	1.580	1.164	1.174	1.204	5.87%
Average Brazil										
soybeans	1.727	1.765	1.565	1.792	1.650	1.800	1.452	1.859	1.722	0.00%
maize	1.779	1.833	1.731	1.750	1.761	1.866	1.649	1.984	1.898	0.78%
rice	1.566	1.349	1.616	1.516	1.687	1.898	1.861	1.738	1.991	3.78%
wheat	0.865	1.151	0.646	1.190	1.139	1.598	1.447	1.765	1.598	9.87%

[a] data for 1987 and 1988 estimated.

Source:World Bank, *Agricultural Sector Review*, op.cit. (original data from: IBGE—*Levantamento Sistematico da Produçao Agricola*).

TABLE 14.7 Brazil—Exports of Grains and Derivatives, 1975—1986 (values in millions of current US$ FOB)

Products	1975	1976	1977	1978	1979	1980	1981	1982	1983	1984	1985	1986	Average Annual Growthrate (%)
Soybeans													
beans	684.9	788.5	709.6	169.9	179.5	393.7	403.7	123.5	308.6	454.1	762.7	243.2	-4.42
meal	465.8	795.0	1150.2	1049.9	1136.9	1449.0	2136.2	1600.3	1793.2	1460.2	1174.9	1180.6	7.06
crude oil	152.4	174.6	274.2	283.2	326.8	411.1	503.3	222.4	155.1	557.2	331.4	71.4	-0.55
refined oil	1.1	21.8	2.7	11.8	7.1	10.1	147.7	150.8	305.9	941.7	271.1	66.7	61.18
Rice	4.2	18.1	1.3	12.0	82.8	38.4	0.1						
Maize	150.9	164.7	135.7	2.2	1.7								
Total grains exp.	1459.4	1962.8	2273.6	1528.9	1734.9	2302.6	3191.0	2097.0	2634.5	3436.7	2540.3	1562.1	3.41
Total agric. exp.	4992.4	6225.8	7549.7	6962.3	7671.6	9809.9	10008.6	8257.3	9451.3	10856.3	9645.4	7136.0	4.35
grains/agr.exp.	29%	32%	30%	22%	23%	23%	32%	25%	28%	32%	26%	22%	
Total exports	8670.0	10128.0	12120.0	12659.0	15244.0	20132.0	23293.0	20175.0	21899.0	27005.0	25639.0	22393.0	10.34
grains/tot.exp.	17%	19%	19%	12%	11%	11%	14%	10%	12%	13%	10%	7%	

Source: World Bank, Brazil—Agricultural Sector Review, op. cit.

TABLE 14.8 Geographic Distribution of EGF and AGF for Major Grains, 1986 (% of value)

	Rice		Maize		Soybeans		Total Applications For All Crops	
	EGF	AGF	EGF	AGF	EGF	AGF	EGF	AGF
Regions								
North	0.13%	4.09%	0.01%	0.12%	0.01%	0.00%	0.56%	1.54%
Northeast	1.62%	5.20%	0.92%	5.77%	2.70%	0.87%	8.03%	5.91%
Southeast	4.77%	7.04%	29.84%	21.08%	24.38%	4.99%	20.79%	13.57%
South	83.14%	11.73%	60.74%	33.80%	46.01%	0.03%	56.99%	23.34%
Center/West	10.31%	71.93%	8.47%	38.93%	26.91%	94.03%	13.62%	55.61%

Source: World Bank—*Agricultural Sector Review,* op.cit.

TABLE 14.9 Distribution of EGF and AGF for Major Grains, 1970-1987, (%)

	rice		maize		soybeans		all other crops	
	EGF	AGF	EGF	AGF	EGF	AGF	EGF	AGF
1970	47.46%	94.99%	12.37%	0.42%	10.99%	0.00%	29.18%	4.59%
1971	33.47%	25.91%	5.83%	6.90%	26.62%	0.00%	34.08%	67.19%
1972	30.43%	0.00%	6.13%	8.54%	27.21%	0.00%	36.23%	91.46%
1973	52.57%	36.74%	13.30%	2.86%	0.28%	0.00%	33.85%	60.40%
1974	19.27%	4.56%	15.48%	74.32%	33.87%	0.00%	31.38%	21.12%
1975	16.53%	0.54%	7.49%	6.10%	41.24%	0.17%	34.74%	93.19%
1976	26.65%	55.25%	9.03%	7.11%	35.16%	0.06%	29.16%	37.58%
1977	16.70%	53.06%	9.92%	38.82%	34.31%	0.00%	39.07%	8.12%
1978	12.27%	23.24%	3.83%	24.53%	24.61%	0.00%	59.29%	52.23%
1979	10.70%	17.69%	6.68%	4.94%	23.31%	0.00%	59.31%	77.37%
1980	12.84%	72.58%	6.84%	0.04%	31.86%	3.35%	48.46%	24.03%
1981	10.39%	48.09%	14.64%	2.29%	28.12%	0.01%	46.85%	49.61%
1982	12.34%	11.34%	11.86%	32.73%	30.60%	0.03%	45.20%	55.90%
1983	16.59%	23.21%	10.06%	46.21%	32.35%	0.00%	41.00%	30.58%
1984	15.31%	42.73%	14.41%	24.60%	21.90%	0.00%	48.38%	32.67%
1985	27.48%	15.57%	12.67%	23.93%	31.24%	17.37%	28.61%	43.13%
1986	25.80%	33.35%	8.14%	40.13%	26.24%	15.27%	39.82%	11.25%
1987	30.64%	39.40%	9.93%	49.82%	25.84%	7.50%	33.59%	3.28%
avg.	23.19%	33.24%	9.92%	21.91%	26.99%	2.43%	39.90%	42.43%
st.dev.	11.96%	24.59%	3.33%	20.71%	9.04%	5.25%	9.30%	27.52%
st.dev./ avg.	51.57%	73.98%	33.61%	94.53%	33.51%	215.93%	23.31%	64.86%

Source: World Bank, *Brazil—Agricultural Sector Review,* op.cit.

TABLE 14.10 Brazil—Share of Cereals and Oilseeds in Value of Crop Production, 1966—1986 [a]

| Category | Percent of Total Value of Principal Crops | | | |
	1966	1973	1977	1986
CEREALS AND OILSEEDS	31.4	38.5	37.7	46.1
Wheat	2.5	3.5	2.8	8.7
Rice (Paddy)	14.2	10.2	7.5	11.1
Maize	13.3	11.9	10.0	13.3
Soybeans	1.4	12.9	17.3	13.0

[a]Other crops included in estimates of national crop production are: manioc, potatoes, peanuts, black beans, tomatoes, oranges, other fruits and vegetables, cocoa, coffee, tobacco, sugarcane, cotton, and other export and industrial crops.
Source: Word Bank, *Brazil—Agricultural Storage and Marketing Review*, op. cit (original data from IBGE).

TABLE 14.11 Brazil—Regional Distribution of Crop Production, 1986

| | Regional Production as a Percentage of Total Production | | | | | |
	North	Northeast	Southeast	South	Center/West	Total [a]
FOODCROPS						
Wheat	0.0	0.0	5.9	85.9	8.2	100.0
Maize	2.0	9.3	32.5	40.0	16.2	100.0
Rice (paddy)	4.9	20.0	16.7	33.4	23.4	100.0
Manioc	15.7	52.2	10.3	17.7	4.1	100.0
Black Beans	0.0	41.8	28.2	19.3	5.1	100.0
EXPORT & INDUSTRIAL CROPS						
Soybeans	0.0	1.6	12.8	47.7	38.2	100.0
Cocoa	10.5	86.2	2.8	0.0	0.3	100.0
Coffee	0.0	2.8	72.7	21.0	0.0	100.0
Sugarcane	0.1	27.6	61.4	5.3	5.5	100.0
Cotton	0.1	18.2	40.7	35.9	5.0	100.0

[a] Rows may not total due to rounding.
Source: World Bank, *Brazil—Agricultural Storage and Marketing Review*, op. cit. (original data from IBGE).

TABLE 14.12 Brazil—Total Fertilizer Use by Selected Crops, 1986/87

Crop	Area Planted ('000 hectares)		Total Fertilizer Use ('000 tons)	
	1986	1987	1986	1987
Wheat	3,893	3,430	845	750
Maize	14,248	13,353	1,440	1,340
Rice	6,272	5,890	890	850
Soybeans	9,164	10,515	1,550	1,700

Source: Anuario Estatistico, Setor do Fertilizantes, ANDA, Sao Paulo, 1988, p. 119.

TABLE 14.13 Brazil—Storage Capacity by Type, 1985/86 (metric tons)

Region	Bag	% of Total	Bulk	% of Total	Total	
North	397,581	85.7	66,560	14.3	464,141	100.0
Northeast	2,365,037	89.8	268,664	10.2	2,633,701	100.0
Southeast	10,686,150	68.0	5,039,904	32.0	15,726,054	100.0
South	12,260,233	38.1	19,911,114	61.9	32,171,347	100.0
Center/West	5,323,483	58.2	3,819,354	41.8	9,142,837	100.0
TOTAL	31,032,484	51.6	29,105,596	48.4	60,138,080	100.0

Source: World Bank, *Brazil—Agricultural Storage and Marketing Review,* op. cit. (original data from CIBRAZEM).

TABLE 14.14 Brazil—Ownership and Use of Agricultural Storage, 1985

	Static Capacity ('000 mt)	(%)	% of Total National Capacity
Public Use [a]			
Public Sector Warehouses [b]	8,508	27.9	14.2
Privately-owned General Warehouse			
Companies	6,352	20.8	10.6
Cooperatives	15,683	51.3	26.1
Subtotal	30,543	100.0	50.9
Private Use [c]			
Public Sector [b]	3,309	11.2	5.5
Commercial and On-Farm	7,544	25.5	12.6
Agro-Industrial	18,702	63.3	31.1
Subtotal	29,555	100.0	49.1
TOTAL	60,098	100.0	100.0

[a] Space available for lease to the general public.
[b] CIBRAZEM, State Storage companies, rail warehouses.
[c] Own use only.
Source: CIBRAZEM, Cadastro Nacional de Unidades Armazenadoras.

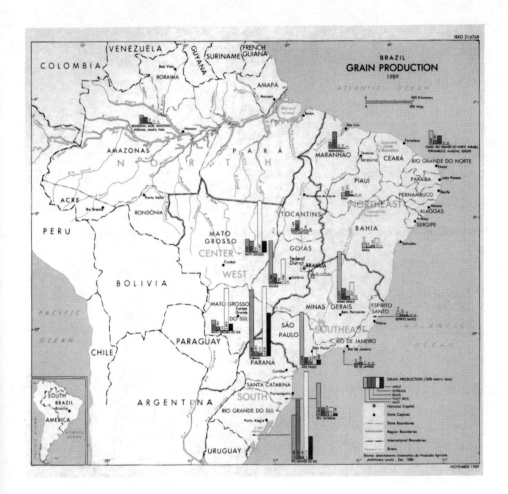

Bibliography

ABIOVE. *Despesas Portuárias no Complexo Soja*. Sao Paulo. Undated (probably 1989).

_____. *Informativo*. All years. Various issues—monthly.

_____. *Planejamento Estratégico para o Setor Soja*. Sao Paulo, May 1989.

Agropec Comercial e Exportadora S.A. Monthly Prices of Soybeans and Soybean Products in 1986/87. Brazil.

Amaral, Cicely Moitinho. *Intervençao Governamental Recente em Mercados Agrícolas no Brasil* (report to the World Bank). Sao Paulo. 1988.

AMEC, Inc. *Brazil—Marketing of Rice, Maize, Wheat, Soya, Coffee, Cocoa, Fertilizer* (report to the World Bank). Arlington, Mass. April 1988.

_____. *Brazil—Agricultural Marketing* (report to the World Bank). Arlington, Mass. September 1988.

_____. *Brazil—Grain Storage; A Preliminary Assessment* (report to the World Bank). Arlington, Mass. May 1987.

Calegar, Geraldo M. and G. Edward Schuh. *The Brazilian Wheat Policy: Its Costs, Benefits, and Effects on Food Consumption*. International Food Policy Research Institute—research report 66. Washington, D.C. May 1988.

Cargill Inc., Brazil. *SETA*. Year 8, No. 45. Jan/Feb 1988.

Carlton, Dennis (1984), "Futures Markets: Their Purpose, Their History, Their Function, Their Successes and Failures," *Journal of Futures Markets*, Vol. 4, no. 3, 237-271.

CFP. *Os Efeitos das Políticas Macroeconômicas Sobre a Agricultura*. Coleçao Análise e Pesquisa, vol.37. Brasilia. 1989.

_____.*A Indústria de Soja no Brasil: Estrutura Econômica e Políticas de Intervençao do Governo no Mercado*. Coleçao Análise e Pesquisa vol.34. Brasilia. 1988.

_____. *Informativo*. All years. Various issues—weekly.

Chalmon, Philippe. *Traders and Merchants Panorama of International Commodity Trading*, Harwood Academic Publishers. 1987.

Code of Federal Regulations. Title 12, Parts 1 to 299, Regulation A, dealing with Bankers Acceptances. 1971.

Dollinger, Noreen L., "Comparative Analysis of Transport and Handling of Export Wheat in Canada and the U.S."Unpublished thesis. University of Illinois, 134 pp. November 1980.

Food and Agriculture Organization. *FAO Fertilizer Yearbook 1988*. Rome: FAO, 1988.

Glaessner, T., W. Todd and L. Aiken. Credit and Marketing Reform Project (Loan 2727-BR), Supervision Report, World Bank. 1989.

Group of Thirty, "Clearance and Settlement Systems in the World's Securities Markets,"New York and London, 1989

Hill, Lowell D. "Effects of Regulation on Efficiency of Grain Marketing", *Journal of International Law*, 17:31 (Summer 1985) 389-419.

Hill, Lowell D., and A. Mustard. "Economic Considerations in Industrial Utilization of Cereals", in Y. Pomeranz and L. Munck (ed)., *Cereals: A Renewable Resource*, St. Paul, Minn: American Association of Cereal Chemists, 1981, pp. 25-53.

Knight, Peter. *Brazilian Agricultural Technology and Tade—A Study of Five Commodities*. New York: Praeger. 1971.

Leath, Mack N., and Lowell D. Hill. *U.S. Corn Industry*. National Economics Division, Economic Research Service, USDA, Agricultural Economic Report No. 479, Washington, D.C., February 1982.

McCalla, Alex, and Andrew Schmitz. "Grain Marketing Systems: The Case of the United States vs. Canada", *American Journal of Agricultural Economics*, 61:2 (May 1979) 199-212.

Office of Technology Assessment, Congress of the United States. *Grain— Quality in International Trade—A Comparison of Major U.S. Competitors*. Washington, D.C., February, 1989.

Ortman, Gerald F. *Analysis of Production and Marketing Costs of Corn, Wheat and Soybeans among Exporting Countries*, Ohio State University—symposium—1986.

Pacheco, Luiz Carlos. *A Comercializaçao da Soja no Brasil*. Curitiba PR., 1987.

Pereira Soares, Ricardo. *Avaliaç&o Econômica da Política Triticola de 1967 a 1977*, Brasilia-CFP, Coleçaô Análise e Pesquisa, vol. 20, 1980.

Ramos de Lins, Everton and José Roberto daSilva. *Avaliaçao Econômica da Opçao de Substituir Importaçoes por Produçao Interna—O Caso do Trigo*. Sao Paulo 1986.

Rask, Norman, Gerald Ortmann, and Walter Stulp. *Comparative Costs Among Major Exporting Countries*. Ohio State University, Dept. of Ag. Economics. Occasional Paper. Columbus, Ohio: Ohio State University. Appendix 3.

Reade, Ryan. "Letters of Credit and Bankers' Acceptances, "*Practicing Law Institute*, Course Handbook Series, No. 450. 1988.

Rezende, Gervásio Castro de. *Estocagem e Variaçao Estacional de Preços: Uma Análise do Papel do EGF*. Brasilia—IPEA/INPES. August 1983.

Silva, José Roberto da. *Oferta de Alimentos no Brasil e a Questao do Trigo: Algumas Consideracoes*. Sao Paulo. 1986.

Silveira, J.M.F.G. da. Progreso Tecnico e Oligopolio: as especificidades da Industria de Sementes no Brasil. [Mimeo]. Campinas, Brazil: UNICAMP. 1985.

Storey, Gary G., Andrew Schmitz, and Alexander H. Sarris (eds). *International Agricultural Trade: Advanced Readings in Price Formation, Market Structure and Price Instability*. Boulder and London: Westview Press. 1988.

Todd, Walker and James Thomson. "An Insider's View of the Political Economy of the Too Big to Let Fail Doctrine,"Western Economics Association, San Diego, California, July 2, 1990.

United States Government. *Estimates of Producer and Consumer Subsidy Equivalents. Government Intervention in Agriculture, 1982-86*. Washington, D.C.: United States Department of Agriculture. 1988.

U.S. Department of Agriculture. *Agricultural Statistics 1988*. Washington, D.C.: USGPO, 1988.

U.S. General Accounting Office. *Grain Marketing Systems in Argentina, Australia, Canada, and the European Community; Soybean Marketing Systems in Brazil*, Washington, D.C.: USGPO, 1976.

Wheat Facts and Trends. Mexico: CIMMYT, 1988.

Williams, Jeffrey. *The Economic Function of Futures Markets*, Cambridge: Cambridge University Press. 1989.

World Bank. *Trade Policy in Brazil—The Case for Reform*. Report no. 7765-BR. Washington, D.C. March 26, 1990.

_____. *Brazil Agricultural Sector Review*, Washington, D.C., July 26, 1990.

_____. *Brazil Agricultural Incentives and Marketing Review*, Washington, D.C., December 1, 1989.

_____. *Brazil Public Expenditures, Subsidy Policies and Budgetary Reform*, Washington D.C., June 1989.

_____. *Brazil Pricing Policies in Agriculture*, Washington, D.C., 1985.

_____. *The Brazilian Maize Trade*, Washington, D.C., 1984.

_____. *The Grain Subsector*, Washington, D.C., June 22, 1990, Volume II: Australia, Canada and the United States.

_____. *The World Bank Atlas*, Washington, D.C., 1989.

PART THREE

Australia

Grain Marketing, Institutions, and Policies

Roley Piggott, Brian Fisher, Julian Alston, and Andrew Schmitz

15

Background

Overview of the Australian Economy

Australia is similar in size to the continental United States but has a population of only 17 million. The population is highly concentrated in the major cities, making Australia one of the most urbanized countries in the world. Gross domestic product (GDP) at market prices was about $A336 billion in 1988/89 or about $A20 thousand per person. During the 1980s, the average annual growth in real GDP has been 2.9 percent (as compared with a 2.4 percent average for all industrialized countries), and the average annual inflation rate has been 9.0 percent (as compared with a 6.5 percent average for all OECD countries). Unemployment has generally been less than 10 percent during this period.

Significant events in the Australian macro-economy over the past several years have included a move to flexible exchange rates, deregulation of the financial system and the pursuit of an incomes policy designed to achieve a fall in real wages and salaries and an increase in real profits. In the early 1980s the share of wages and salaries in national income was about 62 percent while the non-wage share was 38 percent. The respective shares are now 56 percent and 44 percent. Average weekly earnings are currently around $A445.

On average, agriculture has contributed 4.3 percent, mining 6 percent, manufacturing 18 percent and other sectors 71.7 percent, respectively, to GDP during the 1980s. Agriculture's share of Australia's gross domestic product has steadily dwindled over the post-World War II period, falling from about 24 percent in 1949/50 to less than 4 percent in 1988/89. Of particular significance is the decline in the relative importance of agricultural products as export earners. The decline (from about 80 percent of total export earnings in the 1951/52 to 29 percent in 1988/89) came about mainly because of rapid growth in Australia's mineral exports (particularly coal and iron ore but including several others).

Despite the drop in its relative contribution to export earnings, the Australian agricultural economy is still heavily dependent on export markets. The ratio of gross value of agricultural exports to gross value of agricultural production averaged 68 percent during the 1980s (Table 15.1). In the case of grains, export dependency was even higher, averaging 86 percent during the 1980s.

Australia has had a long history of protecting its manufacturing sector, mainly through tariffs, but agricultural industries have also received various forms of assistance. In recent years the government has been reducing assistance levels. It is projected that, by the mid-1990s, the effective rate of assistance to the manufacturing sector will have been reduced by about 30 percent (Industries Assistance Commission, 1988). In general, the levels of assistance that have been afforded the manufacturing sector have exceeded those in the agricultural sector. As of December 31, 1987 the effective rate of assistance to the manufacturing sector as a whole was 19 percent (projected to be 13 percent by the mid-1990s) while the corresponding figure for agriculture was 11 percent (projected to be 9 percent by the mid-1990s). These sector-level figures mask considerable variation in assistance levels across industries. For example, within manufacturing, the effective rate for food, beverages and tobacco was 5 percent while the effective rate for clothing and footwear was 183 percent. Within agriculture, the effective rate was 212 percent for fresh milk but only 2 percent for wool. The effective rate for wheat at the end of 1987 was about 33 percent but this reflects a payout (first and only) under the underwriting provisions of wheat marketing legislation discussed later. The figure for wheat in the absence of the payout would have been about 4 percent (Industries Assistance Commission, 1988a and b).

The major economic problem confronting the government is the trade deficit. During 1988/89 the current account deficit (mainly due to private borrowing) rose by $A5.5 billion to reach $A17.4 billion or 5.2 percent of gross domestic product. The accumulated net foreign debt was $A108.2 billion or 32 percent of GDP in June 1989 ($A6.4 thousand per person). Interest rates reached a record high, with the commercial bank lending rate to prime borrowers averaging 17.2 percent in 1988/89 and projected to average 19.0 percent in 1989/90 (Australian Bureau of Agricultural and Resource Economics, 1989).

Agronomic and Climatic Characteristics

Australia's topography, climate, and soil types vary greatly across the country. Approximately 500 million hectares, or less than two-thirds of

283

TABLE 15.1. Relative Gross Values of Agricultural Production and Exports for Major Commodity Groups, Australia, 1980/81 to 1989/90

Year	Total Farm	Wheat	Barley	Sorghum	Oats	Rice	Maize	Triticale	All grains	Wool	Meat[a]	Dairy
	$A m					Percent						
Production												
1980-81	11,550	14.6	3.3	1.3	1.2	1.2	0.2	b	21.9	14.5	30.2	7.3
1981-82	12,644	20.6	3.7	1.1	1.2	0.8	0.2	0.1	27.8	14.2	26.1	7.8
1982-83	11,627	13.5	2.5	1.1	1.0	0.8	0.2	0.1	19.1	15.1	29.7	9.6
1983-84	15,435	23.4	4.7	1.6	1.3	0.6	0.2	0.1	32.0	13.1	22.8	7.5
1984-85	15,536	20.6	4.9	1.3	0.8	0.8	0.3	0.2	28.9	15.7	24.5	7.1
1985-86	15,491	17.5	3.8	1.2	0.9	0.6	0.3	0.2	24.4	17.4	25.1	7.6
1986-87	17,321	14.6	2.5	0.9	0.9	0.5	0.2	0.1	19.8	19.2	26.4	7.8
1987-88[c]	0,189	10.3	2.3	0.9	0.9	0.6	0.2	0.1	15.3	27.5	24.9	7.4
1988-89[d]	22,296	10.9	2.2	1.0	0.9	0.5	0.1	0.1	15.9	27.0	24.9	7.5
1989-90[e]	23,100	13.0	2.2	0.8	1.1	0.5	0.1	0.1	17.9	22.9	24.9	7.9
Average	16,519	15.9	3.2	1.1	1.0	0.7	0.2	0.1	22.3	18.7	26.0	7.8
Exports												
1980-81	8,204	21.8	4.3	0.7	0.4	1.3	b	b	28.4	24.1	19.5	3.5
1981-82	7,914	21.9	4.3	1.9	0.3	2.5	b	b	30.9	24.2	18.7	3.9
1982-83	7,408	18.8	3.0	0.7	0.2	1.6	b	b	24.4	25.4	22.5	4.6
1983-84	8,380	21.8	7.0	1.3	0.5	1.1	b	b	31.7	24.5	17.8	4.5
1984-85	10,479	27.3	6.3	2.3	0.5	1.1	b	b	37.6	24.8	14.3	4.0
1985-86	11,666	25.6	5.2	1.5	0.2	1.6	b	b	34.1	26.1	14.0	3.7
1986-87	12,183	18.0	2.9	0.7	0.2	1.1	b	b	23.0	31.8	18.7	3.9
1987-88[c]	14,217	13.2	2.7	0.9	0.4	0.9	b	b	15.8	26.9	18.6	3.9
1988-89[d]	14,916	12.1	2.5	0.7	0.3	1.0	b	b	16.7	40.8	15.8	4.2
1989-90[e]	15,280	16.9	2.3	0.4	0.3	1.1	b	b	20.9	32.1	17.6	4.5
Average	11,065	19.7	4.1	1.1	0.3	1.3	b	b	26.4	28.1	17.8	4.1

a   The meat category includes exports of live animals for meat.
b   Denotes a percentage less than 0.1.
c   Preliminary.
d   Estimated by ABARE.
e   Forecast by ABARE.
Source: Australian Bureau of Agricultural and Resource Economics, *Quarterly Review of the Rural Economy,* 10/4 (1988b) and previous volumes.

Australia's land area, is suitable for grazing or cropping. Much of the remaining area is too arid or rugged to support agriculture. In the case of the land that can be farmed, soil moisture patterns and their impact on the length of growing seasons are crucial determinants of land use. About 22 million hectares are used for extensive and intensive crop production, while about 26 million hectares are planted to improved pastures for animal grazing. The remaining 90 percent of agricultural land does not receive enough rainfall to permit cropping and it is used for grazing on native unimproved pastures. Soils are generally phosphorous and nitrogen deficient, and about 25 million hectares of land that are used for cropping or animal grazing regularly receive applications of chemical fertilizer (National Farmers' Federation, 1986).

The principal grain-producing areas are in the so-called "wheat-sheep zone"(Figure 15.1). Depending on location, rainfall ranges from 200 to 800 mm per year, and the length of the growing season varies from five to nine months. Climatic conditions in these areas (periodic droughts and undependable rainfall) make grain production a high-risk activity for most farmers. The fickle export market, another major factor that contributes to risk, is discussed later.

Australian farms are specialized in comparison to European farms because the bulk of revenue comes from just two or three commodities (Davidson, 1982). However, Australian farms tend to be less specialized than their North American counterparts. In the case of grain farms, wool and/or beef enterprises are often complementary. Farms tend to be considerably larger than in most other countries, but yields are low (discussed in the next section). The average wheat farm is about 1,200 hectares in size, but only about 200 hectares are actually sown to wheat each year (National Farmers' Federation, 1987). Other notable features of grain farms in Australia include a high proportion of family-owned farms and a low labor/land ratio (Davidson, 1982).

Income instability has always been a key issue in Australian agriculture, although unfortunately there is little farm-level documentation. Based on an analysis of changes in year-to-year income fluctuations undertaken by the Australian Taxation Department, Harris, et al. (1974) reported that during the late 1960s over 40 percent of farmers experienced year-to-year income fluctuations of 50 percent or more, while other business and wage earner incomes varied by only 23 percent and 17 percent, respectively. Changes in areas, yields and prices, caused by a host of factors, each play a part in contributing to income instability. Most Australian agricultural policies are implemented with the stated goal of improving income stability, although one could argue that "stability"is something of a euphemism for support.

Grains accounted for about 23 percent of the gross value of agricultural production and 27 percent of the gross value of agricultural exports during

FIGURE 15.1 Australia's Principal Grain-Producing Areas

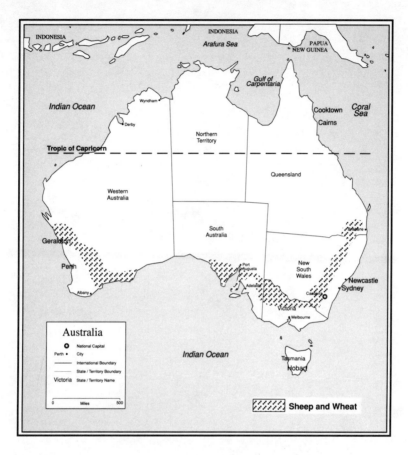

the 1980s (Table 15.1). Grain, wool, and meat production constitute the "big three"of Australian agriculture, and, until very recently, grains dominated in terms of export value. Declines in international grain prices and a buoyant wool market allowed wool to overtake grain as the leading export income earner in 1986/87. Wheat dominates the grain sector.

16

Grain Production and Marketing

Grain Production

Cereal Grains

While all the Australian states except the Northern Territory produce some wheat, New South Wales and Western Australia account for the bulk of production (Figure 15.1). Only spring wheat is grown in Australia's temperate climatic conditions. The variety grown in a given location depends on soils and seasonal conditions. Wheat is sown in the autumn months (usually starting about May) and harvested in early summer after a five- to six-month growing period. The harvest is usually completed by December, but the timings of sowing and harvesting are variable, not only geographically, but also from year to year in any location, depending on the weather. Unlike North American and European winter wheats, there is no dormant growth stage. High-protein, hard wheat (at least 11 percent protein at 11 percent moisture) is planted in the northern part of New South Wales and Queensland, while softer wheats are produced on the southern part of the continent. Australian wheat is classified by protein percentage and other chemical and physical characteristics, with about 70 percent falling into the Australian Standard White category.

Other grain crops primarily include barley, oats, sorghum and rice. Barley is a winter crop grown in the same general areas as wheat, although production is concentrated in the state of South Australia. The majority (about 90 percent) of the crop is two-row barley. Oats for grain is a winter crop planted in southern wheat-producing areas, which sometimes doubles as a temporary source of forage before it is harvested for grain. Yellow field maize is not an important crop in Australia, although it is occasionally planted during the summer season along the eastern coastal area and in

TABLE 16.1. Rice Production and Exports, Australia, 1970/71 to 1988/89

Crop year	Area	Yield (paddy)	Production (paddy)	Exports (milled)
	'000' ha	mt/ha	'000' mt	'000' mt
1970/71	41	7.30	301	96
1971/72	40	6.20	248	169
1972/73	45	6.87	309	152
1973/74	67	6.10	409	125
1974/75	75	5.16	387	160
1975/76	74	5.70	416	215
1976/77	92	5.90	528	254
1977/78	91	5.38	490	273
1978/79	110	6.29	692	239
1979/80	116	5.28	613	450
1980/81	106	7.16	759	276
1981/82	127	6.75	857	558
1982/83	77	6.80	519	384
1983/84	121	5.24	634	241
1984/85	126	6.86	864	327
1985/86	107	6.43	687	499
1986/87	97	6.32	608	359
1987/88[a]	107	7.12	762	297
1988/89[b]	104	7.02	730	430

Note: Production data are based on rice industry figures calculated on a crop year basis.
For example, the 1985/86 data are for the NSW crop harvested in April 1986, the Queensland
summer crop harvested in December 1985, and the Queensland winter crop harvested in
June 1986. Exports are based on Australian Bureau of Statistics data and are generally for exports of the crop harvested in the previous year.
   a   Preliminary.
   b   Estimated by ABARE.
Source: Australian Bureau of Agricultural and Resource Economics, *Commodity Statistical
Bulletin*, December 1988: Table 105.

some irrigated areas. Sorghum is the most important summer cereal and
is grown in northern New South Wales and southern Queensland, although grain sorghum has only become important in the last twenty years.
Most of Australia's rice crop, which (unlike other grains) is generally irrigated, is produced in the southern plains area of New South Wales and in
Queensland. Climatic conditions allow for only one (summer) crop each
year in New South Wales, but two crops per year are possible in tropical
northern Queensland. A mixture of medium- and long-grain varieties are
grown. Sown rice area rose substantially between the late 1970s and 1984/
85 (Table 16.1) because farmers diverted irrigated land away from livestock enterprises to rice in response to relatively high international rice
prices. The reverse occurred in 1985/86 when rice prices fell, and rice area
has remained relatively constant since then. Rice area grew steadily from
1970/71 through 1981/82, while yields increased sporadically, resulting in
a doubling of production since 1971/72.

Wheat yields in Australia are low relative to many other major producers. In recent years, they have ranged from 1.4 to 1.5 metric tons per hectare, averaging 1.3 metric tons per hectare in the 1980s (Figure 16.1). Wheat yields are also very volatile. Note the sharp reductions in 1972, 1977, and 1982 which occurred as a result of drought conditions. Data on world wheat yields are presented in Table 16.2. As can be seen, Australian yields are much lower than those in the United States and the European Community. Also, there have not been any significant increases in Australian wheat yields over the period 1978-1986. This has not been the case for the European Community and the United States. No significant trend was registered for Australia.

Quotas on wheat deliveries and depressed wool returns helped to elevate the importance of coarse grains in the late 1960s and early 1970s. The coarse grain area increased rather markedly in the late 1970s and early 1980s (Table 16.3). As the 1980s progressed, shorter cropping rotations, coupled with year-round production in those areas where the climate permits, helped to maintain the level of coarse grain production (National Farmer's Federation, 1986). However, wheat production also escalated in the late 1970s (Table 16.4), primarily because extra area was sown to wheat in response to depressed beef prices. As a result, wheat has continued to dominate the grains sector.

Cereal Grains and Livestock

A large percentage of Australian wheat land is suited for livestock or grain production. In these areas, wheat farming involves extended rotations with clover and sheep. The rotation varies considerably depending on relative prices. During the early 1970s it was common to plant wheat four years in a row and then pasture for two years. However, due to reduced wheat prices and increased sheep and wool prices, the rotation followed in recent years has tended to include a longer pasture phase. In Australia it is not uncommon to find farms which are highly integrated in three enterprises: sheep, wheat, and beef cattle. Certain farms, however, are completely specialized. In parts of Australia, only cattle ranching occurs; in other parts, only wheat production exists. Farms in the wheat-sheep zone produce a range of products, including wheat, coarse grains, oilseeds, wool, beef, and other livestock products. Crop production in the zone is subject to large fluctuations in area, yield, and output because of price and climatic variability. For example, wheat production fell from 16.4 million metric tons in 1981/82 to 8.9 million metric tons in 1982/83, a drought year. In the following season, it rose to over 22 million metric tons.

FIGURE 16.1 Wheat Yields by Major Exporters in Tons/Hectare

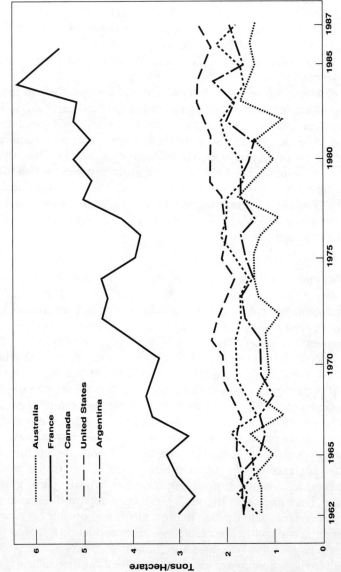

Source: International Wheat Council, *World Wheat Statistics* (London: various issues).

TABLE 16.2. World Wheat Yields, 1978 to 1986 (metric tons per hectare)

	1978 (t/ha)	1979 (t/ha)	1980 (t/ha)	1981 (t/ha)	1982 (t/ha)	1983 (t/ha)	1984 (t/ha)	1985 (t/ha)	1986 (t/ha)
World	2.0	1.9	1.9	1.9	2.0	2.1	2.2	2.2	2.3
Major Exporters									
EC[a]	4.2	4.1	4.4	4.3	4.6	4.5	5.6	5.1	4.6
U.S.	2.1	2.3	2.3	2.3	2.4	2.7	2.6	2.5	2.3
Canada	2.0	1.6	1.7	2.0	2.1	1.9	1.6	1.8	2.2
Australia	1.8	1.5	1.0	1.4	0.8	1.7	1.5	1.4	1.6
Argentina	1.7	1.7	1.6	1.4	2.1	1.8	2.3	1.6	1.8
Major Importers									
U.S.S.R.	1.9	1.6	1.6	1.4	1.5	1.5	1.3	1.6	1.8
China	1.9	2.1	1.9	2.1	2.5	2.8	3.0	2.9	3.0
Brazil	1.0	0.8	0.9	1.1	0.6	1.2	1.0	1.5	1.4
Egypt	3.3	3.2	3.2	3.3	3.5	3.6	3.7	3.8	3.6
Japan	3.3	3.6	3.1	2.6	3.3	3.0	3.2	3.7	3.6
Others									
India	1.5	1.6	1.4	1.6	1.7	1.8	1.8	1.9	2.0
Turkey	1.8	1.9	1.8	1.8	1.9	1.8	1.9	1.5	1.6[f]
Pakistan	1.3	1.5	1.6	1.6	1.6	1.7	1.5	1.6	1.9
Iran	1.0	1.1	1.0	1.1	1.1	1.1	1.0	0.9	1.0
Spain	1.8	1.6	2.2	1.3	1.7	1.6	2.6	2.6	2.0[a]
Mexico	3.7	3.8	3.9	3.7	4.5	4.0	4.4	4.2	4.2
Yugoslavia	3.1	3.0	3.4	3.1	3.3	3.4	3.8	3.6	3.6
Sweden	4.5	4.2	4.1	4.8	5.3	5.1	5.6	4.8	5.6
Bangladesh	1.8	1.9	2.0	1.8	1.8	2.1	2.3	1.9	2.2
Iraq	0.6	0.9	0.9	0.8	0.8	0.8	0.6	0.9	0.8[f]
E. Europe	3.6	3.0	3.6	3.4	3.7	3.6	4.2	3.7	3.9
W. Europe[b]	2.4	2.3	2.9	2.6	2.8	3.1	3.5	3.5	4.1[f]
Africa[e]	0.9	0.9	0.9	0.9	1.1	0.9	1.0	1.1	1.2
Asia & Others[d]	1.2	1.2	1.5	1.5	1.4	1.5	1.5	1.5	1.5
S. America[c]	1.4	1.5	1.4	1.3	1.4	1.6	1.7	1.9	2.0

[a]   EC includes Spain, Portugal for 1986; yield for EC excluding these countries for 1986 is 5.1.

[b]   Excludes EC, Spain, Sweden, Yugoslavia, shown separately.

[c]   Excludes Argentina, Brazil, shown separately.

[d]   Excludes Bangladesh, China, India, Iran, Iraq, Japan, Pakistan, Turkey, shown separately; includes New Zealand.

[e]   Excludes Eygpt, shown separately.

[f]   United States Department of Agriculture figures for Western Europe and Iraq generally are above, and for Turkey generally below, IWC figures.

Sources: International Wheat Council, *World Wheat Statistics*, 1986 and *Market Report*, January 10, 1987. United States Department of Agriculture, Foreign Agricultural Service, *World Grain Situation and Outlook*, FG-2-87, January 1987 (for 1985 and 1986 only).

TABLE 16.3. Coarse Grain Output and Disappearance, Australia, 1970/71 to 1988/89

Crop Year	Area	Yield	Production	Apparent Domestic Disposals[a]	Exports[b]
	'000' ha	mt/ha	'000' mt	'000' mt	'000' mt
1970/71	4,191	1.31	5,473	3,255	2,218
1971/72	4,492	1.29	5,782	2,606	3,176
1972/73	3,891	0.93	3,620	1,957	1,663
1973/74	3,664	1.27	4,671	2,929	1,742
1974/75	3,285	1.35	4,421	1,537	2,884
1975/76	3,868	1.44	5,575	2,427	3,148
1976/77	3,901	1.29	5,019	1,692	3,327
1977/78	4,318	0.98	4,217	2,279	1,938
1978/79	4,663	1.51	7,063	4,496	2,567
1979/80	4,200	1.48	6,210	2,188	4,022
1980/81	4,284	1.22	5,211	2,925	2,286
1981/82	4,847	1.38	6,687	3,672	3,015
1982/83	4,527	0.88	3,966	2,585	1,381
1983/84	5,795	1.64	9,497	5,295	4,202
1984/85	5,525	1.59	8,771	2,584	6,187
1985/86	5,307	1.53	8,144	2,445	5,669
1986/87	4,502	1.58	7,093	3,734	3,256
1987/88[c]	4,626	1.51	6,996	5,177	1,819
1988/89[d]	4,758	1.51	7,205	4,236	2,969

Note: Includes barley, oats and sorghum to 1959/60, maize added in 1960/61, followed by triticale in 1979/80.
[a]   Domestic disposal includes stocks and barley used to produce export malt and is calculated as the residual of production after exports.
[b]   Exports are on a July-June basis and are grain exports only. Sorghum exports prior to 1952/53 are not included.
[c]   Preliminary.
[d]   Estimated by ABARE.
Source: Australian Bureau of Agricultural and Resource Economics, *Commodity Statistical Bulletin*, December 1988: Table 84.

TABLE 16.4. Output and Disappearance of Australian Wheat, 1970/71 to 1988/89

Crop Year[a]	Area	Yield	Production	Domestic Sales[b]	Exports	Closing Stocks
	'000' ha	mt/ha	'000' mt	'000' mt	'000' mt	'000' mt
1970/71	6,478	1.22	7,890	1,681	9,049	3,404
1971/72	7,138	1.19	8,510	1,848	7,760	1,451
1972/73	7,604	0.87	6,590	2,242	4,137	478
1973/74	8,948	1.34	11,987	2,319	7,418	1,882
1974/75	8,308	1.37	11,357	2,394	8,550	1,658
1975/76	8,555	1.40	11,982	1,992	8,233	2,665
1976/77	8,956	1.30	11,667	1,696	9,763	2,137
1977/78	9,955	0.94	9,370	1,750	8,098	780
1978/79	10,249	1.77	18,090	1,966	11,693	4,646
1979/80	11,153	1.45	16,188	2,428	13,197	4,268
1980/81	11,283	0.96	10,856	2,555	9,614	2,044
1981/82	11,885	1.38	16,360	1,705	11,068	4,776
1982/83	11,520	0.77	8,876	3,117	7,280	2,285
1983/84	12,931	1.70	22,016	1,627	14,152	7,518
1984/85	12,078	1.55	18,666	2,150	15,720	8,456
1985/86	11,736	1.38	16,167	2,357	16,162	5,806
1986/87	11,261	1.49	16,778	2,349	14,868	3,772
1987/88[c]	9,063	1.36	12,369	2,460	12,402	2,875
1988/89[d]	8,965	1.60	14,387	1,800	10,562	2,300

Note: Figures do not include on-farm wheat retained for feed or seed.

[a] Years up to and including 1980/81: December 1 to November 30; 1981/82: December 1 to September 30; 1982/83 and subsequent years: October 1 to September 30.

[b] 1984/85 and subsequent years include feed wheat sold privately with permission from the Australian Wheat Board.

[c] Preliminary.

[d] Estimated by the ABARE.

Source: Australian Bureau of Agricultural and Resource Economics, *Commodity Statistical Bulletin*, December 1988: Table 69.

Domestic Markets

Wheat marketed domestically, usually less than 20 percent of each year's crop, is used mainly for human consumption, but also for animal feed and industrial purposes (e.g., starch, gluten, and glucose). The relative importance of different domestic uses varies considerably from year to year. For example, during the last five years, wheat for human consumption fluctuated from 42 percent to 61 percent of domestic sales, stockfeed wheat from 19 percent to 48 percent, and industrial wheat from 10 percent to 21 percent. This can largely be attributed to changes in the demand for stockfeed wheat caused by variations in seasonal conditions and relative prices of alternative feedgrains. The coarse-grain industry is less dependent on export sales than is the wheat industry. Domestic livestock consume a considerable portion of output, and the brewing industries are also an important market outlet (Table 16.3). Rice producers, like wheat farmers, are heavily dependent on export sales, with about 60 percent on average of the total exported during the 1980s (Table 16.1).

Because of Australia's dependence on export markets, domestic consumption trends for wheat and rice are relatively unimportant. However, in the case of coarse grains, the domestic market is growing because of an expanding intensive livestock sector (mainly hogs and poultry) and its requirements for feedgrains. Most Australian beef is produced under extensive conditions, but there are periodic surges in demand for coarse grains to feed cattle when drought affects the availability of pastures.

Export Markets

Australia's Importance as an Exporter

Australia is a major wheat exporter. It follows the United States, the European Community, and Canada in terms of market share. Data on wheat exports as a percentage of production for major exporters are provided in Table 16.5. Wheat exports by the major exporters are given in Table 16.6. There are some interesting comparisons apparent from this table. In the 1960s, the largest wheat exporter in absolute terms was the United States, followed by Canada and then Australia, but Australia had lost that ranking by the late 1980s. In 1986/87 and 1987/88, the European Community exported more wheat than Australia as a consequence of the levels of government support provided to European farmers under the Common Agricultural Policy.

Australian wheat exports are plotted in Figure 16.2. Exports follow production closely and, as a result, export volumes are highly volatile. Data

TABLE 16.5. Wheat Exports as Percent of Production for Major World Exporters, 1961/62 to 1987/88

Year	EC10[a]	United States	Canada	Australia	Argentina
1961/62	13.7	58.4	126.3	72.0	47.7
1962/63	13.5	58.8	58.6	74.5	32.6
1963/64	15.4	74.7	82.2	77.3	39.0
1964/65	19.4	56.5	66.6	72.4	56.9
1965/66	19.1	65.9	90.1	67.3	91.1
1966/67	16.9	57.0	62.3	67.1	35.2
1967/68	15.3	50.5	56.7	75.0	30.7
1968/69	15.8	35.0	47.1	45.2	43.1
1969/70	20.5	42.0	51.6	77.7	32.6
1970/71	10.3	54.6	131.3	114.7	17.2
1971/72	13.2	39.1	95.1	90.2	28.5
1972/73	16.4	76.6	108.1	62.8	39.2
1973/74	13.0	67.4	70.6	61.9	22.8
1974/75	17.5	57.9	81.0	75.3	28.7
1975/76	25.2	55.3	72.2	68.7	36.1
1976/77	12.4	44.4	57.0	82.7	53.0
1977/78	14.7	54.9	80.8	86.4	31.6
1978/79	19.0	67.2	61.9	64.6	49.3
1979/80	23.6	64.4	92.4	81.5	58.3
1980/81	27.4	63.6	84.3	88.6	45.0
1981/82	29.1	63.6	74.4	67.4	45.8
1982/83	25.4	54.6	79.9	82.5	65.3
1983/84	27.7	59.0	82.1	64.3	59.7
1984/85	24.7	54.9	82.7	82.5	68.4
1985/86	24.1	37.5	75.3	97.3	50.6
1986/87	23.4	48.1	66.2	93.3	50.6
1987/88	22.3	76.0	91.0	98.0	42.0

[a] Aggregated for first 10 members of the European Community. It excludes Spain and Portugal.

Source: International Wheat Council, *World Wheat Statistics* (London: various issues); 1986/87 from U.S. Department of Agriculture, Foreign Agricultural Service, FG-9-88, Washington, DC.

TABLE 16.6. Total Wheat Exports By Major World Exporter, 1963/64 to 1987/88
(million metric tons)

Year	EC^a	United States	Canada	Australia	Argentina	Total
1963/64	3.8	23.1	15.1	7.8	2.8	55.8
1964/65	5.4	19.6	11.9	6.5	4.4	50.5
1965/66	5.5	23.4	14.8	5.7	7.9	62.0
1966/67	4.2	20.0	14.8	7.0	3.1	55.8
1967/68	4.4	20.2	8.9	7.0	1.4	51.2
1968/69	5.0	14.7	8.7	5.4	2.8	45.7
1969/70	7.2	16.5	9.0	7.3	2.1	50.7
1970/71	3.1	19.8	11.6	9.5	1.7	54.3
1971/72	4.7	16.9	13.7	8.7	1.3	52.5
1972/73	6.5	32.0	15.6	5.6	3.5	68.3
1973/74	5.5	31.1	11.7	5.5	1.1	63.1
1974/75	7.1	28.3	11.2	8.0	2.2	63.4
1975/76	7.7	31.5	12.1	8.1	3.1	66.5
1976/77	3.9	26.4	12.9	8.4	5.6	61.8
1977/78	4.5	31.5	15.9	11.1	2.7	72.4
1978/79	7.4	32.4	13.5	7.2	3.3	71.7
1979/80	10.3	36.6	15.0	15.4	4.7	86.0
1980/81	12.7	42.1	17.0	11.1	3.9	94.0
1981/82	14.0	49.3	17.8	11.4	4.3	100.7
1982/83	14.1	39.3	21.1	8.5	7.5	96.1
1983/84	14.9	38.3	21.2	11.6	9.6	100.3
1984/85	17.2	38.2	19.1	15.1	8.0	104.1
1985/86	15.0	25.1	17.6	16.1	6.3	87.0
1986/87	15.0	27.3	20.8	14.9	4.3	90.1
1987/88	16.0	43.4	23.6	12.2	3.8	95.8

a   European Community comprised of original member states to 1967/68,
9 member states to 1980/81, 10 member states to December 1985, thereafter
12 members.
Source: International Wheat Council, *World Wheat Statistics* (London: various issues); 1986/87 from U.S. Department of Agriculture, Foreign Agricultural Service,
FG-9-88, Washington, DC.

FIGURE 16.2 Wheat Supply and Disappearance for Australia

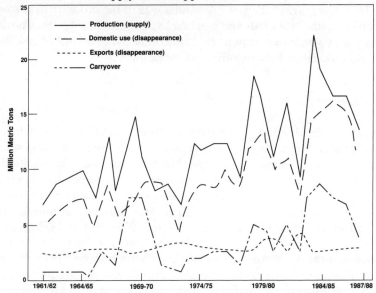

Source: International Wheat Council, *World Wheat Statistics* (London: various issues).

FIGURE 16.3 Market Share of Wheat Exports by Major Exporters

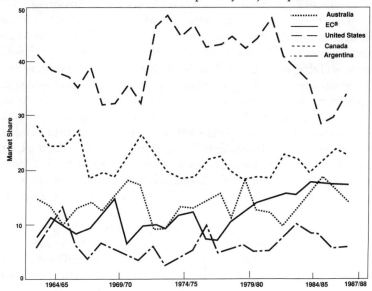

[a] Six original member states in 1967–1968, 9 member states in 1980–1981,
10 member states in December 1965, thereafter 12 members.

Source: International Wheat Council, *World Wheat Statistics* (London: various issues).

on the market share of wheat exports by major exporters are presented in Figure 16.3. The market share for Australia was of the order of 10 percent to 12 percent in the late 1970s and reached 18.5 percent in 1985/86. The dependency on volatile export markets is the other major source of risk (besides yield risk) faced by Australian wheat farmers.

Destinations for Australia's Exports

The various destinations of Australian wheat exports are shown in Table 16.7. Table 16.8 gives imports of wheat by source. By far the largest importers are China, Egypt, and the U.S.S.R. To add an additional perspective to the above, see Figures 16.4 and 16.5. The pattern of world wheat imports is less concentrated than for exports. However, five countries — China, Japan, the U.S.S.R., Egypt, and Brazil — accounted for about half of world wheat imports between 1978/79 and 1985/86. Six countries — the U.S.S.R., China, Egypt, Japan, Iran, and Iraq — accounted for about two-thirds of Australia's wheat exports between 1978/79 and 1985/86. The remaining one-third went to approximately 40 countries.

In 1984/85 Australia provided a substantial share of the total volume of wheat purchased by four of the major wheat importers — Egypt, China, Japan, and the U.S.S.R. When the United Kingdom entered the European Community in 1973, Australia was forced to seek new customers to replace its major wheat market. Australia now depends on sales to centrally planned economies (in particular, the U.S.S.R. and China), the Middle East, Japan (a traditional buyer from Australia), and a number of developing countries (Table 16.9). In the decade ending 1983/84, Australia's share of world wheat exports averaged about 10 percent, but it increased significantly between 1985/86 and 1986/87 in response to increases in Australian production (Table 16.10). It has since declined again and Australia's future market share will be dependent on European Community and U.S. policies with respect to domestic price supports and subsidized exports. During the 1980s, the Middle East was the principal export market for barley, and Japan was the main export market for sorghum (Table 16.9). Australia's share of the world markets for barley and sorghum have fluctuated in a similar pattern as for wheat.

Developing countries close to Australia, together with Hong Kong, constitute the principal export markets for Australian rice (Table 16.9), along with a variety of "short-term" customers. Australia has traditionally controlled only a small share (i.e., about 3 percent) of the world rice market (Table 16.10).

TABLE 16.7. Destinations of Australian Wheat Exports, 1978/79 to 1985/86 (percentage of total Australian wheat exports)[a]

	78/79 %	79/80 %	80/81 %	81/82[b] %	82/83 %	83/84 %	84/85 %	85/86 %
Bangladesh	3.7	0.9	0.9	1.6	0.0	4.6	0.9	0.6
China	23.2	17.8	12.2	15.4	10.6	15.8	9.4	18.3
Egypt	13.1	13.7	17.1	14.5	21.1	13.5	16.7	14.4
Ethiopia	0.5	0.8	0.3	0.2	0.0	0.1	0.7	1.7
India	0.3	0.0	0.0	7.1	0.0	0.0	0.0	0.0
Indonesia	5.5	5.2	6.3	4.1	4.6	3.1	3.8	4.4
Iran	1.5	6.1	6.3	6.7	7.8	13.3	9.4	6.7
Iraq	8.2	7.0	6.3	5.8	7.8	6.8	8.0	4.5
Japan	10.5	7.3	9.7	7.5	13.0	7.6	7.5	5.9
Korea, Republic of	0.0	0.0	0.0	0.2	0.0	3.7	7.0	3.9
Kuwait	1.8	1.2	2.7	1.6	3.3	1.0	1.3	1.0
Malaysia	3.5	3.1	3.9	3.3	4.1	3.3	2.2	2.5
Mexico	0.0	0.0	0.0	0.0	0.0	1.5	2.2	0.0
Pakistan	4.3	0.5	0.8	0.2	0.0	0.2	6.5	0.7
Saudi Arabia	1.9	1.1	0.9	2.1	1.0	0.5	0.0	0.0
Singapore	2.4	2.5	1.6	0.7	0.9	0.6	0.8	0.8
South Africa	0.0	0.0	0.0	0.0	0.0	2.2	0.7	0.6
Sri Lanka	0.4	0.6	1.7	1.4	0.4	0.4	1.4	1.2
United Kingdom	0.0	0.0	0.0	0.0	0.0	0.0	0.0	1.7
USSR[c]	10.7	25.0	18.4	18.8	14.0	12.8	12.2	19.9
Yemen, A.R.	2.4	1.8	3.0	3.2	1.2	2.1	1.8	3.5
Yemen, P.D.R.	0.2	0.1	0.0	0.9	1.4	1.2	1.5	1.0
Others[d]	5.9	5.3	7.9	4.7	8.8	5.7	6.0	6.7
Total	100.0	100.0	100.0	100.0	100.0	100.0	100.0	100.0

[a] Cut-off is 1.5 percent in any year covered by table.
[b] December 1, 1981 to September 30, 1982. Previously December 1 to November 30. Subsequently October 1 to September 30.
[c] Includes Korea (P.D.R.) and Vietnam to 80/81.
[d] Shipments to approximately 40 countries – total varies from year to year.
Sources: Australian Wheat Board, *Annual Reports*, 80/81 to 85/86.

TABLE 16.8. Imports of Wheat by Source, 1984-85 [a] (percentage of total wheat imports)

Importing country (%)	Importer's share of world wheat imports (%)	Exporter's share of each import market						
		Exporting country						
		Australia (%)	U.S. (%)	Canada (%)	EC (%)	Argentina (%)	Others (%)	Total (%)
USSR	28.7	7.2	21.7	27.1	21.6	14.4	8.0	100.0
		(8.1)	(3.7)	(37.3)	(36.0)	(3.7)	(11.2)	(100.0)
China	7.6	19.2	33.0	37.6	1.1	9.1	0	100.0
		(37.5)	(1.1)	(40.9)	(9.1)	(10.2)	(1.2)	(100.0)
Japan	5.9	18.1	57.8	24.1	0	0	0	100.0
Brazil	5.0	0	61.4	24.1	1.0	13.4	0.1	100.0
Egypt	4.4	50.5	37.4	10.5	0	0	1.6	100.0
Korea, Rep of	3.1	32.0	68.0	0	0	0	0	100.0
Iran	2.7	65.8	0	0.9	5.4	21.3	6.6	100.0
Iraq	2.7	45.9	32.8	13.9	0	0	7.4	100.0
EC	2.2	0	38.75	8.7	0	2.6	0	100.0
Asia[e]	13.5	18.8	37.8	7.5	8.5	7.9	5.8	100.0
Africa[f]	11.7	3.9	41.3	7.1	34.1	0.7	12.9	100.0
S. America[d]	4.0	0	71.0	10.1	1.7	17.2	0	100.0
Central America	3.0	13.2	37.9	27.6	12.6	8.3	0.4	100.0
E. Europe	3.0	0	2.2	8.2	35.9	2.0	51.7[c]	100.0
W. Europe[b]	1.3	0	52.5	9.4	31.7	2.3	4.1	100.0
Total	98.8							

[a] Percentages for 1986-87, estimated by the IWC, are shown in parentheses for the USSR and China.
[b] Excludes EC, shown separately.
[c] Includes USSR, 17.3 percent of total.
[d] Excludes Brazil, shown separately.
[e] Excludes China, Japan, Republic of Korea, Iran, Iraq, shown separately.
[f] Excludes Egypt, shown separately.
Sources: International Wheat Council, *World Wheat Statistics*, 1986 and *Market Report*, January 10, 1987.

FIGURE 16.4 Percentage Shares of World Wheat Imports, 1978-79 to 1985-86

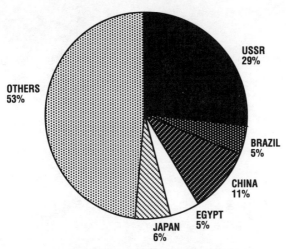

USSR
29%

OTHERS
53%

BRAZIL
5%

CHINA
11%

EGYPT 5%

JAPAN
6%

**Between 1978-79 and 1985-86 five countries
accounted for 47 per cent of world wheat imports.**

FIGURE 16.5 Destinations of Australian Wheat Exports, 1978-79 to 1985-86[a]

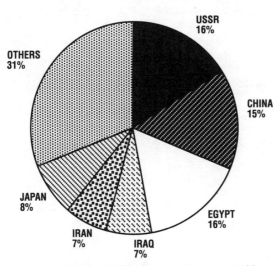

USSR
16%

OTHERS
31%

CHINA
15%

EGYPT
16%

JAPAN
8%

IRAN
7%

IRAQ
7%

**Between 1978-79 and 1985-86 six countries accounted for
69 per cent of Australian wheat exports.**

[a] Percentage of total Australian wheat exports

TABLE 16.9 Percentage Shares and Major Markets for Australian Grain Exports, 1980/81 to 1987/88

Commodity	1980/81	1981/82	1982/83	1983/84	July-June year 1984/85	1985/86	1986/87	1987/88[a]	Average
Wheat and flour									
Egypt	17	14	21	13	17	14	14	18	16
Iran and Iraq	12	12	15	20	17	11	22	27	17
USSR	18	19	14	13	12	20	8	3	13
China	12	15	11	16	9	18	24	10	14
Japan	10	7	13	8	8	6	6	10	95
Others	31	33	26	30	37	31	26	32	31
Total	100	100	100	100	100	100	100	100	100
Barley[b]									
Middle East	24	48	24	37	57	69	45	30	42
Japan	30	12	23	14	15	17	22	36	21
USSR	21	3	0	0	15	0	0	0	5
South America	11	8	25	5	4	6	8	6	9
Others	14	29	28	4	9	8	25	28	23
Total	100	100	100	100	100	100	100	100	100
Sorghum									
Japan	49	51	43	67	78	80	91	92	69
Taiwan	37	36	32	18	19	17	7	0	21
USSR	3	5	0	0	0	0	0	0	1
Others	11	8	25	15	3	3	2	8	9
Total	100	100	100	100	100	100	100	100	100

Rice

Fiji	4	2	4	3	5	—	—	3
Hong Kong	15	10	16	19	16	—	—	15
New Zealand	2	1	1	3	2	—	—	2
Papua New Guinea	32	16	25	32	33	—	—	28
Saudi Arabia	8	4	7	12	7	—	—	8
Other	39	67	47	31	37	—	—	44
Total	100	100	100	100	100	—	—	100

a   Preliminary.
b   Exports include grain for feed and malting purposes.

Sources: Data for wheat and flour, barley, and sorghum are compiled from Australian Bureau of Agricultural and Resource Economics, *Crop Report*, 1988c, No. 45: 51 and *Commodity Statistical Bulletin*, December 1988a: Table 92, 93. Data for rice are from Industries Assistance Commission Report, *The Rice Industry* October 1987: Table D5.

TABLE 16.10. Estimated Shares of Australian Grain Exports In World Trade, 1980/81 to 1988/89 (percentages)

	Wheat	Barley	Sorghum	Rice
1980/81	11.8	10.6	3.5	2.1
1981/82	11.3	13.8	8.1	4.7
1982/83	8.8	4.5	2.6	3.2
1983/84	11.6	21.7	10.8	1.9
1984/85	14.6	26.1	9.2	3.0
1985/86	19.5	20.2	12.8	3.8
1986/87	16.9	11.9	7.5	2.8
1987/88[a]	11.7	12.4	8.5	2.5
1988/89[b]	11.8	12.2	12.0	3.6
Average	13.1	14.8	8.3	3.1

Note: Data are for July-June years for rice and wheat, and October-December years for barley and sorghum.
[a]   Preliminary.
[b]   Estimated by ABARE.
Source: Compiled from data in Australian Bureau of Agricultural and Resource Economics, *Commodity Statistical Bulletin*, December 1988: Tables 79 (wheat); 98 (barley);100 (sorghum); 105 and 107 (rice).

The Marketing System

Institutional Perspective

As in most countries, the marketing of Australia's agricultural commodities is characterized by a complex pattern of regulations and institutional involvement. Few would regard this as an attempt by governments to provide a "level playing field" for market participants to negotiate terms of trade or as a response to market failure. A more widely-held view is that the intervention pattern in agricultural marketing is the outcome of an economic process involving a demand for various regulations on the part of interest groups and a supply of regulations from government in return for political support. Indeed, Australia differs from many developed countries in that one of the three main political parties (the National Party) has an agrarian-based constituency. While there has been considerable division within political parties on matters related to agricultural marketing and policy, the predecessor of the National Party (the Country Party) and, to a lesser extent, the National Party itself, have espoused a philosophy that farmers should have control over how their products are marketed. Moreover, they have been adamant that, where institutions have been set up for this purpose under state or federal legislation, they should operate without ministerial interference.

Marketing boards (or corporations) are common in Australia. These institutions were first created in the 1920s following the failure of voluntary cooperatives to increase and stabilize producer returns. Campbell and Fisher (1982) explain that governments of the day readily provided the necessary legislative backing to establish what were essentially compulsory cooperatives. These eliminated the necessity of expensive support programs and the cost to the government for their creation was minimal. Eleven boards operate at the federal level (each under its own enabling legislation such as the Commonwealth Wheat Marketing Act of 1989 in the case of the Australian Wheat Board) and in excess of fifty at the state level (some constituted under omnibus legislation such as the New South Wales Marketing of Primary Products Act, and others under their own specific enabling legislation).

The Australian Constitution has a profound influence on the pattern of regulatory control over agricultural marketing and the types of activities undertaken by agricultural marketing boards or corporations. The Constitution makes explicit (Section 5.1) the matters on which the Commonwealth Government can legislate, while the states have a general legislative competence. However, the Commonwealth has exclusive powers with respect to the levying of customs duties and excise taxes, and the granting of bounties on the production and export of commodities. The Commonwealth can legislate on export and interstate trade but Section 92 provides for free trade between the states. Powers to set prices and control production belong to the states. Clearly, there is scope for conflict between the states and the Commonwealth in the design of legislation. The Australian Agricultural Council, consisting of the federal minister for Primary Industries and Energy (chairperson) and his state-level counterparts, was established to coordinate federal and state legislation relating to agriculture so as to minimize conflict.

As a result of the Constitutional division of powers, federal-level marketing boards (e.g., the Australian Wheat Board) are engaged mainly in export marketing activities (e.g., licensing of exporters and promotion) and any influence they have over domestic marketing is dependent upon the states enacting legislation complementary to federal legislation. State-level boards and corporations can set prices and control production levels within their borders. However, Section 92 has proved to be most troublesome for the architects of agricultural marketing schemes. In essence, a farmer can endeavor to avoid the strictures of agricultural marketing boards by selling produce across a state border. The High Court has handed down varying interpretations of the legality of such endeavors.

Pressures for change to the highly-regulated nature of agricultural marketing — particularly the operation of marketing boards — began to build in the mid- and late-1970s. Changes to marketing board enabling legislation began to be made in order to: (1) reduce farmer dominance on boards;

(2) provide boards with more marketing expertise and commercial flexibility; and (3) make boards more accountable to parliament. During the 1980s state and federal governments of different political persuasion have initiated steps toward deregulation of several areas of economic activity including agricultural marketing. Substantial changes to wheat marketing legislation came into force on July 1, 1989.

Wheat Marketing pre-July, 1989

Regulation of wheat marketing at the national level began in 1914 when an Australian Wheat Board (AWB) was established under the War Precautions Act and involved acquisition of the wheat crop by the board, price fixing and advance payments to growers on delivery of their crop. There was a period of private trading between the two World Wars and then an Australian Wheat Board was re-established in 1939. The first "wheat marketing plan"was introduced in 1948 and was a development on the wartime marketing arrangements. Since then eight plans have operated, each for a period of five years except for the fifth plan, which ran for six years because the states and the Commonwealth had difficulties reaching agreement on the nature of the required complementary state-federal legislation.

Two policy instruments formed the cornerstones of the first six plans which covered the 31 years from 1948/49 to 1978/79. The first was a "two-price"or "home-consumption"pricing arrangement, in essence a form of price discrimination which kept prices on the domestic market (relatively price-inelastic demand) above prices on the export market (highly elastic demand). Occasionally, when export prices were high, domestic prices were set below export prices. Costs of production figured prominently in the determination of domestic prices. The second instrument was a buffer fund to stabilize returns from the export market. This stabilization fund was the mechanism for maintaining a guaranteed export price. Growers contributed to the fund when the price was above the guaranteed export price and drew from it when the reverse was true. When funds were insufficient, the government (i.e., taxpayers) made up the deficiency. This occurred in 14 of 31 years (Longworth and Knopke, 1982).

In the seventh plan (1979/80 to 1983/84) the price of wheat for domestic milling was fixed using a formula that tracked trends in export prices, but maintained the domestic price at an average of 20 percent above export prices. Domestic stockfeed and industrial wheat were priced at the Australian Wheat Board's discretion. Growers also received a government-guaranteed minimum price, with any deficiency between the market and guaranteed minimum prices being met by the government. This replaced the earlier buffer fund arrangement.

The eighth plan, which ended on June 30, 1989, was similar to the seventh plan. Domestic prices of wheat for flour milling were based on export parity plus a margin of $A15 per metric ton. The guaranteed minimum price was 95 percent of the average of the gross pool returns in the subject year and the two lowest pool returns in the previous three seasons, less pool costs for the subject year. Also, growers were able to sell stockfeed wheat directly to buyers under a permit system.

In all of these plans growers received "advance payments" soon after the (compulsory) delivery of their wheat to the Australian Wheat Board. These advances against the future proceeds from sale of the crop were financed by borrowings from the Reserve Bank (Australia's central bank) prior to 1979, but thereafter the board was given flexibility to borrow from other sources. Subsequent payments were made to growers as the crop was sold, with the wheat grown in any year constituting a separate "pool." Typically it took about three years to close out a pool. The deficiency payments mechanism — where government payments were required to return growers the guaranteed minimum prices — was triggered only in the case of the crop harvested in 1986.

The Australian Wheat Board appointed agencies in each state to receive wheat on its behalf and transport it to port terminals or other designated points. These agencies, commonly referred to as "bulk handling authorities," were usually owned, or at least closely regulated, by state governments. Various marketing channels for wheat are indicated in Figure 16.6. Under the old system, private enterprise seldom participated in the handling, transport or storage. When private agents have been involved they have been regulated closely. Even cooperative marketing has generally occurred as a result of an agreement with a statutory marketing authority. In addition to the institutional framework that governed the transport, storage and handling of wheat, another set of arrangements applied to wheat sales. Because price policy was implemented through these, they are discussed in the public policy section.

Pressures for Change

In a statement on economic and rural policy (Hawke and Kerin, 1986:50), the government drew attention to the high costs of grain storage, handling and transport incurred in the statutory marketing system. For example, wheat storage, handling and transport charges per metric ton rose by more than 50 percent between 1979/80 and 1984/85 while wheat prices increased by less than 20 percent. As a result, the share of net pool returns consumed by these services jumped from 14 percent to 20 percent.

Concern over these costs led to the establishment of a Royal Commission (a government-initiated enquiry headed by a Royal Commissioner with wide investigatory powers) to investigate the matter. The Commission (1988c) discovered a complex set of agreements in operation, which it classified into three categories. First there were agreements designed to "extend market power." An example was an agreement between the Australian Wheat Board and a state bulk handling authority allowing the latter to be the sole receiving agent for the board. Second, there were agreements which the Commission described as "proxies for market forces." These involved "... a structured set of rewards and penalties to promote more efficient operating practices..." For example, agreements between the Australian Wheat Board and the state bulk handling authorities contained various reward and penalty structures. Third, there were agreements to "... improve cooperation and coordination within the system without direct payments." The Commission cited agreements to improve outloading facilities at storage locations as an example.

FIGURE 16.6 Alternative Grain Marketing Channels

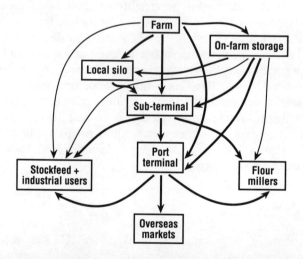

Note: The thinner arrows indicate channels which have become possible as a result of new legislation introduced in July 1989. Prior to the new arrangements some direct sales from growers to stockfeed companies were permitted under a permit system.

Source: Adapted from Royal Commission into Grain Handling, Storage and Transport (1988a).

The Royal Commissioner presented his report in February 1988 in which it was concluded that:

> For the most part, the current system of grain distribution does not meet the criteria of economic efficiency, cost effectiveness and integration, the three assessment criteria in the Commission's terms of reference. (Royal Commission into Grain Storage, Handling and Transport, 1988(a):3).

Its preferred option for change was a "... mixed deregulatory/regulatory approach" in which:

- marketing authorities would interact with a deregulated distribution system, and have a charter to minimize storage, handling and transport costs;
- sole receival rights for storage and handling of grain would be removed, providing marketing authorities with the flexibility and responsibility to employ those agencies able to provide the required services at least cost;
- all road transport restrictions and impediments would be removed, with the result that grain would no longer be one of the few goods that is, to a significant extent, reserved for rail transport;
- port service and sea transport costs would be disaggregated and marketers of grain would pass back to growers and other market participants the actual costs incurred at Australian ports. (Royal Commission into Grain Storage, Handling and Transport, 1988(a):p.8).

In addition to the Royal Commission, the Industries Assistance Commission (now Industries Commission), a federal government agency that makes recommendations to the government on levels of assistance to industries, presented its five-yearly report on the wheat industry in February, 1988 (Industries Assistance Commission, 1988). As in its previous reports on the wheat industry, it argued that the wheat board's statutory monopoly on domestic sales should be removed (to enable greater economic efficiency) although its monopoly status relative to export sales should be retained (mainly on grounds of perceived benefits in obtaining higher export prices). It recommended a number of other changes, including allowing the wheat board to trade in other grains, which would essentially provide the board with greater commercial flexibility.

Current Wheat Marketing Arrangements

Following a lengthy debate after the release of the Royal Commission and Industries Assistance Commission reports, new wheat marketing arrangements (Commonwealth Wheat Marketing Act 1989) came into operation on July 1, 1989. The general thrust of the arrangements is to deregulate handling, storage, transport and pricing on the domestic market (the AWB will have to compete with other agents on the domestic market) but the AWB will retain control (i.e., single-seller status) over export sales. The full effects of these changes will not be felt for some time (the harvest that began around November 1989 was to be the first marketed under the new arrangements) but it has been estimated that the reforms in relation to domestic handling, storage and transport will save the average wheat grower about $10 per metric ton (Royal Commission into Grain Storage, Handling and Transport, 1988(a): 12). As has been the case in the past, the arrangements have required the passage of state-federal complementary legislation to enable a federal-level agency to operate within states. The duration of the new legislation is not limited under the Act and this is a major departure from the past pattern of five-year wheat marketing plans.

Extracts from a release describing the new arrangements are provided in Chapter 18.

The Australian Wheat Board

Membership

Under the new legislation the AWB is comprised of a wheat grower (for the first board constituted under the new legislation) chairperson selected by the Minister for Primary Industries and Energy, eight general members (not necessarily wheat growers) selected for their skills in production and/or marketing and a government member selected by the Minister. Should the chairperson opt to act in a non-executive capacity, the Act provides for the appointment of a Managing Director chosen by the board. The Grains Council of Australia, the producer-representative body, can nominate up to six individuals to serve with a ministerially-appointed Presiding Member on the selection committee for the eight general members of the board. Under previous legislation six of the eight general members had to be wheat growers.

Objectives, Functions and Powers

The objectives, functions and powers of the AWB are set down in the 1989 Wheat Act. The objectives are:

- to maximize the net returns to Australian wheat growers who sell pool-return wheat to the AWB by securing, developing and maintaining markets for wheat and wheat products and by minimizing costs as far as is practicable; and
- by participating, in a commercial manner, in the market for grain and grain products, to provide Australian grain growers, and especially wheat growers, with a choice of marketing options.

The functions of the AWB include:

- to export and control the export and overseas marketing of Australian wheat;
- to promote, fund or undertake research into matters related to wheat marketing; and
- to export and trade in wheat products, other grains and other grain products to the extent that this promotes an objective of the AWB.

The AWB has powers to:

- buy wheat in Australia or overseas;
- import wheat into Australia;
- sell or arrange for sale of wheat in Australia or overseas; and
- provide the same marketing services for grain other than wheat as it does for wheat.

Control over export marketing has been a fundamental feature of the AWB's activities since its establishment. Sometimes this has been rationalized, partly on the grounds that centrally-planned economies prefer to do business with other governments or their statutory agencies rather than with private traders, but this argument has never been given much credence in Australia. The real reason for retaining AWB control over exports has to do with the industry's and government's perception that there are price premiums in certain overseas markets which need to be protected by having a single Australian seller servicing those markets.

The AWB has sold wheat to private multinationals for resale to a third (overseas country) party. Up to 30 percent of total exports have been sold this way in the past, but in recent times the figure has been 10-12 percent. The procedure is initiated by the trader who negotiates with the AWB on price, quality, shipping period and markets. These markets typically include South America and private importers to Asia and the South Pacific (e.g., Malaysia, Indonesia, Thailand, The Republic of Korea, Sri Lanka, Yemen, New Zealand and Fiji).

Trading and Pricing Activities

The following description of the AWB's trading and pricing activities is, of necessity, based on observations of how it operated under the old legislation. No doubt these will change somewhat as it develops its strategies for operating in a more deregulated environment with its greater commercial flexibility.

The AWB establishes competitive daily export prices for all available grades of Australian wheat. Separate prices are set for up to six forward shipment months. The AWB's prices are based on the cash and futures prices of the nearest equivalent grades of U.S. wheat. They are negotiated either as flat figures or as basis contracts. They may be adjusted to take account of freight differentials, quality and marketing factors. Prices are established on a FOB basis and are quoted in U.S. dollars. Periodic cost and freight sales are made.

The AWB has pursued a consistent long-term marketing strategy which has centered on supplying large and regular customers with Australian wheat in the quantities they require. That strategy has been supported by progressive development of marketing support, technical assistance and promotional activities to develop and maintain customer loyalty to the AWB and Australian wheat. Those activities have concentrated in particular on the quality features of Australian wheat, namely its white, clean, dry and insect-free characteristics. Regardless of the international marketing environment, these are perceived to be fundamental to the continuing success of Australia's wheat marketing activities.

The AWB uses other marketing tools as appropriate to compete in the international wheat market. The establishment of long-term agreements is seen as a major objective to ensure long-term market access and to reduce planning uncertainty. A significant percentage of future sales is broadly committed under long-term agreements. The provision of credit is another area where the AWB has had to become involved to remain competitive. Some of Australia's largest markets are supplied on credit terms, although changes to credit insurance arrangements have encouraged a reduction in the supply of credit and shortening of credit terms. Reference has already been made to the fact that the AWB sometimes sells wheat through private multinationals. These "Recognized Exporters" are part of the AWB's marketing strategy in that these exporters have access to, or specialized knowledge of, particular markets. However, the AWB normally reserves certain large and important markets for itself — typically those with government buying agencies or when end use is for nonfeed purposes. These markets include the U.S.S.R., China, Egypt, Iran, Iraq, and those in which long-term agreements are maintained.

The AWB is in constant communication with buyers and sellers around the world. Wheat prices fluctuate daily and, as a result, the AWB changes its asking price at least on a weekly basis. In addition, it is in constant touch with the Chicago Board of Trade and, at times, hedges wheat on the exchange. The precise amount of hedging done by the AWB is confidential. However, it does hedge on the Chicago market — a practice which is in sharp contrast to, for example, the Canadian Wheat Board, which is the sole exporter of wheat for Canada.

In the past, the AWB has sold grain to domestic channels for use in making bread, gluten and the like. Private companies had to purchase wheat from the AWB to process into products for both the domestic and export markets. (These companies are now free to purchase wheat directly from growers.) The exportation of flour from Australia is extremely limited. However, there has been a significant growth in the exportation of gluten from the Australian market.

The marketing problems of the AWB, as in the case of Canada, are exacerbated by uncontrolled variations in quality. As the OTA study points out, seven classes of wheat are produced and marketed in Australia: Prime Hard, Hard, Australian Standard White, Soft, Durum, General Purpose, and Feed. Table 16.11 shows the percentage of AWB receivals of wheat by class and state averages. The largest percentage of wheat grown as a class is Australian Standard White — roughly 68 percent. Roughly 15 percent is hard wheat. The remainder is split among the other classes given in the table. The principal quality difference among classes is the protein level and the end-use performance associated with protein. Australian Standard White wheat generally has protein levels around 10 percent. Table 16.12 illustrates a breakdown of the components that go into wheat grading.

In addition, the AWB has to sell into an extremely volatile market. Figure 16.7 gives international wheat prices for 1978/79 through 1986/87. World wheat prices have fluctuated considerably. Between 1972 and 1974, export prices for wheat measured in U.S.dollars more than doubled from the lower prices which had prevailed for the previous three years. From 1982 to 1987, prices on world wheat markets fell steadily so that, in nominal U.S. dollar terms, international prices in 1987 were at about the same level as 10 years previously (Figure 16.7).

Marketing of Other Grains

As a generalization, it is fair to say that the marketing system for other grains has many of the features that characterized wheat marketing until

TABLE 16.11 Percentage of Wheat Receivals By Class and State Averages, Australia, 1976/77 to 1985/86

Years	Australia Prime Hard (APH)	Australia Hard (AH)	Australia Standard White[a] (ASW)	General Purpose[b] (GP)
1976/77	8.7	18.4	64.5	8.4
1977/78	14.2	17.2	62.8	5.8
1978/79	4.3	15.6	69.2	10.9
1979/80	4.7	16.1	74.3	4.9
1980/81	3.8	14.5	77.6	4.1
1981/82	7.7	19.9	68.1	4.3
1982/83	10.1	13.8	72.3	3.8
1983/84	6.6	12.4	51.5	29.5
1984/85	6.4	13.0	77.5	3.1
1985/86	4.5	13.0	64.7	17.8
Averages over 10 years				
Australia	7.1	15.4	68.3	9.2
New South Wales	15.9	25.7	45.2	13.2
Victoria	--	3.6	90.7	5.7
South Australia	--	24.1	72.7	3.2
Western Australia	--	5.1	87.3	7.6
Queensland	28.6	29.9	26.0	15.5
Tasmania	--	--	--	--

[a]   Includes minor quantities of Durum and soft wheat.
[b]   Includes Australian feed wheat.
Source: Australian Wheat Board, *Annual Reports*.

1989. Statutory agencies abound and the practices of pooling handling, storage and transport costs and pooling receipts from sales on different markets (see discussion under Public Policy in Chapter 17) has been widespread. Given the changes that have been made to wheat marketing one can expect changes to the arrangements for other grains. The Royal Commission that has been discussed had terms of reference relating to grain (as opposed to just wheat) marketing. Moreover, under the new arrangements for wheat the AWB will be able to trade and ship grains other than wheat.

TABLE 16.12 Typical Analysis for the Australian Milling Wheat Classes

	Australian Prime Hard 14%	*Australian Hard*	*Australian Standard White*	*Australian Soft*
Wheat				
Test weight (kg/hl)	79.4	80.0	80.5	78.0
1000 kernel weight (g)	35.2	37.2	35.2	34.8
Grain hardness (P.S.I.)	15	14	17	27
Protein (11% moisture)	14.2	12.2	10.8	8.5
Ash %	1.50	1.50	1.38	1.38
Falling number (sec)	494	460	422	325
Flour extraction %	75	74	75	74
Screening				
Total screenings % (2mm screen)	2.5	2.6	3.1	3.2
Flour				
Protein (3.5% moisture)	13.1	11.0	9.6	7.5
Wet gluten %	40.0	33.7	28.8	2.2
Diastatic activity (mg)	192	237	195	116
Ash %	0.50	0.48	0.47	0.45
Farinogram				
Water absorption %	65.6	65.8	60.8	52.4
Development time (min)	6.0	4.7	3.4	1.9
Extensograph				
Extensibility (cm)	23.2	22.8	20.1	19.6
Maximum height (B.U.)	460	365	320	190
Area (cm)	140	112	95	43

Source: Australian Wheat Board, *Australian Wheat Industry Guide.*

316

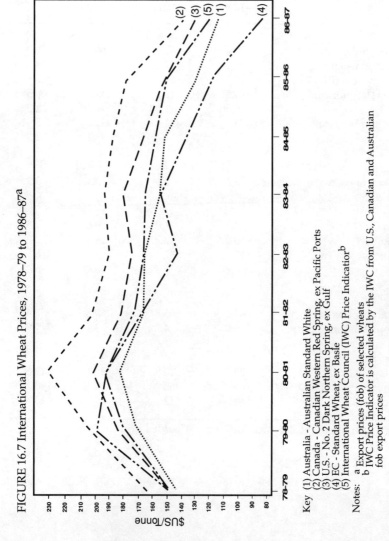

FIGURE 16.7 International Wheat Prices, 1978–79 to 1986–87[a]

Key (1) Australia - Australian Standard White
    (2) Canada - Canadian Western Red Spring, ex Pacific Ports
    (3) U.S. - No. 2 Dark Northern Spring, ex Gulf
    (4) EC - Standard Wheat, ex Basle
    (5) International Wheat Council (IWC) Price Indicator[b]

Notes: [a] Export prices (fob) of selected wheats
       [b] IWC Price Indicator is calculated by the IWC from U.S., Canadian and Australian
           fob export prices

Source: International Wheat Council, World Wheat Statistics (London: various issues).

17

Public Policy

Overview

The remainder of this part focuses on wheat because it is the most important grain crop in Australia and, as mentioned previously, policy with respect to other grains has been similar to that for wheat.

Government interventions that affect the wheat and coarse grains industries are summarized in Table 17.1. This summary table of grain policy instruments is oversimplified in the sense that the grain industry is affected by many policy measures enacted elsewhere in the economy, including in other agricultural industries. One important example is the implicit tax on grain exports that results from protection afforded to domestic manufacturing industries. Another example is the government's contribution to agricultural research.

In the case of wheat, potentially the most powerful instrument affecting producers' returns was the deficiency payment mechanism in the form of the guaranteed minimum price. There is no similar arrangement for other grains. However, the deficiency payment has only been triggered once since 1979 and the main intent of the policy was to reduce growers' perceived price risk. The deficiency payments mechanism has been removed in the new wheat marketing arrangements. Its principal advantage was that its existence allowed the AWB to borrow at favorable interest rates and it allowed for a relatively high first advance payment to be made to growers. The explicit guarantee given to AWB borrowings in the new marketing arrangements is intended to provide these benefits.

Myers, Piggott and MacAulay showed (Table 17.2) that Australian wheat farmers have not been very dependent on government transfers. This fact is also substantiated by more recent Producer Subsidy Equivalent (PSE) measures estimated by the United States Department of Agriculture

TABLE 17.1 Major Grain Policy Instruments In Australia

	Commodity	
	Coarse Grains[a]	*Wheat*
PRODUCTION/CONSUMPTION		
Producer guaranteed price		
Deficiency payments		X[b]
Input subsidies		
• credit		
• fert/pest		
• irrigation		
• machinery/fuel		
• seed		
Crop insurance		
Controlled consumer price	X[c]	X[c]
TRADE		
Imports		
• tariff		
• quota	X	X
• subsidies		
• licensing	X[d]	
• state trading		
Exports		
• taxes		
• restrictions		
• subsidies	X[e]	X[e]
• licensing		
• state trading	X	X
OTHER		
Marketing subsidies		
• storage		
• transport		
• processing		
State marketing	X[f]	X[e,f]

[a] Barley and sorghum.
[b] With pooled returns.
[c] Implicit consumer tax.
[d] Strict quarantine requirements for imported grains effectively eliminate imports.
[e] Export sales insurance.
[f] With pooled marketing costs.

TABLE 17.2 Effects of Wheat Market Intervention on Real Levels and Stability of Some Key Industry Variables, Australia, 1953/54 to 1983/84

Variable	Mean	Percentage Change in Coefficient of Variation
Price of food wheat	7.1	-32.1
Price of stockfeed wheat	2.5	-11.6
Export price	-0.6	2.7
Price to growers	2.0	-16.7
Quantity exported	2.5	-4.7
Production	1.7	-2.5
Revenue from exports	1.7	-2.5
Revenue to growers	3.6	-2.6

Note: The effects were generated by simulating plans 2 to 6 over the period 1953/54 to 1978/79 and plan 7 over the period 1979 to 1983/84.
Source: Myers, Piggott, and MacAulay (1985: Table 4).

(Table 17.3). Roughly speaking, the PSE is the percentage of gross revenue received from government intervention. Australia's wheat PSE averaged 7.7 percent over the 1982-1986 period, which is very small compared to the other developed country exporters (i.e. Canada, EC-10, and the United States). Despite relatively large wheat production subsidies in competing countries, Australian agriculture has managed to continue to produce and export substantial quantities of wheat.

The prevalence of statutory marketing authorities in the grain industry has already been discussed. It has been typical for statutory authorities to have vesting powers in the sense that they could secure ownership of grain to negotiate sales. This mechanism has allowed the implementation of domestic price controls in the form of price discrimination (domestic prices higher than export prices). However, at least in the wheat industry,

TABLE 17.3. Wheat Producer Subsidy Equivalents: 1982-87

Year	Australia	Canada	EC-10	Argentina	U.S.
1982	9.1	14.1	27.0	-14.6	16.8
1983	4.1	20.2	9.9	-29.2	38.1
1984	3.2	34.2	3.8	-41.2	28.5
1985	4.7	33.9	31.1	-0.5	37.9
1986	17.2	49.9	58.8	16.6	63.0
Average	7.7	30.5	26.1	-13.8	36.9

Source: United States Department of Agriculture. Estimates of Producer and Consumer Subsidy Equivalents: Government Intervention in Agriculture, 1982-1987, ATAD-ERS, 1989.

these powers have been eliminated. The only grain that is subject to production controls is rice which is regulated through the implementation of restrictions on irrigation water and land. Wheat delivery quotas were introduced for a short time after the 1969 sowing but they were fairly ineffectual and were annulled in the 1975/76 crop year. Their effect was small because growers were given the opportunity to offset over-quota deliveries with the following season's quota and it was clear in 1972 that international grain shortages were developing.

In the past, grain producers benefited from input subsidies, particularly those for fertilizer and credit. Currently, government intervention in input markets, which includes taxes and tariffs on material and capital inputs, results in a net tax on grain production. It is possible to argue that there is still an implicit subsidy on credit because of the government guarantee on loans to the Australian Wheat Board. Insofar as wheat producers also produce other grains and the AWB can trade in these grains, it could also be argued that the implicit subsidy applies to grains in general.

Domestic markets for grain have been protected through embargoes on imports and strict quarantine restrictions which amount to a prohibitive import barrier.

The Industries Assistance Commission (1988) questioned whether state involvement in storage, handling, and transport of grains results in a net producer tax or subsidy. However, the Royal Commission was of the opinion that it increased producer costs by about $A10 per metric ton despite the fact that state rail authorities, which transport the majority of grain to ports and other destinations, and state-owned bulk handling authorities, did not recover all of their costs from growers.

As will be explained in some detail later, state authorities practice pooling of costs and pooling of revenues. This causes cross-subsidization among producers.

Pricing Practices

Historically, Australian grain marketing bodies have commonly used two pricing practices. The two-price or home consumption pricing policy has already been mentioned. One justification sometimes given for this is that domestic prices can be manipulated counter cyclically to export prices, making the "equalized" or weighted average producer price more stable than the export price. Another is that the scheme protects domestic consumers from extreme fluctuations in export prices. In fact, the home-consumption price for wheat was less than the average export return in some years (Table 17.4). However, in most years the reverse was true, and it seems reasonable to conclude that this pricing arrangement essentially transferred income from domestic consumers to producers. Given the

TABLE 17.4. Australian Wheat Prices, 1970/71 to 1988/89

Crop Year	Unit Value[a]	Guaranteed Min. Price[b]	Home Consumption Price	Feed Wheat[c]	Average Export Return
			--- $A/mt ---		
1970/71	51.15	54.20	63.93	53.28	48.00
1971/72	53.75	55.78	55.40	54.75	48.17
1972/73	54.11	57.61	67.63	56.97	54.35
1973/74	109.44	58.79	71.10	71.10	125.01
1974/75	110.63	73.49	83.40	83.40	122.63
1975/76	104.26	76.55	99.32	99.32	110.13
1976/77	89.05	76.29	105.40	105.40	92.32
1977/78	99.70	80.94	111.16	111.16	103.22
1978/79	126.91	91.96	116.61	116.61	133.63
1979/80	153.08	114.71	130.78	127.30	161.97
1980/81	155.13	131.92	156.12	152.20	154.53
1981/82	156.44	141.55	187.20	147.28	157.65
1982/83	176.45	141.32	203.46	176.67	184.53
1983/84	163.77	150.00	219.41	174.65	170.56
1984/85	171.59	145.35	210.73	201.24	189.01
1985/86	168.21	149.87	213.89	198.20	179.97
1986/87	150.79	139.83	188.92	171.29	136.54
1987/88	163.56	144.29	193.46	177.66	156.70[d]
1988/89[d]	201.00[d]	153.00[d]	210.00[e]	--	189.00[d]

a   Unit value is gross value of production of wheat grown in the crop year divided by quantity sold.
b   The "stabilization price" replaced the guaranteed price in 1974/75, and the guaranteed minimum price replaced the stabilization price in 1979/80.
c   Average Australian standard white stockfeed wheat prices, terminal ports basis, without quantity discounts applied. Figure for 1986/87 is the simple average of July 1986 - October 1986 prices.
d   Preliminary.
e   Estimated by ABARE.
*Source:*Australian Bureau of Agricultural and Resource Economics, *Commodity Statistical Bulletin*, December 1988: Tables 75 and 91; Bureau of Agricultural Economics, Commodity Statistical Bulletin, December 1986: Table 68.

small proportion of wheat sold domestically, this policy has not been very effective in boosting wheat producers' incomes.

The second, and perhaps most distinctive, feature of grain pricing arrangements is pooling — both of prices received and handling and transport costs. This is only possible because of the dominance of statutory authorities in marketing, handling, and transport services for grain. (For a detailed discussion of the nature and extent of price and cost pooling, see Royal Commission into Grain Storage, Handling and Transport (1988d)).

Within broad grades, the AWB used to pool the revenue from all sales of wheat from a particular crop year regardless of where or when it was sold. As already mentioned, export and domestic prices were not the same. Returns from both sources were pooled for particular grades, and all farmers contributing to the pool received an equalized price. Other forms of pooling also occurred at the national level. Port and shipping costs were pooled by the AWB, with the result that export prices were the same for a particular grade regardless of source, except for a small adjustment made to compensate Western Australian growers for the freight advantage they enjoy to Middle East markets. All marketing costs are pooled by the Wheat Board.

Grain handling and storage systems have been characterized by extensive cost pooling, both between handling facilities and among the users of a particular facility within a state. It should be noted that, for many years, the costs associated with the handling and storage of grain were pooled between, as well as within, states. Thus, all growers were charged the same amount per metric ton, regardless of the actual cost of handling their grain. There is also evidence of cost pooling and cross subsidization in the provision of transport services. Almost the entire Australian wheat crop is moved to port by rail. The freight rate that a particular grower is charged may contain elements of subsidy from other growers, other rail users, or taxpayers. Under-pricing of transport services on low-volume branch lines that carry nothing but grain is another source of cross-subsidization.

Grower organizations support pooling even though it involves costs to growers as a whole. Revenue pooling may be seen as a logical concomitant of retaining the AWB role. However, the rationale for grower support for cost pooling is unclear. It is administratively convenient for institutions that essentially replace competitive markets as the determinants of price and cost. Pooling also redistributes income among producers, spreading risk. Perhaps, therefore, some notion of equity provides part of the rationale. It is probably fair to say that grower members of statutory marketing and handling authorities have been under pressure from some quarters to promote policies that are viewed as "equitable,"even if they are not in the interests of all growers. (For a further discussion of pooling, see Quiggin and Fisher, (1988a)).

Pooling of returns from wheat sales will continue under the new wheat marketing legislation but the AWB now has the flexibility to establish a number of pools for each season's crop. Moreover, growers can avoid wheat pools by opting to sell for cash either to the AWB or a private trader. It is difficult to speculate about how much cost pooling will characterize future market arrangements.

Stockholding Policy

Australia does not run a bufferstock policy for grains. Data on world wheat stocks are given in Figure 17.1. Stocks rose significantly during the 1980s. Australia generally does not hold large amounts of stocks. This can be seen from Figure 16.2. Stock levels are directly related to production and exports. When production is high, stock levels are also generally high. In Australia, wheat stocks are generally well below 5 million metric tons. Stock carryover is in the neighborhood of 20 percent to 25 percent of the annual production compared to about 60 percent in the United States and 80 percent in Canada. Additional data on stock levels are given in Table 17.5. (For further discussion of stockholding see Fisher (1978)).

Storage, Handling and Transport Charges

The Royal Commission has estimated marketing costs as shown in Tables 17.6 and 17.7. They found that combined storage, handling and rail freight charges represent 19.1 percent of the expected net pool contribution in 1986/87. The other major cost components are sea freight (16 percent) and AWB administration and interest costs (11 percent).

FIGURE 17.1 World Closing Wheat Stocks, 1978 to 1986

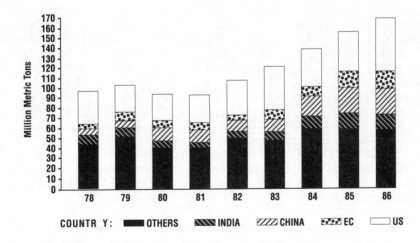

TABLE 17.5. Australian Wheat Supplies and Disappearance, 1961/62 to 1987/88 (million metric tons)

	Supply			Disappearance			
Year	Beginning Stocks	Production	Total	Domestic	Exports	Total	End-of-Year Carryover
1961/62	0.7	6.7	7.4	2.1	4.8	6.9	0.5
1962/63	0.5	8.4	8.8	2.0	6.2	8.2	0.6
1963/64	0.6	8.9	9.6	2.1	6.9	9.0	0.6
1964/65	0.6	10.0	10.6	2.7	7.3	9.9	0.7
1965/66	0.7	7.1	7.7	2.5	4.8	7.3	0.5
1966/67	0.5	12.7	13.2	2.4	8.5	11.0	2.2
1967/68	2.2	7.5	9.7	2.7	5.7	8.3	1.4
1968/69	1.4	14.8	16.2	2.3	6.7	8.9	7.3
1969/70	7.3	10.5	17.8	2.4	8.2	10.6	7.2
1970/71	7.2	7.9	15.1	2.7	9.0	11.7	3.4
1971/72	3.4	8.6	12.0	2.8	7.8	10.6	1.5
1972/73	1.5	6.6	8.0	3.4	4.1	7.6	0.5
1973/74	0.5	12.0	12.5	3.2	7.4	10.6	1.9
1974/75	1.9	11.4	13.2	3.0	8.5	11.6	1.7
1975/76	1.7	12.0	13.6	2.7	8.2	11.0	2.7
1976/77	2.7	11.8	14.5	2.6	9.8	12.3	2.1
1977/78	2.1	9.4	11.5	2.6	8.1	10.7	0.8
1978/79	0.8	18.1	18.9	2.5	11.7	14.2	4.6
1979/80	4.6	16.2	20.8	3.4	13.2	16.6	4.3
1980/81	4.3	10.9	15.1	3.5	9.6	13.1	2.0
1981/82	2.0	16.3	18.4	2.4	11.0	13.4	4.9
1982/83	4.9	8.8	13.8	4.2	7.3	11.5	2.3
1983/84	2.3	22.0	24.3	2.6	14.2	16.7	7.6
1984/85	7.6	18.3	25.9	2.6	15.1	17.3	8.6
1985/86	8.6	16.6	25.1	2.9	16.1	17.9	7.3
1986/87	7.3	16.8	24.1	2.7	14.8	17.5	6.6
1987/88	6.6	12.4	19.0	2.7	12.2	14.9	4.1

Note: 1987/88 data are preliminary.
Source: 1961/62 to 1985/86: *World Wheat Statistics* (London: various years); 1986/87 and 1987/88: IWC Market Report; and U.S. Department of Agriculture, Foreign Agricultural Service, "World Grain Situation Outlook," FG-9-88, Washington, DC.

Storage, handling and transport costs vary widely across states (Table 17.7). In 1986/87 the combined charges for these services ranged from $A23.57 per metric ton in South Australia to $A41.19 per metric ton in New South Wales, which represents between 11 percent and 24 percent of the export customer's CIF price.

TABLE 17.6. Components of the Export Price of Wheat: Australia, 1986/87

	$/tonne	% of Customer Price
Customer Price (c.i.f.)[a]	172.70	
• sea freight	27.70[d]	16.0
Average f.o.b. Port Price	145.00	
• AWB administration & interest cost[b]	19.52	11.3
Net Pool Contribution	125.48	
• Storage and Handling	14.94	8.7
• Rail Freight	17.88[c]	10.4
• Wharfage	1.12	0.6
• Research Levy	0.40	0.2
• Ceres House	0.10	--
Farm Gate Return	91.04	52.7

[a]   Customer price derived from expected BAE f.o.b. return.
[b]   Gross interest cost estimate.
[c]   Weighted State average using 1985-86 charges and 1985-86 receivals.
[d]   Based on freight rate for shipment from east coast of Australia to the Middle-East as at 28 November 1986.
Source: Australian Wheat Board (personal communication), BAE (1986)

Public Policy: Evaluation

For the sake of convenience, the discussion in this section is divided into two parts — pooling and stabilization. In reality, this is somewhat artificial because pooling of returns from different markets could be viewed as part of the overall stabilization package.

Pooling

Until recently, little attention has been paid to assessing efficiency losses associated with pooling costs and returns, although estimates of income transfers were made. In an early paper, Longworth (1967) described both inter-temporal transfers and those likely between growers at a given point in time. Similar observations were made in an Industries Assistance Commission (1983) paper. More recently, a detailed analysis of the transfer effects of pooling was conducted by the Royal Commission into Grain Storage, Handling, and Transport (1988d). In a sample of sites in Western Australia, some growers received income transfers in the order of $A3 per

TABLE 17.7. Average Freight and Storage and Handling Charges Deducted From Grower's First Advance, Australia, 1984/85 to 1986/87, ($/tonne)

	NSW	VIC	OLD	SA	WA
Freight[a]					
1984/85	23.67	21.20	16.24	10.48	16.49
1985/86	24.59	22.52	15.84	11.18	16.02
1986/87	24.49	22.09	16.21	11.13	16.01
Storage and handling					
1984/85	17.20	13.75	20.00	12.74	13.05
1985/86	16.70	13.80	19.00	11.93	13.05
1986/87	16.70	14.63	17.00	12.44	13.05
Combined cost					
1984/85	40.87	34.95	36.24	23.22	29.54
1985/86	41.29	36.32	34.84	23.11	29.07
1986/87	41.19	36.72	33.21	23.57	29.06

a   These figures are calculated on the basis of total freight deductions made from growers divided by (D – N – B), where
D = total receivals by AWB in a State.
N = tonnes delivered that did not incur a freight deduction (e.g., direct farm to port deliveries)
B = tonnes sold under grower to buyer arrangements.
Average freight deductions, although primarily for rail transport, in some cases may be for road transport, where the road transport is organized by the bulk handler or rail authority (for example in Western Australia and Victoria).
Source: AWB, personal communication, May 20, 1987.

metric ton, while others paid an additional $A2 per metric ton for grain handling.

As previously mentioned, the Australian Wheat Board pools across wheat types, which they classify as broad grades, such as Australian Standard White. This masks price signals to growers with respect to grain quality, which has a distinct and obvious effect on farm-level investment decisions. However, it also camouflages signals to the plant breeding industry, distorting research investments. Little inquiry has been conducted into the trade-offs involved in choosing the optimum number of grain classifications, given the demand characteristics for Australian wheat and the costs of handling, transporting, and storing additional grades.

The effects on efficiency can be considered, first, in terms of cost pooling among sites and, second, in terms of pooling among users at a given site. In the case of the first question, researchers usually postulate that transport cost differentials outweigh those for handling. Thus, it is always cheapest to deliver grain to the nearest receiving point. For example, Spriggs, Geldard, Gerardi, and Treadwell (1987) make this assumption, which would apply, in particular, to pooling among widely separated regions when crop acquisition is compulsory. Assuming that the elasticity of grain supply is 0.8 and that handling charges represent approximately 15 percent of the grain price, Quiggin and Fisher (1988a) showed that the welfare loss associated with a 10 percent distortion in grain handling charges is approximately 0.02 percent of the value of the crop. If the farm-gate price is $A100 per metric ton, the welfare loss would be $A0.02 per metric ton. Given such small efficiency losses for widely separated sites, it is noteworthy that cost pooling among states was discontinued well before serious consideration was given to suspension of pooling within states. This was likely because of the associated large income transfers among states.

In the case of cost pooling among nearby receiving points, welfare losses arise when deliveries are made to points with high marginal costs rather than to those with low marginal costs. If, for example, the transport costs to two sites are equal, then the social loss is equal to the difference in marginal costs. If the high-cost site is closer, then the loss from pooling is equal to the difference in marginal costs less the savings in transport costs. Effects of this kind are only relevant within regions where variations in marginal costs are greater than the additional transport costs. In the case of typical cost differences among sites in New South Wales, Quiggin and Fisher (1988a) showed that welfare losses in the order of $A1.50 per metric ton may not be uncommon when handling costs are pooled among nearby sites.

Pooling among the users of a particular elevator may also give rise to efficiency losses. Some deliveries will always be more costly to process than others because of differences in delivery time, amount unloaded (both in total and per truck), moisture content, and qualitative factors, such as the number of segregations required. For any given period of operating time, a convex, short-run operating cost function implies that a uniform pattern of delivery is preferable to an uneven one, and, hence, some degree of peak-load pricing improves efficiency (Quiggin and Fisher 1988a). In some cases, however, the gains from peak-load pricing may be small and are probably not worth the transaction costs.

Other factors also affect system costs. For example, there are likely some benefits, such as reduced administrative charges, associated with large deliveries, which may be reflected in bulk discounts. It is also probable that variations in the size of delivery trucks will affect the cost at the receiving point. Since some overhead is associated with unloading, costs per metric ton will decline as truck size increases up to some limit set by the design of the receiving facilities. The absence of bulk discounts, together with cost pooling among users at a particular site, discourages the use of larger-than-average size trucks. It is also worth noting that the resource costs per metric ton per kilometer associated with a typical double-axle farm truck used for delivery from farm to elevator is some three times greater than that of six-axle articulated units (Bureau of Transport Economics, 1987). If full resource costs are not recovered from truck operators, pooling across users at a particular receiving point causes losses, not only to the handling facility, but also to society at large.

The discussion in this section centered around the direct effects of pooling on the grain handling industry. At least in theory, the efficiency costs of pooling at this level can be measured. What cannot be easily assessed, however, are indirect costs of pooling. These may arise, for example, because the performance of the grain handling system is not adequately evaluated as a result of a lack of publicly available information in a system dominated by statutory authorities. Deregulation of the grain handling system should give rise to improvements on this front.

Stabilization

A number of evaluations of wheat stabilization policy have been made, including Longworth (1967), Industries Assistance Commission (1978; 1988), and Longworth and Knopke (1982). Changes in wheat marketing plans over the years do, however, mean that some of the earlier analyses are somewhat redundant.

Two elements of wheat marketing plans that remained intact over the years are home-consumption pricing and the payment of an equalized return to producers from domestic and export sales. It has been argued that this is done to stabilize producer returns.

Longworth and Knopke (1982) took into account other stabilization instruments besides home-consumption pricing and equalization. They calculated that the cumulative effects of wheat price policy over the period 1948/49 to 1978/79 constituted a net social welfare loss of $A677 million in constant (1979) dollar terms. They estimated this to be about equivalent to the returns from one average-size crop. They found that producers were net losers over the period, while domestic consumers were net gainers. This occurred because the authors included the first five-year marketing plan in their analysis. During that plan, domestic prices were held below world prices, resulting in substantial transfers to domestic consumers. If the first plan is omitted from the analysis, producers (consumers) become net gainers (losers), and the cumulative social costs are substantially lower at $A159 million.

A more recent study by Myers, Piggott, and MacAulay (1985) used econometric simulation to analyze the effects of Australian wheat price policy on the level and stability of key market variables, such as producer prices and industry revenues (Table 17.2). The authors concluded that only modest progress was made in achieving the stated objectives. In part, this is because home-consumption pricing proved to be a relatively ineffective instrument, given proportionately small domestic sales. In addition, in some stabilization plans, the buffer fund rules limited the extent to which stabilization funds could be used to moderate producer prices (see, for example, Campbell and Fisher, (1982: 130)). In a later study, Alston, Freebairn and Arch (1988) found a more significant stabilizing effect from hypothetical home-consumption pricing schemes for Australian wheat.

Policy Issues

Three key issues are discussed in this section. The first, and the one that has received most attention in recent times, concerns deregulation of marketing and sale arrangements. The second concerns the effects of protection for the manufacturing sector on the grain industry. While this issue has not received much scrutiny recently, it has in the past, and could come to the forefront again depending on movements in the exchange rate and the income levels of grain farmers. Finally, attention is devoted to Australia's international competitiveness in wheat markets in light of other countries' policies and Australian policy choices.

Deregulation

The most fundamental issue that has faced the Australian grain industries in recent years has been deregulation of storage, handling and transport as well as the selling of the crops. It has caused major debate within the industry and was responsible partly for a change in leadership of the agrarian-based Australian National party. It remains to be seen what will happen in the grains industries other than wheat, now that the latter has been led down a deregulatory path.

There have been calls for the deregulation of agricultural markets for many years (see, for example, Campbell (1977)), and, in the case of the grains industries, some deregulation was advocated by agencies such as the Industries Assistance Commission (1983), the Bureau of Agricultural Economics (1983), and others (for example, Hussey, 1986; Spriggs, et al., (1978)). The main arguments that have been advanced supporting deregulation revolve around the inefficiencies that stem from pooling costs and returns and the lack of a competitive atmosphere among marketing and handling authorities. The maintenance of domestic prices above export prices has also been criticized, not only because of the welfare effects on domestic consumers, but also because returns to producers are unlikely to increase much given the small domestic market.

The main arguments advanced in favor of the retention of monopoly grain marketing authorities are that such agencies enjoy a better relationship with government-controlled buying organizations in importing countries than do private firms; that the Australian monopolies can exercise some market power in export markets; and that competition in the grain markets would lead to the exploitation of producers.

In this debate, one has to recognize that there is little empirical evidence to support or refute any of these contentions. In essence, the questions are empirical ones. What is not clear is whether the benefits from enhanced export returns under the AWB system are greater than the associated costs of pooling. More work is needed to enable these questions to be resolved. Continued grower support for the AWB system suggests that growers believe they are benefiting.

There is strong evidence suggesting that the grain handling and rail transport industries are spatial natural monopolies (Quiggin and Fisher, 1988b). In terms of the traditional market failure criteria, the technical conditions underlying the grain handling and transport industry yield a prima facie case for intervention. Given the natural monopoly characteristics of the industry, three main policy alternatives are available: non-intervention; public ownership (perhaps with tendering out of some services); or regulation of the resulting monopoly. All approaches have significant disadvantages. The statutory monopoly solution, which was in place for almost half a century, has demonstrably significant costs. Because an unregulated environment will maintain some element of competition in both the handling and transport industries, owing to the use of on-farm storage and road transport and competition from interstate ports, it is conceivable that overall system costs could be lowered — perhaps in the order of $A10 per metric ton, as suggested by the Royal Commission.

Assistance to the Manufacturing Sector

Australia's agricultural export industries have long been concerned about the extensive protection offered to the manufacturing sector through tariffs and quantitative restrictions. These policies cause adverse effects on rural export industries, such as the wheat sector, for two important reasons. First, there are direct costs, because imported inputs used in agriculture are simply made more expensive. Further, tariffs can increase the cost of living, which results in higher labor costs for agriculture, owing to Australia's centralized wage-fixing system. As a result of Australia's more or less price-taker status on agricultural export markets, these cost increases cannot be passed on.

Tariffs also affect Australian export industries in a more indirect, but probably more important, manner. Tariffs on manufactured goods result in a higher value for the Australian dollar, with the net effect that a hidden tax is placed on exports. In the case of grains, whose world prices are set in U.S. dollars, farmers receive fewer Australian dollars per metric ton than would otherwise be the case. For commodities such as wool, which have prices established in Australian dollars, the tariffs make the product more expensive in terms of foreign currency, with the effect of "pricing it out of the market."

Recently, the adverse consequences of the tariffs were used as an argument in support of increased levels of assistance to rural industries, including the wheat industry. In other cases, they were used as a justification for the continuation of existing forms of assistance, such as the two-price scheme. These forms of aid came to be known as "tariff compensation"or "compensatory assistance."

Claims for compensatory assistance were made on the basis of efficiency as well as on equity grounds. Proponents argued that because of the tariffs, fewer resources were employed in relatively low-cost industries, such as wheat production, than would otherwise be the case. However, the efficiency argument for compensatory assistance to industries such as wheat may be more strongly grounded on the differential levels of assistance across Australia's rural industries, rather than between the agricultural and manufacturing sectors. This is based on the assumption that there are small resource flows between agriculture and the rest of the economy as noted in Pandey, Piggott, and MacAulay (1982), and possibly substantial resource flows among particular rural industries. For the wheat industry to justify continued or increased assistance, supporters need to demonstrate that aid would result in resource flows from higher cost (compared with wheat) industries rather than lower-cost industries. (For a summary of these issues, see Fisher (1985, 1986).)

In May, 1988 the Federal Government announced that a general program of tariff reductions would be implemented over the next few years. In particular, by July, 1992 industries with tariffs currently exceeding 15 percent (nominal rate) are to have their rates reduced to 15 percent or less, while those with tariffs currently between 10 percent and 15 percent will have their rates reduced to 10 percent or less. Tariffs protecting some agricultural industries (e.g., dairy, dried vine fruits, wine and citrus) will be among those reduced to 15 percent by July, 1992.

International Competitiveness

In recent years, farmers and policy advisors have been interested in the question of Australia's international competitiveness in world wheat markets, especially in light of European Community and U.S. trade policies. The Industries Assistance Commission (1988a) found that interventions by other countries in their wheat markets in 1986 reduced international prices by about 15 percent, causing a decline in average net farm income in the wheat-sheep zone of about 5 percent. More recently, Roberts and Love (1989) estimated the effects of the U.S. Export Enhancement Program on Australian wheat growers. They found that the program cost Australian wheat growers between $A215 million and $A337 million in 1987 and 1988 combined.

However, domestic policy choices also have an impact on Australia's international competitiveness. For example, the Industries Assistance Commission estimated that real net farm income in the wheat-sheep zone could be raised by 5 percent over about a two-year period if any of the following occurred: (1) real wages declined by 2 percent on an economy-wide basis; (2) on-farm productivity improved by 6 percent; (3) off-farm costs for grain storage, handling, and transport decreased by 45 percent; (4) real government expenditure on goods and services dropped by 54 percent; or (5) nominal rates of assistance for manufacturing were reduced by 70 percent. The Commission (1988a: 62) stated that "these domestic measures are not presented as alternatives. Their purpose is to illustrate the extent to which some domestic factors within the control of Australian policy makers could influence farm incomes in Australia's wheat-sheep zone to an extent similar to relaxation of other countries' agricultural interventions."The implication is that, while much attention is currently focused on the negative impact of other countries' policies on Australia's competitiveness as a wheat exporter, domestic policy choice has also had a negative effect.

Since the May, 1988 Economic Statement by the Federal Government, a package of microeconomic reforms has been pursued, particularly reforms to a number of work practices, including those on the waterfront. In implementing these reforms the government has had to proceed cautiously to avoid serious confrontation with the strong labor trade union movement.

18

The Australian Wheat Board[1]

The following material has been extracted with only very minor modification from a release on the new marketing arrangements (Department of Primary Industries and Energy 1989) made in August, 1989.

Operation of the Australian Wheat Board

The AWB will continue to accept into a pool all wheat offered to it, provided it meets certain classification and quality standards and is not seed wheat or wheat that has been already sold by the Australian Wheat Board.

Each season the Board determines the classification and quality standards to be used in classifying and pooling of wheat and these standards must be widely publicized to allow growers to make informed choices in regard to marketing their wheat.

A major change under the new arrangements is that the AWB is able to establish a number of pools for each season's wheat, if it so chooses. For example, it could run separate pools for categories of wheat of different qualities or it could differentiate pools according to time of delivery of wheat. These are matters for the AWB to decide in consultation with the Grains Council of Australia (GCA).

The new legislation continues the major functions of the AWB including control over the export marketing of wheat, payment to growers and borrowing monies to make those payments, domestic sales and arranging for receival, storage and transport of wheat; and growers still have full access to these services through the Australian Wheat Board. However, the AWB now has increased functions including contracting for, marketing and trading of wheat and grains other than wheat (produced in Australia or in other countries), trading and exporting wheat and grain products, and the power to store, handle and transport, wheat or grain and their products, or arrange for such services, or own or operate such facilities.

AWB Exempt from State Restrictions

The AWB is required to use the most cost effective methods available for storage, handling and transport of wheat and to account separately for these costs to growers as appropriate. The legislation enables the Commonwealth, after notifying the state concerned, to exempt the AWB (and other grain trading corporations) from certain restrictive state regulations which have the potential to increase the cost of marketing wheat and other grains destined for the inter-state or export trade.

The state legislation in question includes that requiring grain to be carried by rail, monopoly control over grain handling and port operations and state vesting of wheat in state authorities.

Delivery or Marketing of Wheat by a Grower

From July 1, 1989, growers have a choice of delivering their wheat to the AWB (via an authorized receiver) for inclusion in a pool or of selling it for cash to the AWB or to a merchant, miller or any other person. Wheat which is not delivered to a pool will be traded in the domestic market except that some cash purchases by the AWB may be exported through a commercial arrangement between its pools and its cash trading operation.

Permits or grower-to-buyer authorizations are no longer required for sales outside the pool except in Queensland where the compulsory acquisition powers of the State Wheat Board have been retained under state legislation.

Warehousing, contract growing, or other marketing arrangements may be put in place by various participants in the industry at their commercial discretion. The AWB has set up a cash trading facility to compete for wheat not delivered to a pool and to purchase other grain. This operation must be kept separate from its pooling operation. The AWB is offering contracts for specialized lines of wheat as well as forward selling contracts.

Payment for Wheat and Underwriting

(a) Underwritten Advance Payments

The Minister is required to make at least two determinations of the aggregated net return for each season, the first before October 1, and the second around March 1, as the basis for the government's guarantee of borrowings. The percentage figure applicable to the guarantee will phase down to 80 percent in the fifth season of these arrangements.

The legislation does not set the level or timing of advance payments. The AWB is currently offering the following options:

- advance payment plus subsequent payments;
- advance payment plus cashout;
- advance payment plus equity certificate; and
- deferred payments.

The AWB continues to determine allowances for quality, variety, storage, transport and other relevant factors, including port services and blue water costs on a port zone basis, to be accounted for in the advance or later payments.

(b) Final Payments

Pools may be closed at any time when the AWB considers that the returns will not be significantly changed by continuing the pool. At that time the Board will determine the actual net return per ton for the pool and make a final payment to growers if that return exceeds the total of advance payments already made.

Growers may apply for a payment in lieu of a final payment or request a certificate of entitlement stating a grower's equity in one or more open pools.

Payment for Nonpool Wheat. The price for wheat purchased by the AWB cash trading facility, merchants, millers or others is determined by negotiation between the grower and the buyer on the basis of the market conditions at the time of sale and the provision of services such as storage and transport.

Grower Levies

All wheat sold or processed for or by a grower is subject to a levy which is divided between the Wheat Research Trust Fund and the Wheat Industry Fund at levels determined each season by the GCA. From July 1, 1989, the total rate of levy has been set at 2.5 percent of the net value of wheat sold or processed. Of this, 2 percent is apportioned to the Wheat Industry Fund and the remaining 0.5 percent to research.

Exemptions from the levy apply for small quantities of production and where processed products are used for domestic purposes only. The purchaser or receiver is required to pay the levy on behalf of growers and may deduct the amount from the price paid for wheat.

Wheat Industry Fund

The Wheat Industry Fund is a source of finance to be used as collateral or security to enable the AWB to develop and undertake commercial activities outside of the conduct of pools. The fund may also be used with approval of the GCA for other purposes such as to increase the level of advance payments above the level provided by the government borrowing guarantee, or to provide insurance for credit sales. Growers hold equity in the fund.

The Government is providing a five year guarantee of borrowings up to $100 million to assist in the establishment of the fund.

Notes

1. Extracts from a press release on Australia's new wheat marketing arrangements operative from July 1, 1989.

Bibliography

Alston, J.M., J.W. Freebairn and A.M.J. Arch. *Effects of Producer Price Equalization in Victorian Agriculture*. Research Report Series No.63, Melbourne: Department of Agriculture and Rural Affairs, June 1988.

Australian Bureau of Agricultural and Resource Economics, *Agriculture and Resources Quarterly*, 1/3 (1989).

_____. *Commodity Statistical Bulletin* (December 1988a).

_____. *Quarterly Review of the Rural Economy* 10/4 (1988b).

_____. *Crop Report* (November 1988c).

Bureau of Agricultural Economics. "Wheat Marketing in Australia: An Economic Evaluation". BAE Occasional Paper No. 86. Canberra: Australian Government Publishing Service, 1983.

_____. *Commodity Statistical Bulletin* (December 1986).

Bureau of Transport Economics. "Competition and Regulation in Grain Transport: Submission to Royal Commission". Occasional Paper 82. Canberra: Australian Government Publishing Service, 1987.

Campbell, K.O. "Are Statutory Marketing Boards Really Necessary? Some Relevant Issues". *Commodity Bulletin* 5(1977): 1-15.

Campbell, K.O. and B.S. Fisher. *Agricultural Marketing and Prices*. Melbourne: Longman Cheshire, 1982.

Davidson, B.R. "The Economic Structure of Australian Farms". In D.B. Williams (ed.), *Agriculture in the Australian Economy*. Sydney: Sydney University Press, 1982: 29-54.

Department of Primary Industries and Energy. "Wheat Marketing Arrangements From July 1, 1989". Mimeo, Canberra, August 1989.

Fisher, B.S. "An Analysis of Storage Policies for the Australian Wheat Industry". Research Report No. 7, Department of Agricultural Economics, University of Sydney, 1978.

_____. "Frontiers in Agricultural Policy Research". *Review of Marketing and Agricultural Economics* 53(1985): 74-84.

_____. "The Rural Recession: An Assessment and an Analysis of Some Policy Options". *Australian Quarterly* 58(1986): 146-52.

Harris, S., J.G. Crawford, F.H. Gruen, and N.D. Honan. "The Principles of Rural Policy in Australia: A Discussion Paper". Canberra: Australian Government Publishing Service, 1974.

Hawke, R.J.L. and J. Kerin. *Economic and Rural Policy.* Canberra: Australian Government Publishing Service, 1986.

Hussey, D. "Australian Grain Marketing: Its Economics and Politics". Policy Paper No. 6, Australian Institute for Public Policy, Perth, 1986.

Industries Assistance Commission. *Wheat Stabilization.* Canberra: Australian Government Publishing Service, 1978.

_____. *The Wheat Industry.* Report No. 329. Canberra: Australian Government Publishing Service, 1983.

_____. *The Rice Industry.* Report No. 407. Canberra: Australian Government Publishing Service, 1987.

_____. *The Wheat Industry.* Report No. 411. Canberra: Australian Government Publishing Service, 1988.(a).

_____. *Annual Report 1987/88.* Canberra: Australian Government Publishing Service, 1988(b).

Longworth, J.W. "The Stabilization and Distribution Effects of the Australian Wheat Industry Stabilization Scheme". *Australian Journal of Agricultural Economics* 11(1967): 20-35.

Longworth, J.W. and Phillip Knopke. "Australian Wheat Policy, 1948-79: A Welfare Evaluation". *American Journal of Agricultural Economics* 64(1982): 642-54.

Myers, R.J., R.R. Piggott, and T.G. MacAulay. "Effects of Past Australian Wheat Price Policies on Key Industry Variables". *Australian Journal of Agricultural Economics* 29(1985): 1-15.

National Farmers' Federation. *Australian Agricultural Yearbook 1986.* Melbourne: Publishing and Marketing Australia, 1986.

_____. *Australian Agriculture: The Complete Reference on Rural Industry.* Melbourne: Morescope Pty. Ltd. 1987.

Office of Technology Assessment — U.S. Congress. *Grain Quality in International Trade: A Comparison of Major U.S. Competitors*, OTA-F-402 (Washington, DC: U.S.Government Printing Office, February 1989).

Pandey. S., R.R. Piggott, and T.G. MacAulay. "The Elasticity of Aggregate Australian Agricultural Supply: Estimates and Policy Implications". *Australian Journal of Agricultural Economics* 26(1982): 202-19.

Quiggin, J. and B.S. Fisher. "Pooling in the Australian Grain Handling and Transport Industries". In Brian S. Fisher and John Quiggin (eds.), *The Australian Grain Storage, Handling and Transport Industries: An Economic*

Analysis. Research Report No. 13, Department of Agricultural Economics, University of Sydney, 1988(a): 47-53.

_____. "Market and Institutional Structures in the Grain Handling Industry: An Application of Contestability Theory". In Brian S. Fisher and John Quiggin (eds.), *The Australian Grain Storage, Handling and Transport Industries: An Economic Analysis*. Research Report No. 13, Department of Agricultural Economics, University of Sydney, 1988(b): 33-46.

Roberts, I. and G. Love. "Some International Effects of the U.S.Export Enhancement Program". *Agriculture and Resources Quarterly* 1 (1989): 170-81.

Royal Commission into Grain Storage, Handling and Transport. *Executive Summary and Recommendations*. Canberra: Australian Government Publishing Service, February, 1988(a).

_____. *Volume 1 Report*. Canberra: Australian Government Publishing Service, February, 1988(b).

_____. "Institutional Arrangements". In *Volume 2 Supporting Papers*. Canberra: Australian Government Publishing Service, February, 1988(c): Supporting Paper 2.

_____. "Pricing Practices". In *Volume 3 Supporting Papers*. Canberra: Australian Government Publishing Service, February, 1988(d): Supporting Paper 6.

Sanderson, B.A., J.J. Quilkey, and J.W. Freebairn. "Supply Response of Australian Wheat Growers". *Australian Journal of Agricultural Economics* 24(1980): 129-40.

Scobie, G.M. and P.R. Johnson. "The Price Elasticity of Demand for Exports: A Comment on Throsby and Rutledge". *Australian Journal of Agricultural Economics* 23(1979): 62-66.

Smith, A.W. and J.P. Brennan. "The Effects of Wheat Delivery Quotas on Wheat Supply". Paper presented to the 22nd Annual Conference of the Australian Agricultural Economics Society, Sydney, February 1978.

Sniekers, P. and G. Wong. "The Causality Between U.S.A. and Australian Wheat Prices". *Review of Marketing and Agricultural Economics* 55(1987): 37-50.

Spriggs, J. *An Econometric Analysis of Report Supply of Grains in Australia*. Foreign Agricultural Economics Report No. 150, United States Department of Agriculture, Washington D.C., 1978.

_____., J. Geldard, W. Gerardi, and R. Treadwell. "Institutional Arrangements in the Wheat Distribution System". BAE Occasional Paper 99, Australian Government Publishing Service, Canberra, 1987.

PART FOUR

Canada

Grain Marketing, Institutions, and Policies

Andrew Schmitz and W.H. Furtan

19

Background

Canada has a small, open economy. The 27 million Canadians enjoy a relatively high standard of living by international standards. This high living standard has been dependent upon a productive labor force, an abundance of natural resources and favorable trading relations.

The Canadian system of government is a parliamentary democracy. The power of government is invested in the Prime Minister and the Cabinet who are members of The House of Commons. The highest judicial body, the Supreme Court, is appointed by the Prime Minister. The Senate, Canada's upper house, is also appointed by the Prime Minister. Canada has 10 provinces and two territories, all with a parliamentary system of government. Canada has two official languages, French and English.

The political parties in Canada are the Conservatives, the Liberals, and the New Democratic Party. These three parties are all national, and they contest elections in all provinces and territories. A number of regional parties represent regional or special interests. However, only the Conservatives and Liberals have ever held government at the national level. Canada is a member of the British Commonwealth of Nations.

General Economic Conditions

The Canadian economy is market-driven. Government intervention has taken the form of state-owned corporations, regulation, and extensive social welfare programs. In recent years, the number of state-owned corporations has declined as many of these companies (e.g., Air Canada) have been sold to the private sector. The government has used regulation in the financial sector, the transport sector, the agriculture sector, and others; however, the current Conservative government has moved to reduce the

level of regulation. Social programs, such as the nationally funded medi-
care program and unemployment insurance program, have remained intact.

Economic Factors

While the Canadian economy has grown by about 3 percent per year, it
has experienced high rates of inflation and unemployment as well as high
interest rates (Table 19.1). Consumer prices rose fourfold between 1967
and 1986 (Table 19.1). Inflation was particularly high from 1975 to 1983
when prices increased at approximately 7 percent per year. Unemployment in Canada is currently around 10 percent, double the level of the early 1970s. The 10 percent unemployment rate is partly the result of a rapid
expansion in the size of the labor force in the the mid 1970s, and a subsequent decline in numbers from the peak reached during that period. Currently, unemployment is low (approximately 3 percent) in Ontario and

TABLE 19.1.   General Statistics on the Canadian Economy, 1967-88

Year	Consumer price index, all items	Unemployment rate	Gov't of Canada average bond yield 10 yr.	Gov't annual deficit
		%	%	($'000,000)
1967	36.5	3.8	5.94	(108)
1968	37.9	4.5	6.75	(36)
1969	39.7	4.4	7.58	994
1970	41.0	5.7	7.9	247
1971	42.2	6.2	6.95	(139)
1972	44.2	6.2	7.23	(530)
1973	47.6	5.5	7.56	434
1974	52.8	5.3	8.90	1,268
1975	58.5	6.9	9.04	(3,823)
1976	62.9	7.1	9.18	(3,337)
1977	67.9	8.1	8.70	(7.343)
1978	73.9	8.3	9.27	(10,854)
1979	80.7	7.4	10.21	(9,383)
1980	88.9	7.5	12.48	(10,663)
1981	100.0	7.5	15.22	(7,315)
1982	110.8	11.0	14.26	(20,281)
1983	117.3	11.9	11.79	(24,993)
1984	122.3	11.3	12.75	(30,024)
1985	127.2	10.5	11.04	(31,685)
1986	132.4	9.6	9.52	(24,069)
1987	138.2	8.8	9.20	(23,373)
1988	143.8	7.8	10.22	(19,688)

Source: Canadian Statistical Review, 11-003E. Monthly Statistics Canada, National Income and Expenditure Accounts , Cat. no. 13-001 .

Quebec, but it is much higher in the rest of the country. The government bond rate is currently 10 percent, up from 7 percent in the 1970s. In the early 1980s, the government raised the rate on 10-year bonds to over 15 percent to reduce the rate of inflation and to slow the economy.

In discussing Canadian interest rates, it must be remembered that much of the Canadian economy is based on resource commodities such as forestry, agriculture, mining, and petroleum. These sectors of the economy are particularly hard hit by high interest rates. The western provinces of Canada, which produce mainly resource products, are affected more than other areas. The interest rate policy used to control inflation is not evenly distributed across the country because of the regional nature of production.

Budget Deficit

The Canadian government has run a budget deficit since 1970, and reached an annual deficit of over Can$30 billion in 1984 (Table 19.1). This deficit is financed by borrowing from both within and outside of Canada, and by the printing of money. The most rapidly growing sector of the government budget is interest payments (currently about 5 percent). Unfortunately, all 10 provincial governments, and a number of state-owned companies, are also running budget deficits. The total public debt is thus underestimated if only the federal deficit is considered.

Market Size

Canada has a small domestic market and thus has to depend upon international trade. The United States is Canada's most important market, followed by the European Community and Japan. Merchandise exports and imports (Table 19.2) have provided a favorable trade balance on the current account in a few years. The recent slowdown in the growth of exports has led to concern about Canada's international competitiveness. To increase exports, Canada has been an active supporter of more liberalized world markets at the General Agreement on Tariffs and Trade (GATT). The U.S.-Canada Free Trade Agreement (FTA), signed in 1989, will open the U.S. and Canadian markets to each other and thus stimulate trade and economic growth.

The exchange rate between the U.S. dollar and the Canadian dollar is shown in Table 19.2. Between 1967 and 1971, the Canadian dollar traded at a premium to the United States; however, in most years the U.S. dollar has been at a premium to the Canadian dollar. The two economies are closely linked; the Canadian monetary authorities usually follow U.S. monetary policy.

TABLE 19.2 General Trade and Exchange Statistics for Canada, 1967-1986

Year	Merchandise Exports (including re-exports)	Merchandise Inputs $ '000,000	Current Account	Net capital Account	US-Canada Exchange Rate
1967	11,419.9	10,872.6	-499	1,020	1.077
1968	13,679.0	12,358.1	-97	1,230	1.077
1969	14,871.1	14,130.4	-917	1,201	1.077
1970	16,820.1	13,951.9	1,106	811	1.044
1971	17,782.4	15,313.7	431	1,694	1.010
1972	20,222.0	18,271.5	-386	2,060	0.991
1973	25,648.8	22,725.7	180	75	1.000
1974	32,738.2	30,903.0	-1,460	2,351	0.978
1975	33,616.4	33,961.6	-4,757	5,555	1.017
1976	38,166.4	36,607.5	-3,842	8,076	0.986
1977	44,495.1	41,523.0	-4,301	4,885	1.063
1978	53,361.0	49,047.9	-4,935	4,318	1.141
1979	65,581.5	61,157.0	-4,894	8,851	1.171
1980	76,680.5	67,902.6	-1,904	2,418	1.169
1981	84,432.1	77,139.9	-6,131	14,587	1.199
1982	84,392.7	66,738.5	2,824	-713	1.234
1983	90,555.6	73,098.3	3,066	2,626	1.232
1984	111,729.8	91,492.5	3,437	3,615	1.295
1985	119,566.4	102,640.7	-1,186	6,991	1.366
1986	120,593.4	110,205.2	-9,268	13,219	1.389

Source: Canadian Statistical Review 11-003E Monthly Statistics Canada, National Income and Expenditure Accounts, Cat no. 13-001 .

Education

The educational system in Canada is a provincial responsibility according to the constitution (i.e., The British North America Act). There are substantial differences in the educational system among provinces; however, in all provinces education from grades 1 to 12 (13 in some provinces) is supported by public funds. Postsecondary education is funded partly from public monies and partly through tuition fees. The proportion of public to private funding varies from one institution to another and from one province to another.

The level of education for farm workers in the rural areas of Canada is generally rather low (Table 19.3). While the number of years of schooling in the rural areas has increased, only a small percentage of farm workers have a college education.

TABLE 19.3. Number of Agriculture Graduates Per 10,000 Male Farm Workers In Canada

1960	1970	1980
9.35	17.78	75.81

Source: Hayami Y., and V.W.Ruttan. *Agriculture Development:An International Perspective*, John Hopkins Press, 1985.

The Agricultural Sector

Since the late 1970s, agriculture's share of Canada's gross domestic product (GDP) has varied between a low of 2.2 percent in 1978 and a high of 3.0 percent in 1981 and 1982. Agriculture's share of GDP is currently approximately 2.5 percent (Table 19.4), an increase of about 1 percent over the average for the 1960s. The agricultural sector as a whole is growing at about the same rate as the overall economy.

TABLE 19.4. Gross Domestic Product for Canada and Canadian Agriculture: 1965-1987.

Year	GDP for Canada	GDP for Agriculture	GDP for Agriculture as % of total GDP
	(millions of dollars, 1981=100)		
1965	175,359	2,258	1.3
1966	187,263	2,886	1.5
1967	192,752	2,259	1.2
1968	203,072	2,423	1.2
1969	213,946	2,607	1.2
1970	219,498	2,472	1.1
1971	232,137	2,791	1.2
1972	245,441	2,967	1.2
1973	264,369	4,602	1.7
1974	276,006	5,780	2.1
1975	283,187	6,147	2.2
1976	300,638	5,905	2.0
1977	311,504	3,070	1.0
1978	325,751	7,012	2.2
1979	338,362	8,267	2.4
1980	343,384	9,116	2.7
1981	355,994	10,611	3.0
1982	344,543	10,267	3.0
1983	355,445	9,168	2.6
1984	377,865	10,106	2.6
1985	395,217	9,508	2.4
1986	407,736	11,580	2.8
1987	424,136	11,767	2.7

Source: Statistics Canada. *Quarterly Economic Review*, Annual Reference Tables,Cat 15-001, Ottawa.

TABLE 19.5. Number of Farms and Land Area Classified by Tenure of Operator, Canada, 1971 and 1981.

	1971		1981	
Item	*Number*	*%*	*Number*	*%*
Total No. of Farms	366,128	100	318,361	100
Owner	251,066	69	201,657	63
Tenant	19,200	5	19,847	6
Part owner, Part				
tenant	95,862	26	96,847	30
Total acres	169,668,614	100	162,815,073	100
Owner (acres)	75,852,208	45	63,784,275	39
Tenant (acres)	9,487,446	6	11,135,054	7
Part owner, Part				
tenant	84,328,960	49	87,895,629	55
Share owned	46,116,736	55	48,782,858	56
Share rented	38,212,224	45	39,112,771	44
Total acres				
owned	121,968,994	72	112,567,113	69
Total acres				
rented	47,699,670	28	50,247,825	31
Average farm				
size (acres)	463		511	
Owner (acres)	302		316	
Tenant (acres)	494		561	
Part owned, Part				
tenant (acres)	880		907	
Share owned (acres)	481		504	
Share rented (acres)	399		404	

Source: Statistics Canada. *Census of Agriculture, 1971 and 1981*, Ottawa.

The Canadian farm structure is divided into three ownership classes: owners, tenants, and part owners/tenants. The majority of land is owned by owner-operators (Table 19.5). Approximately 69 percent of the farmers own all their land and only 5 percent are full-time tenants. This ratio has remained constant over the last 20 years.

Farms are becoming larger and more mechanized. As a result, the percentage of the total labor force working in agriculture has dropped, from 6.89 percent in 1968 to 3.55 percent in 1988 (Table 19.6), and the number of people working directly in agriculture has also decreased. The agricultural labor force as a percentage of the total labor force differs among provinces;

TABLE 19.6. Total National and Agricultural Labor Force, Canada: 1965-1988

Year	Total Labor Force ('000 persons)	Agricultural Labor Force	Agric. Labor as % of Total Labor Force
1965	7,141	594	8.32
1966	7,420	544	7.33
1967	7,694	559	7.27
1968	7,919	546	6.89
1969	8,162	535	6.55
1970	8,374	511	6.10
1971	8,631	510	5.90
1972	8,891	481	5.41
1973	9,279	467	5.03
1974	9,662	473	4.90
1975	9,974	483	4.84
1976	10,203	472	4.63
1977	10,500	464	4.42
1978	10,895	474	4.35
1979	11,231	484	4.31
1980	11,573	479	4.14
1982	11,92	511	4.29
1983	12,109	516	4.26
1984	12,316	514	4.17
1985	12,532	511	4.08
1986	12,746	498	3.91
1987	13,011	452	3.47
1988	13,275	471	3.55

Source: Statistics Canada. *Labour Force Annual Averages*, Cat.71-529 (occasional), Ottawa.

Saskatchewan has the highest ratio and British Columbia and Ontario, the lowest.

A number of land lease arrangements are used in Canada. In the private market, the two lease arrangements are crop share and cash rent. Under the crop share agreement, usually one-third of gross revenue goes to the landlord and two-thirds to the tenant. The landlord pays taxes on the farm property and may pay part of the fertilizer and other cash costs. The amount of cash rent paid depends on the productive value of the land and the crops grown. Management decisions, such as what crops to grow, are usually made by the tenant in consultation with the landlord.

The government owns a large amount of farm land in Canada. Most of the government land is used for grazing and various grazing leases exist. Government land is rented on a cash basis; the rent is usually based on some fraction of the assessed value of the land. The government lease rates are always below private rates. In some provinces, such as Saskatchewan, the government has purchased land and then rented it back to younger farmers in order to help them get established in farming. These types of programs have proven controversial as they have had limited success.

20

Grain Production and Marketing

The Grain Sub-Sector

The major crops grown on the prairies are wheat, barley and canola (rapeseed), in that order. Other crops, such as flax, rye, and lentils, are also produced, but they do not contribute as much to gross cash receipts. In this report we focus on the three most important crops.

Production

The production of wheat in Canada has increased marginally in the past 20 years (Table 20.1). Any increase that has occurred has resulted from an increase in area rather than an increase in yields (see Table 20.2). Wheat yields in Canada have lagged behind those of Europe and the United States (Furtan et al.) and have been a constraint to increased production. Barley production has increased through increases in both area and yields (Table 20.3). Barley yields have increased through variety improvements, something not accomplished in wheat. The largest increase in production has been in canola (Table 20.4). The area of canola planted has increased fivefold in the past 25 years; and yields have increased as well. In 1988, the area planted in canola was about equal to that of barley for the first time in Canadian history (Table 20.2). The need to improve the oil quality of canola has been a constraint on increased yields, as the goal of improved oil quality often required a sacrifice in yield.

Exports

The production of agricultural commodities is very different between Eastern and Western Canada as shown in Figure 20.1. In the Prairie region,

TABLE 20.1. Canadian Wheat Supplies and Disposition: Crop Years 1968/69 to 1987/88 (in '000 metric tons)

Crop Year	Total Supplies	Domestic Dis. Farm	Comm.	Exports Wheat & Flour	Carryover July 31
1968-69	35,991	2,479	2,006	8,323	23,183
1969-70	41,450	2,166	2,402	9,430	27,452
1970-71	36,476	2,355	2,295	11,846	19,980
1971-72	34,393	2,435	2,351	3,720	15,887
1972-73	30,402	2,384	2,381	15,692	9,945
1973-74	26,107	2,280	2,292	11,446	10,089
1974-75	23,393	2,016	2,560	10,779	8,038
1975-76	25,119	2,396	2,408	12,336	7,979
1976-77	31,566	2,523	2,289	13,436	13,318
1977-78	33,176	2,460	2,561	16,040	12,115
1978-79	33,251	2,466	2,790	13,084	14,911
1979-80	32,107	2,688	2,809	15,889	10,721
1980-81	30,013	2,732	2,509	16,262	8,510
1981-82	33,313	2,831	2,322	18,447	9,713
1982-83	36,449	2,602	2,492	21,368	9,983
1983-84	36,488	3,191	2,342	21,765	9,190
1984-85	30,389	2,914	2,294	17,583	7,598
1985-86	31,850	2,958	2,583	17,725	8,584
1986-87	39,951	3,956	2,481	20,783	12,731
1987-88	38,681	5,128	2,836	23,515	7,202

Source: Canadian Wheat Board (1989). *Annual Report 1988*, Canadian Wheat Board,Winnipeg.

production is concentrated in grain and livestock while, in the east, dairy and poultry make up a larger part of net farm income. This difference in production mix is important for at least three reasons. First, the farm programs that support dairy and poultry are production and import quotas; grain and livestock programs are supported by deficiency-type programs. Second, grain and livestock producers stand to gain from freer world trade, but dairy and poultry producers would likely lose from more liberalized world markets. This difference has led to a policy conflict in Canada. Canada's position on freer trade (GATT) reflects this dichotomy within Canadian agriculture. Third, the Prairie region, with its export orientation, is subject to a high degree of uncertainty due to changing world market forces caused by droughts, policy changes and the like.

Export markets for wheat have changed significantly in the past 20 years (Table 20.5). In the past, Canada's wheat exports were almost exclusively high protein wheats. World wheat trade in low protein wheats has expanded much faster than the demand for high protein wheats (Table

TABLE 20.2. Acreage and Yield of Canadian Wheat, Barley and Canola, 1965-1988.

Year	Wheat		Barley		Canola	
	Acres	Yield	Acres	Yield	Acres	Yield
		kg/ac.		kg/ac.		kg/ac.
1965	27,892	621	5,893	766	1,435	356
1966	29,293	754	7,160	862	1,525	383
1967	29,671	528	7,780	666	1,635	344
1968	29,018	593	8,500	786	1,056	418
1969	24,550	727	8,970	856	2,022	376
1970	12,075	708	9,480	890	4,074	404
1971	18,994	735	13,508	930	5,341	406
1972	20,915	672	12,050	897	3,318	397
1973	23,215	678	11,520	851	3,205	382
1974	21,570	588	11,370	738	3,160	368
1975	22,855	716	10,590	856	4,520	407
1976	27,165	841	10,302	980	1,778	471
1977	24,275	778	11,330	1004	3,590	550
1978	25,670	803	10,060	979	6,980	501
1979	25,380	644	8,730	973	8,420	405
1980	27,060	680	10,950	971	5,140	483
1981	30,056	709	12,730	1006	3,463	530
1982	30,600	859	11,875	1085	4,390	512
1983	33,160	772	9,830	945	5,750	453
1984	31,860	635	10,395	885	7,610	447
1985	33,220	694	10,800	976	6,875	504
1986	34,310	878	10,810	817	6,430	577
1987	32,630	769	11,240	1119	6,560	582
1988	13,025	464	9,210	981	8,960	450

Source: Canadian Wheat Board (1989). *Annual Report 1988*, Canadian Wheat Board, Winnipeg.

20.6) and offers a larger potential growth market which should be exploited. Recently, Canada licensed some low protein wheats and this will lead to higher returns to Canadian farmers (Ulrich, Furtan and Schmitz 1986).

When Canada first exported wheat, the United Kingdom and other Western European countries were the primary markets. Today the Soviet Union and China import approximately 50 percent of Canada's wheat exports. With both China and the Soviet Union attempting to increase grain production through policy reform, Canada is in a vulnerable position with regard to future export markets (Carter, McCalla, and Schmitz 1989).

Barley exports from Canada are much smaller than wheat; in 1987/88 approximately 4.5 million metric tons were exported with 1.7 million metric tons going to Saudi Arabia (Canadian Wheat Board Annual Report). Over one-half of the barley produced in Canada is used in the domestic

TABLE 20.3. Canadian Barley Supplies and Disposition: Crop Years 1968/69 to 1987/88 (in '000 metric tons)

Crop Year	Total Supplies	Domestic Dis.		Exports Wheat & Flour	Carryover July 31
		Farm	Comm.		
1968/69	9,949	3,876	1,200	575	4,298
1969/70	12,381	4,602	1,391	1,923	4,465
1970/71	13,354	5,158	1,145	3,910	3,141
1971/72	16,239	6,121	1,270	5,020	3,828
1972/73	15,112	6,081	1,231	3,598	4,202
1973/74	14,420	5,576	1,531	2,776	4,537
1974/75	13,327	4,557	1,653	3,013	4,104
1975/76	13,614	4,837	1,688	4,326	2,763
1976/77	13,276	4,634	1,641	3,783	3,218
1977/78	15,020	4,582	1,690	3,540	5,208
1978/79	15,605	4,943	1,967	3,800	4,895
1979/80	13,373	5,139	2,142	4,086	2,006
1980/81	13,400	4,899	1,777	3,521	3,203
1981/82	16,927	5,370	1,424	6,002	4,131
1982/83	18,096	5,631	1,688	5,648	5,129
1983/84	15,338	5,656	2,174	5,537	1,971
1984/85	12,267	5,434	1,896	2,781	2,156
1985/86	14,543	5,713	1,731	3,795	3,304
1986/87	17,878	6,643	1,342	6,718	3,172
1987/88	17,129	7,285	1,555	4,594	3,695

Source: Canadian Wheat Board (1989). *Annual Report 1988*, Canadian Wheat Board,Winnipeg.

livestock feeding industry (Rosaasen and Schmitz); thus, barley is not as important an export crop as wheat. Malt barley varieties are grown by 50 percent of the farmers even though only 10 percent of the barley is selected for malting. The price of malt barley is about twice that of feed barley, though it yields only two-thirds that of feed barley. Because farmers cannot be sure if their barley will be selected by the maltsters, they plant the malt varieties and hope to get their barley selected. This gives the maltsters an advantage over the producers because it leads to an excess supply of malt barley (Ulrich, Furtan, and Schmitz, 1985).

Canola exports go almost entirely to the Japanese crushers. The Japanese crushers are protected with a favorable canola oil import tariff that provides the Canadian producers of canola a captive market (Furtan, Nagy and Storey). As a result of this distortion, the Canadian crushing industry has never become well developed. However, while the U.S. market for canola oil appears very promising because of the lower level of cholesterol in canola oil, the Canadian crushers are struggling in this market be-

TABLE 20.4. Canadian Supply and Disposition of Canola/Rapeseed: Crop Years
1966/67 to 1986/87 (metric tons)

Year	Stocks July 31	Production	Domestic Crush	Exports
1966/67	76,340	585,134	112,599	313,387
1967/68	131,292	560,186	117,004	279,163
1968/69	225,050	439,984	157,260	324,568
1969/70	111,325	624,929	170,129	132,039
1970/71	83,529	1,637,467	194,478	1,061,654
1971/72	254,216	2,154,561	273,289	966,219
1972/73	978,386	1,299,555	353,170	1,226,050
1973/74	468,974	1,206,568	334,414	888,664
1974/75	280,912	1,163,476	275,968	592,987
1975/76	399,648	1,748,616	347,160	683,026
1976/77	1,048,648	836,886	549,714	1,017,871
1977/78	199,000	1,973,100	630,300	1,013,600
1978/79	325,000	3,497,100	725,100	1,642,295
1979/80	1,068,100	3,411,100	897,300	1,742,600
1980/81	1,476,900	2,483,400	1,003,300	1,372,600
1981/82	1,327,900	1,936,700	945,400	1,359,300
1982/83	692,400	2,246,000	904,100	1,271,300
1983/84	486,400	2,609,300	1,159,300	1,497,600
1984/85	119,600	3,427,900	1,290,400	1,455,900
1985/86	470,100	3,507,000	1,211,100	1,455,700
1986/87	950,000	3,798,000	1,991,000	2,126,000

Source: Agriculture Canada (1989). Market Commentary. Agriculture Canada, Ottawa.

cause they cannot purchase enough canola. The canola crushed
domestically in Canada is consumed in the Canadian market with only
small amounts of canola oil exported from Canada, and much of this is as
food aid. Canola meal, which is used for livestock and poultry, has not
been developed to the extent it can compete with soymeal in livestock rations.

Location

The location of production and cropping patterns in Canada's major
grain producing region, the Prairie region, depends largely upon the soil
zones (Figure 20.2). In the brown soil zone, wheat is produced in a wheat-
fallow rotation. The brown soil zone is short of moisture in most years and
farmers fallow as a method to conserve moisture. The dark brown soil
zone, which receives more moisture than the brown zone but less than the

FIGURE 20.1 Cash Receipts from the Sale of Selected Farm Commodities, Ontario and Quebec, and Prairie Provinces, 1986

Prairie Provinces

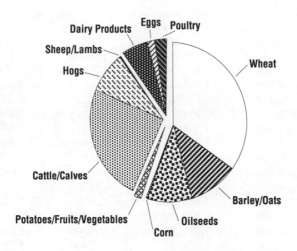

Ontario and Quebec

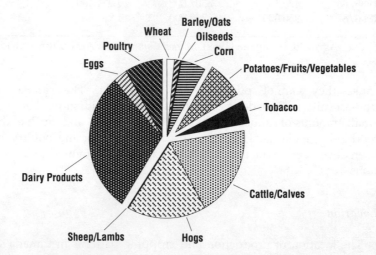

Source: Fulton, Rosaasen and Schmitz.

TABLE 20.5. Canadian Wheat Exports (Including Durum) by Select Countries: Crop Years 1978/79 to 1987/88. (in metric tons and percent)

Crop year	Country						
	EEC	% of Total	USSR	% of Total	China	% of Total	Total Wheat
1972/73	2,156	14.3	4,440	29.5	4,172	27.7	15,056
1973/74	2,323	21.3	979	9.0	1,462	13.4	10.902
1974/75	2,582	25.2	352	3.4	2,244	21.9	10,228
1975/76	1,999	17.2	3,236	27.8	1,207	10.4	11,637
1976/77	2,598	20.4	1,043	8.2	2,074	16.3	12,711
1977/78	2,859	18.7	2,146	14.0	3,469	22.8	15,246
1978/79	2,032	16.5	1,429	11.6	3,102	25.2	12,299
1979/80	2,183	14.3	2,579	16.9	2,516	16.5	15,212
1980/81	2,209	14.1	3,971	25.5	2,879	18.4	15,567
1981/82	2,042	11.3	5,019	27.9	3,101	17.2	17,972
1982/83	1,881	8.9	6,959	33.2	4,424	21.1	20,956
1983/84	2,037	9.5	6,761	31.7	3,428	16.1	21,283
1984/85	1,216	7.1	6,019	35.2	2,780	16.2	17,072
1985/86	1,458	8.4	5,219	30.1	2,558	14.7	17,311
1986/87	1,627	8.0	5,391	26.6	4,065	20.1	20,193
1987/88	1,321	5.7	4,500	19.4	7,586	32.7	23,172

Source: Canadian Wheat Board(1989). *Annual Report 1988.*—Wheat Board, Winnipeg.

black soil zone, is the transition area. All three crops — wheat, barley and canola — are produced in the dark brown zone. The cropping pattern in the dark brown zone is wheat-wheat-fallow, wheat-barley-fallow or wheat-canola-fallow; however, some farmers have adopted a much longer all-crop rotation through extensive use of farm chemicals.

Most of the canola and barley is produced in the black soil zone, largely because these crops respond to higher rainfall than wheat. Many farmers in the black soil zone do not incorporate fallow into their rotation except when required to control weeds. Wheat produced in the black soil zone is often of lower quality than that produced on the brown and dark brown soil because of the shorter growing season.

Production Costs

The cost of production for wheat, barley, and canola on fallow and stubble for each soil zone is shown in Table 20.7. To cover the total cost of production, farmers would need to receive at least Can$110 per metric ton. In the past few years, wheat prices have been below this level, which necessitated the introduction of government programs. As yields increase mov-

TABLE 20.6. Profile of Total Canadian Wheat Exports, 1957-1964 to 1974-1981 (Averages of Periods)

Period	H Class	M Class	S Class	D Class	Total
Canadian Exports (thousand metric tons)					
1957-1964	8,086	0	213	603	8,902
1965-1973	10,431	1	128	987	11,547
1974-1981	10,498	345	1,267	1,801	13,911
World Trade (thousand metric tons)					
1957-1964	8,834	15,924	5,977	1,008	31,743
1965-1973	13,678	20,790	7,679	2,532	44,679
1974-1981	14,992	29,779	17,082	3,585	65,438
Canadian Market Share (%)					
1957-1964	91.3	0	3.0	67.1	28.5
1965-1973	76.4	0	1.7	36.7	25.8
1974-1981	69.8	1.2	7.5	50.2	21.3

Notes: The H class includes all No. 1, No. 2, and No. 3 Canadian Western Red Spring Wheat from Canada and all Hard Red Spring Wheat from the United States. The M class includes all No. 1 and No. 2 Canadian Utility Wheat from Canada, Hard Red Winter Wheat from the United States, and all exports from Australia and Argentina. The S class includes all Canadian Western Red Winter Wheat, No. 3 Canadian Utility and Eastern Wheat from Canada, all soft wheat exports from France, and Soft Red Winter and White Wheat from the United States. The D class is all durum wheat.
Source: Ulrich A.,W.H. Furtan, and A. Schmitz (1986). "The Effects of Licensing Regulation on Canadian Agriculture: The Example of HY320," *Journal of Political Economy,* 95,1, (Jan.), pp.10-32.

ing from the brown soil zone to the black soil zone, so do the costs of production. In the black soil zone, farmers have more weeds to contend with than in the brown soil zone, thus pushing up the costs (they also use more fertilizer). Also, harvest costs are higher in the black soil zone because of the need to dry the harvested grain before it can be stored to prevent losses in quality.

Farm Size and Age Structure

In Canada, farm sizes vary depending upon the region and on the commodities produced. The data in Table 20.8 is for the Prairie region of Canada where most of the grain is produced. The majority of farms (60 percent) are commercial, and they account for 77.5 percent of gross farm sales. Large corporate farms are only 2 percent of the total number of farms, but they produce 13.9 percent of gross farm sales. Marginal farms (those that are not economically viable) make up 38 percent of all farms in the Prairie region. The high percentage of marginal farms in the region is of major concern to policymakers.

Policymakers are also concerned about the age distribution of the farming population. Relatively few farmers are under the age of 35, and a significant number are over 64 (Table 20.8). There are at least two reasons for the aging of the farm population. First, farming has recently witnessed some difficult economic times. In 1985, the Prairie region experienced a major drought and a grasshopper infestation, both of which reduced crop yields. In 1986 and 1987, farmers faced the lowest prices for grains in history (Figure 20.3). In 1988, a drought in the Prairie region cut production by over 35 percent. In 1989, unprecedented rains at harvest reduced grain quality to record low levels. Young people are not ready to invest their lives in farming under conditions of uncertainty such as these. Many of them prefer to work in the off-farm labor market where the returns are perceived to be greater. The present situation contrasts sharply with the situation in the 1970s when farm incomes were high and producers had a high degree of optimism about farming.

The second reason for the aging of the farm population is that farming is virtually impossible to enter unless a parent/relative starts a son/relative in the business. The human and monetary capital requirements for a new entrant are too great to meet without sizable financial support.

For the above reasons, and because the available technology allows a single-farm family to farm larger areas, the number of farmers on the prairies will continue to decline, and those farms remaining will get larger.

The economic health of the Prairie grain economy is not strong. While government financial support (to be discussed later) has helped, the industry has experienced a general economic decline. This is best illustrated by the drop in capital values of farm assets by over 50 percent in nominal terms in the past seven years.

Market Structure

The pricing and marketing of wheat and barley differ substantially from that of canola. However, the transportation, storage, and delivery system is the same. In this report, we will first deal with wheat and barley, which come under the control of the Canadian Wheat Board (CWB or board). Canola is marketed through the Winnipeg Commodity Exchange (WCE).

The CWB is the sole marketing agency for the wheat and barley grown in the Prairie Region which is sold into the export market or on the domestic market for human consumption. The CWB may also market these grains into the domestic feed grain markets when additional supplies are required, but producer sales to the domestic feed grain market, handled by the private trade, are usually adequate.

Historical Background and Current Objectives

The CWB, established as a Crown Agency by the Canadian Wheat Board Act of 1935, was preceded by two earlier federal government marketing boards set up to market wheat during World War I. During World War II, the CWB was empowered to market all Canadian grains. After the war, it returned to marketing wheat, barley and oats. Prior to 1966, the CWB's statutory authority had to be renewed every 5 years. In 1966 this requirement was dropped and the Canadian Wheat Board Act became permanent legislation.

In 1974, the sale of western grains for use in animal feeds within Canada was removed from the CWB's jurisdiction and returned to the private sector. The CWB is the residual supplier in the domestic feed grain market as discussed in detail later. Oats was removed from the board's control and returned to the private sector in August, 1989.

FIGURE 20.2 Soil Zones of Western Canada

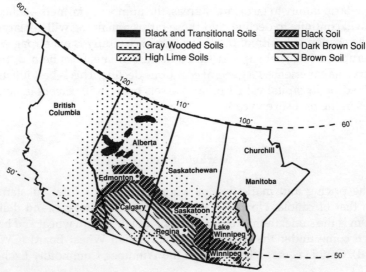

Source: Office of Technology Assessment.

TABLE 20.7. Production Costs for Wheat, Barley and Canola: The Prairie Provinces of Canada, 1988. (dollars per acre)

Crops	Soil Zone Rotation	Total Variable Costs[a]	Total Fixed Costs[b]	Total Costs
	brown zone			
wheat	fallow	40.67	54.23	94.91
barley	fallow	47.08	62.82	109.91
canola	fallow	47.86	66.87	114.73
wheat	stubble	46.51	53.91	100.42
barley	stubble	45.91	56.83	102.74
	dark brown zone			
wheat	fallow	39.50	59.36	98.86
barley	fallow	50.54	65.55	116.09
canola	fallow	45.43	61.65	107.28
wheat	stubble	48.46	56.15	104.61
barley	stubble	50.84	64.72	115.55
canola	stubble	62.63	59.57	122.20
	black zone			
wheat	fallow	51.20	68.42	119.62
barley	fallow	48.21	72.71	120.92
canola	fallow	54.26	67.90	122.16
wheat	stubble	61.37	67.23	128.60
barley	stubble	62.03	70.62	132.65
canola	stubble	68.24	67.74	135.98

a   Total variable costs include: seed, fertilizer, herbicides, insecticides, fuel, repairs, custom work, crop insurance, labor and operating capital charge.
b   Total fixed costs include: management costs, land costs, and equipment and building capital recovery costs.
Source: R.A.Schoney, Tom Thorson and Ward P. Wiesensel. *1988 Results of the Saskatchewan Top Management Workshops.* Dept. of Agricultural Economics, University of Saskatchewan, Bulletin: FLB 88-01, 1988.

The Canadian Wheat Board Act gives the CWB three major responsibilities:

- to market wheat and barley grown in the Prairie region to the best advantage of grain producers,
- to provide price stability to producers through an annual "pooling"or price-averaging system,
- to ensure that each producer obtains a fair share of the available grain market.

The CWB is a government sales agency as it owns no physical facilities for the handling of grain (Schmitz and McCalla). It employs the services of both private and cooperative elevator companies to carry out the logistics of the physical handling of grain. The CWB's responsibility is to bring the highest possible returns to producers and give them equitable access to the export market. Consumer welfare is not an overriding concern of the board.

The CWB is the world's largest single grain marketing agency. As Mc-Calla and Schmitz (1982) point out, state trading, especially in wheat, is common in many parts of the world.

First, the proportion of wheat trade which involves only private traders is small and declining, involving about 5 percent of the trade in 1973-1977. The reciprocal of course is that 95 percent of world trade in wheat involves a state trader on at least

TABLE 20.8   Number of Farms and Sales, By Farm Size and By Farm Operator, Prairie Region, Canada, 1985

Type of Farm	Number of farms		Farm sales	
	Number	Distribution (percent)	Average Amount (dollars)	Distribution (percent)
Small marginal farms	50,775	38.0	16,592	8.6
Part-time farmers				
Age<35	7,580	5.7	14,556	1.1
age 35 to 64	14,835	11.1	13,169	2.0
Full-time farmers				
age<35	5,890	4.4	19,566	1.2
age 35 to 64	22,465	16.8	18,759	4.3
Commercial farms	80,125	60.0	94,807	77.5
Part-time farmers				
Age<35	2,925	2.2	77,875	2.3
Age 35 to 64	5,010	3.7	84,469	4.3
Full-time partners				
Age<35	9,965	7.5	95,079	9.7
Full-time farmers				
Age<35	2,730	2.0	158,403	4.4
Age 35 to 64	43,260	32.4	111,747	49.3
Elderly farmers				
Age>64	16,235	12.2	45,062	7.5
Large corporate farms	2,610	2.0	522,498	13.9
Mostly family owned	1,955	1.5	471,717	9.4
Mostly owned by others	655	0.5	674,068	4.5
ALL FARMS	133,510	100.0	73,424	100.0

Source: Auer, L. *Canadian Prairie Farming, 1960-2000,* Economic Council of Canada, Ottawa.

FIGURE 20.3 Real Farm Price of Wheat, Prairie Provinces, 1926–1987

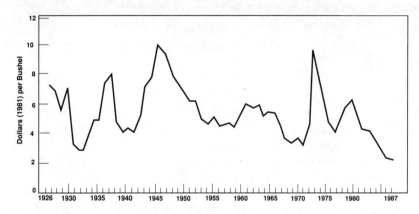

Source: Fulton, Rosaasen, and Schmitz.

one side of the transaction.... Secondly, state-trader to state-trader transactions account for about one-third of the trade and seem to be stable. Thirdly, the importance of state trading on the export side of the market is declining, reflecting (1) the rising importance of the United States as an exporter and (2) the discontinuance of state trading by Argentina. Fourthly, the rise in importance of state-trading importers is substantial, reflecting the rising importance of centrally planned and LDC importers in the wheat market, most of whom pursue state trading.

The CWB has three to five commissioners, who are appointed by the government and a staff of about 525. The commissioners periodically seek advice from an advisory committee elected by farmers, but the advisory group has no real control over the commissioners. Unlike the Australian Wheat Board, the Canadian Board is not directly responsible to producers. It answers to the federal government.

Most of the wheat produced in Western Canada is marketed through the CWB. About 95 percent of the wheat produced enters the primary elevator system; the remainder is used on-farm for feed or seed or is sold locally. Of the wheat that does enter the elevator system, 97 percent is delivered to the CWB and 3 percent is delivered to the private trade. Private traders are permitted only to buy feed wheat, which they subsequently sell on the Winnipeg cash and futures market. The CWB does not trade on the futures market.

Half of the barley produced in western Canada is fed to livestock. Most of this grain is either fed on farms or sold directly to feedlots and does not enter the country elevator system. The CWB, therefore, markets approximately 50 percent of the barley produced. The marketing process for barley is identical to that for wheat.

Marketing

The CWB markets grain in two basic ways. The largest portion of sales is made under contracts negotiated annually between the CWB and buyers. This is in contrast to earlier years when most sales were made through accredited exporters. Although this type of sale no longer dominates, most sales made by the CWB still involve a degree of participation by private trading companies operating as accredited exporters for the CWB.

Section 5(1) of the CWB Act states, "the Board shall sell and dispose of grain acquired by it pursuant to its operations under this Act for such prices as it considers reasonable with the objective of promoting the sale of grain produced in Canada in world markets." The board also has broad discretion in deciding the most appropriate selling methods for carrying out its mandate. The board, of course, is always subject to possible direction from the Federal Cabinet as has been the experience regarding domestic wheat pricing in recent years and regarding export wheat pricing in earlier years when the International Wheat Agreement (IWA) had pricing and supply obligations.

The world grain markets have changed significantly since the present CWB came into existence in 1935 (Carter et al.). The CWB's methods of export selling and pricing have changed also. Up to the early 1960s, the board posted and maintained stable offering prices at Canadian port positions. CWB stocks were sold for cash only, for the most part, to authorized exporters (the trade) who acquired ownership at those points and undertook the export merchandising of Canadian grain along with grains they acquired from other origins. Most of the export business was done in parcel or very small cargo lots. The primary customers were a large number of United Kingdom and European millers and merchandisers. Stable offering prices were possible due to the IWA price range negotiated periodically between major importing and exporting countries and cooperative action by the United States and Canada in carrying stocks and/or curbing production if necessary.

By the 1960s, many changes had occurred or were underway in the world grain industry. Import buying was concentrating into fewer and highly integrated organizations, private as well as state. Media attention focused on the large purchasing, shipping, and financing requirements of the Soviet Union and the People's Republic of China, which were initially served by the Australian and Canadian Wheat Boards due, in no small part, to U.S. foreign policy. Nevertheless, agricultural policymakers in the

United States resented growing commercial wheat markets for other exporting countries while they struggled to find aid outlets abroad and to curb production at home. U.S. support for IWA and export pricing cooperation evaporated. A new era in the international wheat trade had begun.

In response, the CWB moved to a higher volume, more market-oriented approach designed to meet the specific requirements of customers. This included new features such as direct selling, long-term supply assurances, 18- to 36-month credit, and more feed grain exports. The CWB continued its traditional emphasis on quality control and continued to strive to achieve the best return for western producers.

Direct Sales Contracts

Most sales of prairie wheat and barley into export markets are made by CWB teams negotiating directly with overseas buyers, both state and private. The specific sales contracts include the usual conditions such as price, quantity, grade, port, and shipping period. They might involve only one small cargo (less than 20,000 metric tons) or many large cargoes (over 35,000 metric tons) shipped over a few months. Such sales are usually made on the basis of an in-store or FOB position at a Canadian port, with cash payment at time of delivery of each cargo. The CWB negotiates and executes such contracts in much the same manner as private multinational firms. Prices reflect competitive factors at the time of negotiation. When export prices are volatile, the volumes and shipping periods covered by a particular negotiation tend to be reduced in order to minimize the risk for both the buyer and the seller. When export prices are stable, the volume and shipping periods under a single contract may cover the requirements for several months.

The CWB still makes indirect sales via authorized exporters (Canadian and multinational firms), depending on the availability of supplies relative to the board's forward commitments. These firms will respond to the board's posted offering prices at port positions with bids that may be over or under those levels and for quantities of less than cargo size or of several cargoes. In times of tight supplies, authorized exporters may be required to name the final destination before the board will consider countering their bids.

Long-Term Agreements

The CWB negotiates many of its export sales with a government agency responsible for essential food supplies and/or imported supplies for the client country. Government agencies of this type may want the security of an assured minimum supply and/or the availability of credit when making purchases over the coming few years. Similarly, the CWB will want certain minimum purchase assurances if they are to make a supply commitment for the future. The Canadian government will also seek minimum purchase assurances if credit is required and if the government will need to guarantee credit sales under subsequent contracts.

A long-term agreement (LTA) normally covers a three- to five-year period. The CWB undertakes to supply a specified quantity, and the buyer undertakes to purchase that quantity, over that period. Minimum annual quantities for particular types of grain and preferences regarding grades, ports, and shipping periods are usually indicated. These are to be specifically fixed in subsequent contract negotiations as are prices in relation to competitive factors at the time of each contract negotiation. The LTA will indicate if, and to what extent and under what terms, credit may be available for subsequent purchases.

Sales on Credit Terms

The availability of credit for periods of 18 to 36 months was a significant competitive factor in many grain markets in the 1960s. Even though commercial interest rates were relatively low, exporters (including Canada, in some cases) frequently provided subsidized interest rates. This latter practice ceased, at least for Canada, during the seller's market of the 1970s. Today the CWB offers credit only at the commercial rate at which it must borrow; it does not try to profit or lose if a customer needs credit. However, Canadian interest rates are now relatively high and most buyers, if they need credit, can find it cheaper from other sources.

The Canadian government guarantees the bank loans that the CWB makes to finance credit sales. Thus, if principal or interest payments due on a credit sale cannot be obtained from the buying country, the board's loan at the bank will be paid by the federal government. The federal government establishes maximum limits on the credit that can be extended to particular countries, and this may apply on commodities other than grains. The board will check to see if the limit is sufficiently high before negotiating a particular grain deal on credit terms.

Domestic Sales

As stated earlier, the CWB is the exclusive seller of Prairie wheat and barley sold domestically for human consumption; this basically means sales to millers and maltsters. In the case of wheat, the board has been directed on prices by the Cabinet since the early 1970s. Various two-price wheat plans, with domestic prices independent of export price levels, were employed, generally with the objectives of giving maximum price protection for consumers and minimum price protection for producers. Given present and expected developments resulting from the 1989 U.S.-Canada Free Trade Agreement, the board now is pricing wheat to domestic millers at prices that reflect North American values. They are also working with the milling industry to establish new shipping and invoicing procedures. Currently, domestic wheat prices for an upcoming 60- or 90-day period are established on the 15th day of the month preceding the start of the designated period. Mills then book wheat as they need it, drawing out of CWB-owned stocks and paying the established price for the period plus a surcharge for CWB carrying costs.

Market Development

The CWB is heavily involved in market development programs. Programs for particular countries frequently involve milling and baking tests and, in the case of feed grains, feeding trials to determine if grains available in Canada are suitable for the country's needs. If Canadian grains lack particular necessary qualities, the CWB holds discussions with plant breeders to determine whether new varieties possessing the desired qualities can be developed.

Producer Pricing and Policy

The CWB achieves price equalization for grain producers through a price pooling system. Receipts received by the CWB from sales of a particular grain are "pooled"in a single fund.

At the beginning of each crop year (August 1), the government establishes initial producer prices for grain sold to the CWB. These prices are announced in advance, normally in April, to allow farmers to adjust their seeding intentions. Separate prices are established for each grade of wheat. Receipts from CWB sales into the domestic and export markets are then pooled. Producers receive the initial payment at the time of delivery. In some years they receive an interim payment during the crop year (if prices strengthen considerably) and a final payment once the crop year is over. The pool is then closed and the CWB deducts its administrative expenses, interest costs, etc., from the pool. Each producer receives the price initially set (before freight deductions) no matter what date the grain was sold to the CWB within a particular crop year.

When selling to the CWB, producers' marketing costs are deducted in two stages. Freight costs and primary elevator handling costs are deducted from the initial payment at the time of delivery. Other costs (interest, insurance, storage, etc., and the board's operating costs) are later charged against the "pool"before the final payment is made to the farmers.

In addition to pooling, the CWB regulates producer deliveries to primary elevators through marketing quotas. There are no production quotas. The quota system is used to ensure that the kinds and quantities of grain needed to meet sales are delivered when required and that each producer receives a fair share of available markets. Marketing quotas are allocated on the basis of a farmer's seeded area. At the time of harvest, farmers generally have to store grain at their own cost since the CWB sets the initial quotes at levels significantly below production. However, through the 1980s, producers have sold their saleable quantities prior to the close of a given crop year.

In terms of pricing, the initial price for wheat set by the CWB is generally below the final pooled price. Also, the CWB at times makes an interim payment during a given crop year. Over the years, the amount the federal government has paid because of a deficit in the pooled account (i.e., when the initial price has been set above the final price actually received) has been small.

The Winnipeg Commodity Exchange

Canola is priced on the Winnipeg Commodity Exchange (WCE). The WCE was first established in Winnipeg in 1887 as the Western Grain and Produce Exchange. Forward contracting was introduced in 1889 and futures contracting in 1904. A futures contract for canola was started in 1963 and continues today. The exchange received its present name in 1972.

Farmers may hedge a portion of their expected canola crop at time of planting and then lift the hedge when they deliver the grain. However, most farmers sell their canola to an elevator company and do not deliver against a futures contract.

When elevator companies purchase canola from farmers, they either hedge it on the WCE or hold it for speculation. The canola is then moved into export markets or to domestic crushers. As most of the canola is exported to Japan it moves through either the port of Vancouver or the port of Prince Rupert.

Farmers may deliver their canola directly to domestic crushers. The CWB, which controls the quotas for canola, usually provides the crushers with a larger quota than the elevator companies. This is done to allow the domestic crushers continuous access to the canola supply.

Futures contracts for rye and flax are also traded on the WCE. Farmers, elevator companies, and processors use the WCE for these commodities in the same manner as for canola. Smaller, less important commodities, such as lentils, are produced under contract between farmers and processors.

Grain Handling and Transportation

Canadian farmers deliver most of their grain to country elevators. Canadian elevator numbers have declined from 5,800 in 1933 to 3,000 today. About half of them are owned by wheat pools (a form of producer cooperative) or cooperatives. The three wheat pools (Alberta, Saskatchewan, and Manitoba) own about 1,800 elevators and handle about 65 percent of the grain produced. The United Grain Growers (a grain cooperative) owns about 500 elevators and handles 20 percent of the grain produced. Private elevator companies own the remainder of the elevators. Because of costs, these elevators are capable of storing only a portion of farmers' grain at harvest. As a consequence, on-farm storage is substantial. The CWB's delivery quota system controls the flow of grain to the elevator system.

Canada has two transcontinental railway companies, Canadian Pacific Rail and Canadian National Rail, that move grain from elevators to export sites. Because of the location of production, grain has to be carried long distances overland before it can be exported. Canadian grain moves essentially in only two directions from point of production, either east or west.

Most of the grain produced in Western Canada is moved by rail rather than by truck or barge. The farmers deliver their grain to primary elevators located on rail lines. The government regulates rail freight rates, which have changed little in the last 90 years. The rates were fixed in 1900 and were not changed until 1983. Under the 1983 Western Grain Transportation Act, rates were allowed to increase. However, the freight rate at which railway companies can increase/decrease rates at particular points is still controlled by government legislation.

The private costs (i.e., those paid by farmers) of moving grain from country elevators to tidewater are given in Table 20.9. The farmers essentially receive a subsidy for rail transportation. The railway freight charge of Can$6.65 per metric ton represents only part of the full cost of moving grain. Under the Western Grain Transportation Act, the government contributes an additional Can$20 to Can$30 per metric ton to the cost of moving grain. Grain shipped from Saskatchewan receives a larger subsidy per metric ton than grain from Western Alberta because it is further from tidewater. The Canadian Grain Commission controls the costs of moving grain, such as elevator and terminal charges. The Commission determines the maximum tariff (or rates) companies can charge. The actual tariff is then set by the elevator companies.

The allocation of rail cars to particular elevators is made by the Western Grain Transportation Authority (WGTA) and the CWB on behalf of the Canadian Grain Commission. The CWB owns no marketing or transportation facilities. Rather, it contracts for these services with the national railroads and with the cooperative and private elevators. The CWB controls the wheat and barley delivered by farmers to country elevators by the

TABLE 20.9. Estimated Costs of Moving Wheat From a Mid-point to Export Position: 1978/79 to 1987/88 (dollars per metric ton)

	1978/79	1980/81	1982/83	1984/85	1986/87	1987/88
Primary Elevator						
Elevation	4.50	5.30	6.18	6.40	6.27	6.27
Removal of						
dockage	.92	1.00	1.06	1.13	1.67	1.67
Shrinkage	.40	.56	.48	.47	.32	.34
Carrying costs	3.53	7.21	4.36	6.25	2.49	2.30
Railway Freight	5.07	5.07	5.07	8.08	6.27	6.65
Marketing						
Interest,banking						
etc.	1.97	3.43	.32	1.76	.78	.20
CWB admin. costs	1.20	.80	.78	1.34	.86	1.34
Terminal Position						
Storage	.54	.55	.89	1.99	.83	1.00
Fobbing charges						
• via St. Lawrence						
ports	3.01	3.77	4.67	5.01	5.24	5.24
• via Pacific						
seaboard	3.12	3.82	4.74	5.08	5.23	5.23
Lake Trans.	10.21	14.40	16.90	18.32	17.55	17.88
Transfer Position						
Storage	.43	.40	.69	1.03	.99	1.19
Fobbing charges	1.10	1.30	1.47	1.60	2.27	2.15
Total						
via St.Lawrence	32.88	43.79	42.90	53.38	45.54	45.73
via Pacific	21.25	27.74	23.91	32.50	24.72	24.50

Source: Canadian Grain Commission.

quota system discussed earlier, and coordinates logistics with national railroads. Grain cars are allocated to country elevators under a block shipping system whereby Western Canada is divided into 49 shipping blocks.

Railway grain cars are allocated to both the CWB grains and non-board grains by the WGTA. In administrating railway car allocation, the WGTA follows a set of guidelines set out in the Western Grain Transportation Act. The WGTA works closely with the CWB and the sectors which deal with non-board grains, and handles producer requests for cars. In allocating hopper cars among the various sectors, the WGTA considers many factors including major sale commitments by the CWB. The non-board sectors and the CWB are treated equally in hopper car allocations. At times, all cars are allocated to CWB grains but there are times when non-board grains have priority. In terms of board grains, the CWB allocates cars to country elevator points upon request from elevator companies according to the shipments made by that elevator company within that particular shipping block. In the past the CWB has received about 80 percent and the non-board sector 20 percent of the available railway grain cars.

Quality Control. The federal government has enacted rigid regulations to control the quality of grain. These regulations have two major areas of importance: licensing of new varieties; and the establishment of the Canadian Grain Commission, which supervises the handling of grain. The Commission's quality control system involves all facets of the grain industry from the breeding of new varieties to the delivery of grown products to consumers. Of equal importance, however, is the system that establishes the criteria for the release of new varieties where quality control actually begins. The Commission has the legislative authority for licensing grain-handling facilities, setting grade standards, providing official inspection and weighing services, handling foreign complaints, and ensuring that the quality is maintained on grain moving through the system. The Commission is totally fee supported; it assesses fees to recover its operating costs.

Feed Grain Policies

The CWB had complete control of all interprovincial grain movements from 1949 to 1973. This included not only trade between the western and central regions of Canada but also movements between Manitoba, Saskatchewan, and Alberta. Some Central and Eastern Canadian feed users bitterly alleged that the CWB sold to export customers at a price below that to domestic users. The CWB also controlled movements to Western Canadian feed mills until 1961. In the late 1960s, grain gluts depressed prices in the non-board market, but Eastern Canadian feed users could not legally purchase Western Canadian grain except from the CWB. Trucking of grain between the Western Canadian provinces was also illegal, though some farmers may have circumvented the law by moving it across provincial boundaries. Western Canadian producers criticized the lack of grain movement, particularly because of their reduced cash flow.

In 1974, in response to these problems, the government implemented a feed grain policy which created a non-board domestic market. The CWB was still responsible for export markets, industrial grains, and grains for human consumption, and it eventually became an ongoing residual supplier to the domestic feed grain market.

This dual market creates some problems. A producer delivering to the CWB receives an initial payment and then, if the final sales price minus handling and administrative costs exceeds the initial payment, a final payment. Thus, a producer does not know how much he will receive for the grain he has sold until 6 to 18 months after delivery. This makes it difficult for producers to compare prices between the CWB and non-board delivery options.

Farmers' Opinions of the Grain Marketing System

Canada's grain marketing system is complex because of the many types of organizations involved (e.g., cooperatives, state trading companies, private companies, and futures markets). Certain farm groups have challenged the usefulness of the CWB and have suggested a return to marketing through the private grain trade. However, a survey of over 600 Prairie farmers in 1989 indicated that farmers do support the CWB. Farmers prefer the CWB as a market organization over the private trade for all major grains, including canola (Table 20.10). Even though the federal government removed oats from the CWB in 1989, the majority of farmers did not support this decision.

TABLE 20.10   Preferred Marketing System By Crop, Canada

Crop	Canadian Wheat Board	Grain Company	Other
Wheat-domestic use	82	15	1
Wheat-export	83	14	1
Barley-domestic use	68	27	2
Barley-export	75	20	1
Oats-domestic use	59	32	4
Oats-export	64	28	2
Canola	47	40	2
Rye	51	29	4
Flax	47	38	3

Source: Decima Research (1989). "Assessment of the Attitudes of the Prairie Farming Community." Personal communication with Alberta Wheat Pool.

21

Programs Affecting Prairie Agriculture

Government programs have a major impact on the structure of the Canadian grain sector. This chapter provides a brief description of the major programs that have influenced the development of Prairie agriculture. Over time, some programs have been eliminated and new ones have been introduced. At least three major farm programs that had been in place for several years were still in place as of January 1, 1989. These are the Western Grain Stabilization Act (WGSA), Crop Insurance (CI), and the Western Grain Transportation Act (WGTA). However, effective 1991, programs such as WGSA were replaced by the Gross Revenue Insurance Program (GRIP).

Temporary Wheat Reserve Act (TWRA)

Between 1955 and 1970, Canada accumulated large grain stocks. At times, wheat stocks in Canada exceeded those in the United States. The CWB's policy was to impose tight marketing quotas on farmers in order to achieve certain price targets. Given the tight delivery or marketing quotas, stocks accumulated at the farm level. To mitigate the costs of on-farm storage, the government authorized a major program under the Temporary Wheat Reserve Act (TWRA) designed to provide partial compensation to farmers for on-farm storage of their grains. This program resulted in a major income transfer to Prairie farmers.

Lower Inventory for Tomorrow (LIFT)

During the late 1960s, there was a glut in the world wheat economy. Large wheat stocks had accumulated in both the United States and in Canada. The Canadian government attempted to reduce stocks through the LIFT program. Under this program, producers were provided incentives to reduce wheat plantings. The LIFT program lasted for only one year, 1970. During that year, seeded area fell to 4.9 million hectares compared with the 9.9 million hectares seeded in 1969. This led to a drop in production from 17.7 million to 8.5 million metric tons. The cost to the federal government was Can$63 million. This amount was paid directly to farmers.

Two-Price Wheat (TPW)

The Canadian wheat economy has always been subject to large price swings, to a great extent as a result of international market forces. The majority of the wheat grown in the Prairie region is exported. To deal with the instability and stabilize the price of wheat to Canadian producers and consumers, the federal government introduced a Two-Price Wheat (TPW) program in 1967. Under the program, if Canadian wheat export prices fell below a specified domestic floor price, Canadian consumers paid the floor price. In effect, this provided a subsidy to domestic producers. However, if the export price rose above a specified ceiling price, then consumers paid only the ceiling price. In that case, the subsidy flowed from producers to consumers. When the export price fell between the ceiling price and the floor price, then the domestic and export prices were equal; hence, no group would receive a subsidy. In the late 1960s and early 1970s, producers gained at the expense of consumers but, in the late 1970s, the opposite was the case as consumers gained at the expense of the producers.

Until 1986, essentially under the TPW, domestic prices were allowed to move within a band. However, in August, 1986, a revised domestic wheat policy allowed the CWB to establish a domestic price at anywhere between Can$220 and Can$404 per metric ton. The domestic price was set at Can$257 per metric ton. This effectively terminated the price band concept.

The wheat market fell drastically in 1987. Domestic wheat consumption of hard red spring wheat in Canada is somewhere between 10 percent and 15 percent of total Prairie production. Given the domestic price of Can$257 per metric ton on a world price of approximately Can$96 per metric ton, Prairie farmers received a benefit of roughly Can$162 per metric ton on 15 percent of the wheat they grew. Thus, for a farmer producing 490 metric tons in Saskatchewan, the TPW program provided an additional revenue of roughly Can$11,000.

Because of the Free-Trade Agreement (FTA) signed between Canada and the United States, the TPW program was abolished on August 1, 1988. Prices of wheat sold for human consumption in Canada are now based on the world market price. To compensate for the loss caused by the price drop, producers received the full benefit of the TPW program for the 1988/89 crop year. The payment was determined by calculating the difference between Can$257 per metric ton and the average domestic selling price of wheat stored in Thunder Bay.

Special Canadian Grains Program (SCGP)

World grain prices are heavily dependent on the U.S. farm programs. The cornerstone of U.S. farm policy is the farm bill, which is passed at intervals of between five to six years in the United States. Under the 1985 Farm Bill, the loan rate for wheat and corn and other major commodities was lowered significantly. As a result, the world market price also dropped. The intent on the part of the United States was to lower the loan rate, thus lowering export prices to encourage expanded export sales, at least in terms of quantity, and reduce U.S. stocks of wheat and corn. In response to the drop in the U.S. loan rate, the Canadian government introduced the Special Canadian Grains Program (SCGP) in December, 1986. The program made a payout of Can$1 billion to Canadian grain and oilseed producers. The payments were made in 1987 in two installments (Can$300 million and Can$700 million, respectively). Payments were based on the area that farmers seeded to the designated crops in 1986, the regional crop insurance yield, and the relative price decline in each commodity due to the trade war between the United States and the European Community (EC). The crops covered under the program included wheat and durum, barley, oats, rye, mixed grains, corn, soybeans, canola, flax, and sunflower seeds. The maximum payment to any one individual was Can$25,000.

In December 1987, in response to continuing low prices for grain, the government again announced SCGP. The payouts under the second SCGP totaled Can$1.1 billion, and the limit was set at Can$27,500 per producer. This amount was paid out in 1988. The basis of the payment was somewhat different than the previous program in that the summer-fallow area was considered along with additional crops.

Drought Payment (DP)

The 1988 Prairie grain crop was greatly reduced following one of the worst droughts in Canada's history. In certain areas, yields were less than during the drought of the 1930s. In 1988, the government announced a drought-assistance program for Canadian farmers. The Prairie region was to receive payments in excess of Can$500 million. Payments were made in April and May, 1989, and again in August and September, 1989. These payments were based on 1988 yields and were assessed on a township basis. In the payments, factors such as land quality were included. Everything else equal, the higher the land quality, the greater was the payout.

Western Grain Stabilization Act

The Western Grain Stabilization Act (WGSA) was introduced in 1976 to deal with large swings in the income of Prairie farmers. The program was put in place to avoid a repetition of the economic downturn of the late 1960s which was due to declining international grain prices and sales. The program is voluntary. Farmers who join the program contribute a percentage of their gross sales — up to a maximum of gross sales of Can$60,000 — to the stabilization fund. The federal government also contributes to the fund. Over time, the levy paid by farmers has increased from 1 percent to 4 percent of the Can$60,000 maximum. The seven major grains grown on the prairies (wheat and durum, oats, barley, rye, flax, canola, and mustard) are eligible under the program. The one exception is grain fed to livestock on the farm where it is produced.

The federal government's average annual contribution to the program for the five-year period 1979-1983 was Can$107.6 million. Payouts from the program are triggered when the net cash flow from the seven grains grown in the Prairie region falls below 90 percent of the previous five-year average net cash flow. The payout to an individual producer is determined by the level of his/her levy for the current and previous two years. The above formula is designed to stabilize regional net cash flow. However, the income of the individual producer may not be stabilized if, for instance, the farmer has a poor crop in a year when the Prairie region as a whole has had a good crop and relatively high prices.

Since 1983/84, the Prairie grain economy has worsened significantly, largely because of sharply falling grain prices. This situation has led to a significant increase in payouts under WGSA. Farmers who contributed a maximum levy received roughly Can$25,000 under WGSA for the 1987 calendar year. In December, 1987, the federal government added Can$750 million dollars to the stabilization fund. As this amount was still insufficient to cover the deficit created by WGSA, the levies to producers were increased to 4 percent of a maximum of Can$60,000 grain sales for a given year. Though the WGSA is a voluntary program, it has a high level of participation. As of December 30, 1987, over 80 percent of Prairie grain producers participated in the program.

Crop Insurance

Yields of crops grown on the Prairies have fluctuated dramatically as has grain quality. Farmers have always been interested in having governments implement a crop insurance program. The first Crop Insurance program (CI) was introduced in 1961. Crop insurance provides for losses caused by natural hazards such as frost, fire, floods, hail, insects, plant diseases, and drought. Many crops are covered, including wheat and durum, oats, barley, flax, canola, rye, sunflowers, mustard, utility wheat, and canary seed. The program is funded by producer premiums, which are matched by the federal government. However, each provincial government pays the administrative costs of the program.

Crop insurance programs differ among provinces. In Saskatchewan, the program offers farmers the choice of 60 percent or 70 percent coverage for their risk area and soil class. For example, in a given area, if the long-term yields are 2.0 metric tons per hectare on summer fallow, then, if the farmer chooses 70 percent coverage, he or she is insured for 1.4 metric tons per hectare. The producer can select various prices associated with yield levels. In 1989, for example, under crop insurance, the farmer could select one of three prices and the premiums varied accordingly. Many farmers chose the high-price option. In this case farmers who chose the high-price option were covered for roughly Can$257 per metric ton of red spring wheat. However, when the cutoff date arrived, the federal government forecast the price of wheat as roughly Can$162 per metric ton. As a result, farmers who collect crop insurance would receive Can$162 per metric ton provided they chose the high-price option. Prior to 1989, there were essentially only two prices farmers could choose from. Final prices were identical to the prices agreed to at the time crop insurance was taken out. A producer who took out the high price (say, Can$147 per metric ton) in the spring of the year was automatically guaranteed Can$147 per metric ton if crop insurance was collected in the fall. The high-price option was introduced in 1989. Which crop insurance premium is used depends on many factors including soil type, type of crop grown, type of crop rotation, and the level of coverage selected. In addition, coverage levels increase and premiums decrease if the farmer has a high-performance record (i.e., no claims or small claims).

The average annual federal government cost for crop insurance for the Prairie provinces over the five-year period 1981/82 to 1985/86 was Can$120 million. However, in both 1988 and 1989, the amounts far exceeded this level. Because of the severe drought in 1988, the payout exceeded Can$500 million dollars. Given the flooding conditions that existed in 1989, crop insurance payouts could easily have approached Can$500 million dollars.

Not all farmers use high-price option crop insurance. Those who do not are generally the ones who are well financed and have landholdings scattered in different locations. Also, certain producers move in and out of the program depending on factors such as spring climatic conditions. Some producers only insure against possible hail losses, through private companies. However, over 80 percent of the producers in the Prairie region have some form of crop insurance, and a record number participated in 1989. The percentage varies from year to year since, at low market prices, the coverage is less and, hence, a certain percentage of the farmers will not take out insurance.

The yield base for insurance is the 10-year moving average of the yield on each class of soil. The farmer can insure his crop at 60 percent or 70 percent of this yield base. What changes from year to year is the price at which these yields can be insured. For example, those farmers in 1989 who opted for the high-price option would receive, under crop insurance, a price coverage of at least Can$37 per metric ton higher than the maximum price coverage available during the 1988 season. The low-price coverage in 1988 partly explains why the federal government responded to the severe drought by introducing a drought-assistance payment.

Western Grains Transportation Act (WGTA)

The Western Grains Transportation Act (WGTA) was passed in November, 1983. Under the act, the federal government provides railways with an annual payment of up to Can$658.6 million (plus an inflation index) to cover the transportation of eligible grain from Prairie shipping points to Thunder Bay, Churchill, Vancouver, and Prince Rupert. The Can$658.6 million is referred to as the Crow Gap. It was an estimate of the shortfall in revenues experienced by the railways in moving grain at the statutory rates at the time the legislation was introduced. In years when exports are low, the payout under WGTA may be less than Can$658.6 million because of small volumes. The amount paid out each year is calculated on a dollar per metric ton moving basis and varies with the distance to port.

The 1983 Act essentially replaced what is known as the Crow Rate. The Crow's Nest Pass Agreement was introduced in 1897 as an instrument for economic development in Canada. The federal government provided a subsidy to the Canadian Pacific Railway Co. (CPR) for the construction of a railroad in southeastern British Columbia. The CPR agreed to reduce the freight rates on grain and flour moving eastward out of the Prairies that, for certain settlers, affected moving into the west. The Crow Agreement was used as an instrument to integrate economically the Canadian economy, to encourage the development of the West, and to provide inexpensive food for the population of Central Canada. Under the Crow, the rate at which grains and oilseeds could be shipped out of Western Canada (known as the statutory rate) remained fixed. However, as time passed, the railways contended that those rates were too low to cover the cost of an adequate and efficient transportation system. In response, the federal government paid for branch line maintenance, while both the federal and provincial governments undertook the purchase of the railway cars. It was the rising cost of the federal government and branch line maintenance and the increased demands for policy change from the railways, livestock producers, processors, and selected grain groups that finally led the federal government to enact the new legislation in 1983.

The Gross Revenue Insurance Program and the Net Income Stabilization Account

The federal government put into effect in 1991 the Gross Revenue Income Program (GRIP) and the Net Income Stabilization Act (NISA). These programs essentially replace both CI and WGSA and represent marked changes in Canadian agricultural policy.

GRIP is one component of the new saftey net program made available for farmers. This program goes a step beyond crop insurance by offering both yield and price protection. In 1991/92, GRIP will be delivered through CI. Revenue protection premiums are shared by farmers, provinces, and the federal government. The price guarantee on crops is based on a 15-year moving average of past prices. In addition, farmers are guaranteed a minimum yield based on an area average. However, farmers can use an individual farm coverage yield if this is above the area average.

NISA enables farmers to set aside money which they can draw on in difficult times. It allows him/her to set up a personal income stabilization account. Subject to a cap limitation, the farmer can set aside up to 2 percent in an individual account to be matched by the federal and provincial governments. The program triggers payments when a farm's gross margin falls below a five-year average or the net income falls below $10,000.

Net Transfers

Figure 21.1 gives the net transfer to Prairie producers from the selected government programs discussed above. This covers the period 1953-1987. As the data indicate, for many years the largest transfer from the government to the producers was through the WGTA. However, in 1987 the SCGP and WGSA also became major farm programs. In addition, in the years following 1987, another special grains program and a drought program provided major transfers to producers.

It is interesting that, during the course of the development of Prairie agriculture, much of the debate centered on how commodities should be marketed rather than the extent to which the farmer's income should be supported through farm programs. For most of the period prior to 1950, there were few transfers to Prairie producers from either the provincial or federal governments. Prior to that time, generally, the Crow Rate was not a subsidy to producers. No other program of any major significance was in place. It is generally safe to assert that the accumulative total of government transfers to producers in the Prairie region from 1984 to 1989 will far exceed the accumulative total for all of the years prior to that period. During the latter part of the 1980s, droughts coupled with the lowest real price of wheat in history brought forth a government response by income transfers to Prairie grain producers.

Other Programs

Agricultural Credit

Prairie agricultural producers are large users of credit. As Figure 21.2 indicates, between 1971 and 1986 Prairie's outstanding farm debt increased significantly. This is especially true for lending by the provincial government. The federal lending institution for farmers is the Farm Credit Corporation (FCC). It is generally directed at new farmers wishing to enter into agriculture or small farmers wishing to expand. The interest rates associated with the FCC loans are not significantly different from those charged by many chartered banks. However, the FCC loans are much narrower in scope than commercial bank loans, since they focus primarily on land purchases whereas private banks provide credit for many aspects of the farm business, including operating loans, machinery loans, and the like.

Provincial governments have their own financial institution for lending to farmers. One example is the Agricultural Credit Corporation of Saskatchewan (ACC). Credit has been available for many purposes, including livestock loans and loans for building and construction. Through ACC, the government of Saskatchewan instituted a Can$25 break or production loan at 6 percent interest in Spring, 1986, to provide operating capital for spring seeding. In addition, the provincial government of Saskatchewan introduced a credit policy whereby young farmers could borrow up to Can$250,000 at a subsidized interest rate in order to acquire farm land. As of January 1, 1989, the production loan and the farm purchase loan have been terminated. We have not made an assessment of the extent to which incomes have been transferred to farmers through subsi-

FIGURE 21.1 Net Transfers to Prairie Producers under Selected
Government Programs

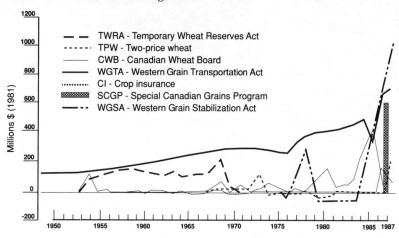

Source: Fulton, Rosaasen, and Schmitz.

FIGURE 21.2 Real Prairie Farm Debt Outstanding, by Major Financial
Institution, 1971-1986

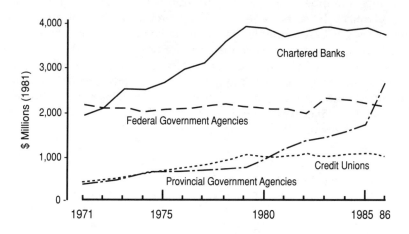

Source: Fulton, Rosaasen and Schmitz.

dized interest rates. However, these amounts appear to be small, relative to those transferred under such programs as the SCGP and the WGTA.

Others

There are other programs of much smaller significance that also impact the Prairie grain economy. Many of these are directed at the livestock sector. These programs include feed grain policy disaster programs in response to droughts that impact the livestock sector, feed-freight programs, and grazing leases on community pastures. Also, there are programs dealing with venture and equity capital which impact the livestock sector.

Two programs worth noting revolve around the actual selling of wheat. The CWB is the sole exporter of wheat from the Prairie region. It sets an initial price for wheat and barley and regulates farmer deliveries through quotas. The CWB uses the concept of pooling where all producers receive the same price for a given grade of grain regardless of time of sale during any given crop year. Once the CWB announces the initial prices, they, in essence, become price supports. If, at the close of the crop year, the pool revenues from CWB sales are insufficient to cover these prices, differences are made up by the federal government. Until the 1985/86 and 1986/87 crop years, the federal government seldom made a payment into the pool account. When it did, it was for a small amount and largely for barley. However, in the latter half of the 1980s, the federal payments were significant and exceeded Can$300 million dollars. In these years as in previous years, the barley pool experienced the greatest deficit.

Because a number of producers are strapped for cash at harvest time, cash advances have been available for producers and administered through the CWB. A farmer holding a quota book can obtain a maximum of a Can$30,000 cash advance interest free for one year after the grain is in farm storage; the stored grain serves as collateral. At the time of delivery, a specified amount is deducted to pay for the cash advance. To illustrate, in 1987/88, a producer with a crop advance had Can$73 per metric ton deducted on Canadian western red spring wheat at the time of delivery to country elevators until the advance was paid off. Given an initial price of Can$96 per metric ton, the producer only receives Can$22 per metric ton net of the cash advance. Beginning the 1989/90 crop year, cash advances were still available but an interest charge on the advance was imposed. Prior to that time, cash advances held for one year were interest-free. The total interest on cash advances through the 1980s was very small relative to the size of transfers through such programs as WGSA and others.

Stocks

The Canadian government has no explicit policy dealing with grain stocks. Stock levels are largely a function of the selling strategy used by the CWB. When marketing quotas (which are set by the CWB) are tight, stocks build up at the farm level. The cost of storage is borne by the farmers.

Stock levels in Canada declined markedly during the 1970s (Figure 21.3). In 1970 stock levels in Canada were above those in the United States and the European Community. As the data suggest, in the late 1980s the United States is by far the largest stockholder of wheat. Stocks held by the European Community approximate those held in Canada.

There are many possible explanations why the United States holds much larger stocks than Canada where at one time stock levels were roughly equal. It has often been suggested that, during the late 1960s and early 1970s, Canada and the United States acted as duopolists in setting wheat prices. Thus, both countries held relatively large volumes of stocks

FIGURE 21.3 Year-End Stocks of Wheat in the World, Selected Countries, 1960-1987

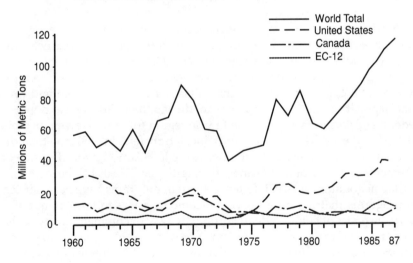

Source: Carter, McCalla, and Schmitz.

in order to keep up prices. However, with the growth of the European Community in the world grain economy, both in terms of production and in terms of exports, Canada felt that it could no longer significantly influence world prices by holding large volumes of stocks. One could hypothesize that in the 1980s Canada became a price taker. It appears that the CWB pricing strategy is one where the volume of stocks held at the end of a given year is kept to a minimum.

In the above context, Canada's policy relating to wheat stocks is significantly different from the United States, which uses stocks as a significant policy instrument in order to stabilize and raise farm income. As suggested earlier, Canada uses stabilization policies for this purpose rather than stock policies. For example, the WGSA is essentially a buffer fund program, where money is collected from farmers in good times and paid out in bad times. This is in sharp contrast to the United States where stocks are part of government policy. Because of the nature of the U.S. farm programs, sizable stocks are often accumulated where the cost of storage is paid for by the government.

An Overview of Programs

Prairie grain farmers have received sizable transfers from governments through various programs. Figure 21.4 illustrates how realized net-farm income is influenced by these transfers. In 1987, for example, net farm income in the prairie region would have been negative had it not been for government transfers Government transfers from both federal and provincial governments became significant during the 1980s. A breakdown of federal support to farmers in the Prairie region for 1987 is given in Table 21.1. The figures do not include provincial transfers. In 1987, the total direct payments, rebates, and liabilities exceeded Can$4 billion. The size of payments given to prairie farmers during the late 1980s has been of concern to negotiators at GATT. Often, Canadian programs have been labeled as programs that not only provide income transfers to farmers but also have the effect of being trade distorting. By way of comment, many of the programs in Canada, even though they provide sizable transfers to prairie farmers, are not trade distorting (Furtan et al. and Fulton et al.), as many of the government payments to farmers are made after the crop has been planted. This is especially true for the drought payments and those payments made under the Special Grains Program. Studies indicate that programs of this type have not had major impacts on Canadian grain production. It is generally felt that the CWB pricing strategy is essentially one which keeps the United States and other countries in line concerning competitive conditions. When the United States lowered its loan rate under the 1985 Farm Bill, Canada followed suit by lowering its export price.

In this regard, Figure 21.5 shows some interesting comparisons. It has often been alleged that Canada priced in the international market such that it sold wheat below the U.S. loan rate, thus making the United States no longer competitive in international markets. The data in Figure 21.5 indicate that this certainly was not the case. Prices received by producers in the Prairie region were generally above the U.S. loan rate.

TABLE 21.1 Federal Support to Farmers in the Prairie Provinces, Canada, 1987

	Manitoba	*Saskatchewan*	*Alberta*	*Total*
(millions of dollars)				
Direct payments [a]				
Western Grain Stab. Act(net)	255	743	360	1,358
Agricultural Stab. Act(net)	---	-3	-6	-9
Crop Insurance(net)	17	41	54	112
Dairy subsidy	11	7	19	37
Other direct payments [b]	1	7	23	31
Rebates	26	53	116	195
Special Canadian Grains Payment	155	408	252	815
Total(net)	465	1,256	818	2,539
Other major liabilities				
Farm Credit Corp. (cumulative deficit) [c]	107	340	153	600
Payments to railways [d] under the WGTA	136	483	251	870
Total direct payments, rebates and liabilities	708	2,079	1,222	4,009
(thousands of dollars)				
Payment per farmer (self-employed)	28	36	26	31

[a]   Net payments are total payments less farmers' contributions.
[b]   Includes such payments as compensation for animal losses and damage to waterfowl.
[c]   The cumulative deficit is attributed to the Prairie provinces on the basis of their respective shares of FCC loan arrears during 1987/88.
[d]   Net of the $71.7 million that was to be refunded to the federal government in 1987/88.

Source: Fulton, M.,K.A. Rosaasen, and A. Schmitz (1989). Canadian Agricultural Policy and Prairie Agriculture. Economic Council of Canada, Ottawa.

FIGURE 21.4 Direct Government Payments and Realized Net Income, Prairie Region, 1971–1987

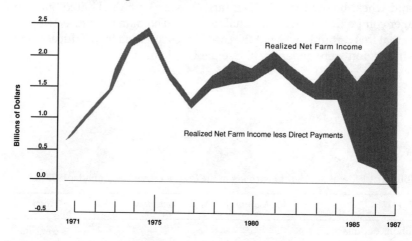

Source: Fulton, Rosaasen, and Schmitz.

FIGURE 21.5 Comparison of U.S. Loan Rate with Canadian Wheat Prices (at Farm Level), Crop Years 1980/81 to 1985/86

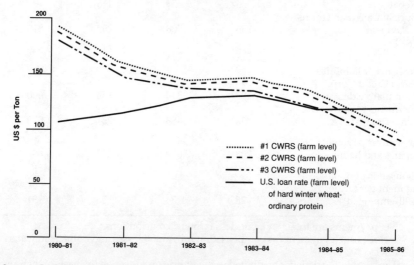

Source: Carter, McCalla and Schmitz.

22

Conclusions

The CWB continues to receive strong support from producer groups. However, it is not necessarily transferable as a model to other countries. As McCalla and Schmitz (1979) point out,

> It is important to stress that, although the Board is supported by most growers and agricultural groups in Canada, it does not mean that it is necessarily a model that has transferability. The character of the Board is a unique function of Canadian history and institutions and, as such, it should be judged only in that context.

In spite of farmer support for the CWB system, the Canadian government appears to be moving in the direction of increased deregulation. Some may view this trend as a serious threat to the future of the CWB as a central selling agency. This concern has been strengthened as a result of pressures from the United States to weaken the powers of the CWB and as a result of the U.S.-Canada Free Trade Agreement (FTA) signed in 1989. Considerable controversy exists over whether or not the CWB can market and price effectively under the FTA.

There have been other concerns raised about the CWB's effectiveness in achieving the highest possible returns to Canadian grain producers. One of these issues focuses on hedging strategies and whether or not the CWB should hedge a certain quantity in order to minimize producers' risk. For example, Carter (1989) argues that the CWB should hedge part of expected output at the time of farmers' plantings. He contends that, even though prices are pooled, farmers are still exposed to significant risks. Currently, the CWB does not hedge on major futures markets, including Chicago.

In this report, we have not analyzed in detail the efficiency of the Canadian system relative to those of other countries. However, several available studies provide some evidence. As argued above, it is generally not true that the CWB underprices the United States by pricing below the U.S. loan rate. This has been a contentious issue and seems to have been a factor in the U.S. government's decision to lower the loan rate under the 1985 Farm Act. Also, in terms of pricing, there is no evidence to suggest that U.S. prices are consistently above Canadian prices even though the U.S marketing system is essentially one made up of the private sector. In a study by McCalla and Schmitz (1979) comparing the grain marketing systems of Canada and the United States, they concluded that, once farm programs were taken into account, prices received by producers in the two countries were similar.

The international cooperation of exporters in determining pricing and stocks policies for wheat broke down in the 1960s. In response to the new trade environment, the CWB increased exports and reduced stocks. In contrast to the period of cooperation where Canada and the United States acted as duopolists in the setting of wheat prices, the United States has become the price setter in the market and Canada a price taker.

As our report shows, agricultural policy in Canada plays a significant role in the income received by Prairie grain farmers. This has been especially true during the 1980s. The CWB is involved in the pricing of Canadian wheat. However, it does not set agricultural policy in Canada. The farm programs put in place in Canada in the 1980s have propped up farm income but have in no way significantly interfered with the operations of the CWB.

Interestingly, prior to the 1980s there were relatively small transfers from both provincial and federal governments to the Prairie grain economy. Much of the debate centered not on whether or not farmers should receive government transfers, but on how commodities should be marketed. For example, a debate in the 1930s and 1940s centered on marketing issues much more than on government transfers. In the 1930s, farmers in western Canada received little financial support from the federal government. Sizable transfers were made to Prairie producers from both federal and provincial governments in the 1980s. In evaluating these transfers, however, it is important to stress that government transfers are not identical to economic subsidies. Transfers can be large while the corresponding true economic subsidies can be quite small. In fact, generally, Canadian farm programs pertaining to the western region are relatively efficient. Even though large transfers have occurred in recent years, these programs have not been output-increasing to any degree and as a result have not been trade distorting. The evidence of this is the stability of Canada's share of the wheat trade-approximately 18 percent to 22 percent. In the international trade context, Canadian government programs appear to be trade distorting and output increasing as measured by standard producer subsidy equivalent measures. However, these are the wrong measures to use when evaluating Canadian farm programs. Most of the programs in Canada are of an ad hoc nature, and payments made through these programs are generally made after harvest. As several studies show, these programs are relatively efficient in that they affect the distribution of income without causing a major misallocation of resources.

Bibliography

Agriculture Canada. *Market Commentary*. Agriculture Canada, Ottawa, 1989.

Auer, L. *Canadian Prairie Farming, 1960-2000*, Economic Council of Canada, Ottawa, 1989.

Canadian Grain Commission. *Canadian Grain Exports, 1987-88*. Canadian Grain Commission, Winnipeg, 1988.

Canadian Wheat Board. *Annual Report 1988*. Canadian Wheat Board, Winnipeg, 1989.

Carter, C., A. F. McCalla, and A. Schmitz. *Canada and International Grain Markets: Trends, Policies, and Prospects*. Economic Council of Canada, Ottawa, 1989.

Carter, Colin A., Andrew Schmitz, David M. Orr, and Robert A. Zortman. "The Canadian Grain System". Chapter 4 in *Grain Quality in International Trade: A Comparison of Major U.S. Competitors*. Office of Technology Assessment, Congress of the United States, Washington, D.C., February 1989.

Decima Research. "Assessment of the Attitudes of the Prairie Farming Community". Personal communication with Alberta Wheat Pool, 1989.

Fulton, M., K. A. Rosaasen, and A. Schmitz. *Canadian Agriculture Policy and Prairie Agriculture*. Economic Council of Canada, Ottawa, 1989.

Furtan, W. H., J. G. Nagy, and G. G. Storey. "The Impact on the Canadian Rapeseed Industry from Changes in Transport and Tariff Rates," *American Journal of Agricultural Economics*, 61:2 (May 1979) 238-48.

Furtan, W.H., T. Y. Bayri, R. S. Gray, and G. G. Storey. *Grain Market Outlook*. Economic Council of Canada, Ottawa, 1989.

Hayami, Y., and V. W. Ruttan. *Agriculture Development: An International Perspective*, Johns Hopkins Press, 1985.

McCalla, Alex F., and Andrew Schmitz. "Grain Marketing Systems: The Case of the United States Versus Canada", *American Journal of Agricultural Economics*, 61:2 (May 1979) 199-212.

_____. "State Trading in Grain". Chapter 3 in *State Trading in International Markets: Theory and Practice of Industrialized and Developing Countries*, ed. M. M.Kostecki. London and Basingstoke, The Macmillan Press, Ltd., 1982.

Rosaasen, K. A., and A. Schmitz. *The Influence of Feed Grain Freight Rates on the Red Meats Industry in the Prairie Provinces*, Department of Agricultural Economics, University of Saskatchewan, Saskatoon, 1985.

Schmitz, Andrew, and Alex McCalla. "The Canadian Wheat Board". Chapter 4 in *Agricultural Marketing Boards: Prices, Profits, and Patterns*, ed. Sidney Hoos. Cambridge, Massachusetts: Ballinger Publishing Company, 1979.

Schoney, R.A., Tom Thorson, and Ward P. Wiesensel. *1988 Results of the Saskatchewan Top Management Workshops*. Bulletin FLB88-01, Department of Agricultural Economics, University of Saskatchewan, Saskatoon, 1988.

Statistics Canada. *Canadian Statistical Review* 11-003E Monthly. National Income and Expenditure Accounts, Cat. No. 13-001, various years.

_____. *Census of Agriculture 1971 and 1981*. Ottawa.

_____. *Labour Force Annual Averages*, Cat. No. 71-529, Ottawa, occasional.

_____. *Quarterly Economic Review*. Annual Reference Tables. Cat. No. 15-001, Ottawa.

Ulrich, A., W. H. Furtan, and A. Schmitz. "The Effects of Licensing Regulation on Canadian Agriculture: The Example of HY320", *Journal of Political Economy*, 95:1 (January 1986) 10-32.

_____. "Public and Private Returns from Joint Venture Research: An Example from Agriculture", *Quarterly Journal of Economics*, 51:1 (February 1985) 103-131.

PART FIVE

The United States

Grain Marketing, Institutions, and Policies

Lowell D. Hill

Karen Bender, Philip Garcia, Harold Guither, Robert Hauser, David Lins, Martin Patterson, Kostas Stamoulis, and Sarahelen Thompson cooperated in preparing this part. The views expressed here are solely those of the author and should not be attributed to the institution with which he is affiliated.

23

Social and Economic Environment

The U.S. Economy

The grain industry in the United States operates within an open, market-driven economy. The United States is the wealthiest of the industrial market economies. In terms of current (1987) dollars, in 1987 the U.S. Gross National Product (GNP) was $4,448.5 billion. By comparison, the GNP for Japan was $2,359.0 billion, and that for West Germany was $1,129.9 billion (Figure 23.1). However, in per capita income, the United States ranked third, after Switzerland and Japan, tying West Germany. (Figure 23.2).

Growth in real GNP (1982 dollars) has been erratic, and has ranged from -2.5 percent in 1981/82 to 6.8 percent during the peak of the most recent recovery in 1983/84. Average annual growth in 1980-1987 was 2.8 percent compared to 3.8 percent in 1960-1970, 2.2 percent in 1970-1975 and 3.4 percent in 1975-1980.

The breakdown of GNP by major types of products (goods, services and structures) shows the increasing share of services in the GNP over time. Services as a share of GNP rose from 43.4 percent in 1970 to almost 51 percent in 1987. The share of goods (other than structures) fell from 46.1 percent in 1970 to 43.0 percent in 1980 to 39.6 percent in 1987. The share of structures remained around 10 percent during the 1970-1986 period. The increasing share of services in the National Product has caused economists some concern about the growth potential of the economy.[1]

The Federal Deficit

A question of current debate is whether the growth of the federal deficit and the need to borrow to finance it could inhibit economic growth by crowding out private credit demand for investment. The effects of the fed-

FIGURE 23.1 Gross National Product of Selected Countries, 1985 vs. 1987

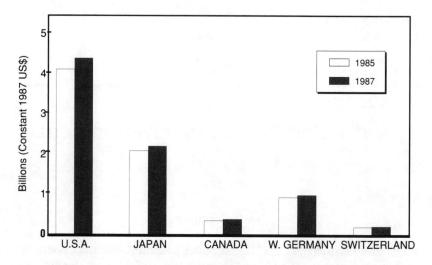

Source: USDC, Bureau of the Census, 1989.

FIGURE 23.2 Per capita GNP for Selected Countries, 1985 vs. 1987

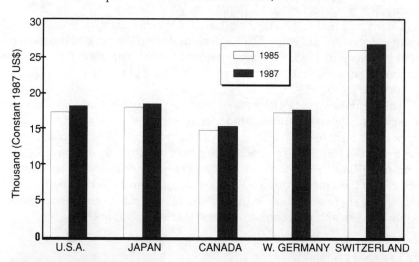

Source: USDC, Bureau of the Census, 1989.

eral deficit on the value of the dollar and on imports, exports and the trade balance are also important issues.

As the federal deficit has evolved over the last 30 years, the budget has been in surplus only twice (in 1960 and 1969)[2]. Between 1970 and 1976 the deficit was extremely volatile, ranging between $2.8 billion and $53.2 billion. In 1981 it increased by 62.1 percent, from $78.9 billion to $127.9 billion. It increased again by almost 63 percent in 1982, and reached a peak of $221.2 billion in 1986. For each of the fiscal years 1988 and 1989, the deficit is estimated to be around $145 billion (*Economic Report of the President,* 1989).

The accumulation of federal deficits has contributed to a sharp increase in the federal government's interest-bearing debt, from $382.6 billion in 1970 to $1,147 billion in 1980 to an estimated $2,590 billion in 1988.[3] While it took 30 years (1945 to 1975) for the debt to increase from $260.1 billion to $544.1 billion (a 109.2 percent change), it has taken only 14 years to increase from $544.1 billion to $2,590 billion (a 376 percent change).

The government's total debt has increased from 41.6 percent of GNP in 1983 to 46.4 percent in 1985 to a projected 54.6 percent in 1988. However, a closer examination of the trend for the most recent years (1986: 51.2 percent, 1987: 53.7 percent, 1988 54.6 percent) shows a reduction in the *rate* of increase of the debt. Meanwhile, the federal deficit as a percentage of GNP has declined from a peak of 6.3 percent in 1983 to 5.4 percent in 1985 and an estimated 3.5 percent in 1988.

There is no uniform agreement on the effects of the government deficit on economic activity and on major measures of performance of the economy (growth and inflation). High federal deficits have been present during recessions and recoveries, and can be found during periods of low as well as high real interest rates.

The argument that the federal deficit crowds out private demand for investment depends on the desirability of government securities to domestic and foreign investors. This desirability is directly related to the credibility of the issuing authority and the expected strength of the economy. The massive influx of foreign capital into the United States in recent years has helped finance the increasing federal debt, keeping real interest rates lower than they would otherwise have been. The effects of a hypothetical crowding out are a function of the relative productivities of private vs. public investment.

An institutional system for reducing the deficit has been in place for three years. The Gramm-Rudman-Hollings Act sets deficit targets and "sequester triggers" which, if and when reached, automatically cancel budget authority on a wide range of programs and impose mandatory budget cuts except in the event of a recession. To increase the President's ability to control the deficit, recent U.S. administrations have been trying to obtain a line-item veto authority that would empower the President to veto budget items considered to be wasteful or purely responses to special interest groups.

Inflation and Unemployment

Inflation. In the five years before 1974, inflation for all items, as measured by the Consumer Price Index (CPI), was modest, about 5 percent a year. Sharp increases in energy prices of 15.3 percent in 1973 and 57.3 percent in 1974 contributed to a rapid increase in the CPI. In 1973, the annual rate of increase in the CPI was 6.2 percent, and by 1974 it had jumped to 11.0 percent. From 1975 to 1978, the average inflation rate was 7.3 percent a year. Further dramatic increases in the rate of change of energy prices in 1979, 1980 and 1981 (by 35.1 percent, 38.0 percent and 21.5 percent, respectively) pushed inflation even higher, to 11.3 percent in 1979 and 13.5 percent in 1980. In 1981, inflation dropped back to 10.3 percent.

The recession of the early 1980s led to a drop in the inflation rate to 6.2 percent in 1982 and 3.2 percent in 1983. From 1985 to 1987 the inflation rate averaged 3.03 percent, a substantial drop from the level it reached in 1980 (*Monthly Labor Review*, 1989). The lowest percentage annual change in the inflation rate (1.9 percent) was in 1986. This coincided with a substantial (19.1 percent) decrease in oil and home fuel prices.

The Consumer Price Index for Food increased by an average of 5.1 percent a year during the period 1960-1987, compared with 5.2 percent for all items. Since 1960, retail food prices have almost kept up with general inflation. During the same period, oil and household fuel commodities increased by an average of 7.6 percent a year. This picture has changed recently. By 1986/87 the overall annual average increase in the CPI had slowed to 2.8 percent, outpaced by a 3.7 percent average gain in food prices. In contrast to food prices, household fuel prices dropped by an average of 9.4 percent per year.

Unlike food prices, from 1975 to 1987 producer prices for farm commodities did not keep up with general inflation. Processing and marketing margins play an increasingly important role in the retail value of food. From 1970 to 1987, the average growth in the Producer Price Index (PPI) for Farm Products was 1.1 percent a year in contrast to 3.1 percent a year for the PPI for Processed Foods and Feeds, 6.6 percent for the overall Consumer Price Index and 6.5 percent for the CPI for retail food. The rates of change of producer prices for farm commodities were unstable, fluctuating from an average of 11.0 percent a year during the early 1970s (1970-1975) to -4.9 percent in 1981/82 and -9.9 percent in 1984/85. Both real and monetary factors contributed to these wide swings, as did special characteristics of farm level prices.

Unemployment. Unemployment has become a major criterion in gauging economic performance. Unemployment rates adjust slowly to major changes in real GNP. During the last two decades, the United States has experienced a wide range of unemployment rates, from approximately 5.5 percent in the early 1970s to 9.7 percent in 1982 and 1983 (civilian workers). The steady increase in unemployment rates from 1979 through 1982 led to a leveling off in 1983, and a steady decrease to 5.4 percent in November of 1988.

In 1987, the industry with the highest unemployment rate (11.6) was construction, followed by agriculture (10.5 percent) and mining (10 percent). The rate for manufacturing was 6.0 percent. In 1983, the year of peak unemployment, construction still had the highest unemployment rate (18.4 percent), followed by mining (17 percent), agriculture (16 percent), and manufacturing (11.2 percent). Since then, unemployment has decreased steadily in all industries except mining, which has followed a more unstable path.

External Sector

The external sector of the U.S. economy has increased in importance recently due to the increasing importance of exports and imports in the total economic activity, and a significant increase in the trade deficit and in the net holdings of U.S. assets by foreigners.

Related issues include fluctuations in the value of the dollar, the United States' change from a creditor to a debtor nation, and the long-run economic implications of that indebtedness.

Exports and Imports. From 1970 to the last quarter of 1988, exports of goods and services increased in real terms by 189 percent, from a value of $178.3 billion to $514.0 billion.[4] Imports of goods and services increased by 192 percent, from $208.3 billion to $607.9 billion. The shift to flexible exchange rates and the abolition of the Bretton Woods system in 1973 was followed by a significant drop in the value of the dollar vis a vis other currencies. The excess supply of dollars during the late 1960s is the primary reason for the devaluation. This devaluation shows up in the increase in net exports and in the current U.S. account balance. Net merchandise exports increased from -$6.4 billion in 1972 to $.9 billion in 1973, while the balance on current account jumped from -$1.7 billion to $11.2 billion.

The current deficit on the current account results from both the effects on exports of the tight monetary policy of the early 1980s and the effects on imports of the recent recovery. The tight monetary policy of the early 1980s and the federal government's increased demand for credit to finance the domestic budget deficit increased real returns on U.S. dollar-denominated assets, raising the value of the dollar and worsening the competitiveness of the U.S. economy. In 1982, total exports of goods and services (in 1982 dollars) fell 8 percent, from $392.7 billion to $361.9 billion, breaking an almost uninterrupted increase since 1970. The level of exports did not regain the 1981 value until mid-1986. Imports, on the other hand, increased (in 1982 dollars) by 10 percent in 1983 and by almost 24 percent in 1984.

Although exports have been increasing almost steadily since 1984, the increase in the real value of imports continues unabated, even though the dollar has depreciated dramatically since its peak value in February of 1985. The result is that the deficit on the current account balance reached $128.9 billion in 1987. While the dollar depreciation made U.S. goods more competitive, an increase in disposable income following the recovery of the 1980s served to fuel the increased demand for imports.

In terms of accumulated claims, the U.S. net international asset position climbed from $58.5 billion in 1970 to a peak of $141.1 billion in 1981, turned negative in 1985 (-$269.2 billion), and declined further to -$368.2 billion in 1987. Western Europe was the primary creditor for the United States (-$387.2 billion in 1987), followed by Japan (-$80.7 billion). Canada and Latin America were net debtors of the United States, with debts of $50.7 and $22.4 billion respectively.

Two issues relate to the trade deficits and debt accumulation, the sacrifices to future U.S. consumption levels that will be needed to pay the debt, and the structure of the debt—what percentage is in liquid assets and what is in the form of direct capital investment. For the first, the low savings and investment rates in the United States indicate that much of the inflow of foreign capital is used for consumption rather than investment purposes. For the second, the data show that the major part of the capital inflow is invested in U.S. government treasury bills. This could lead to a rapid outflow of foreign investment if investment conditions in the United States become unfavorable.

Income Distribution

The United States has a large middle class, and a growing minority of people whose incomes fall below the poverty level. A high percentage of women are employed. In 1987, there were 90.3 million males aged 15 or older, and 85.6 million were reported as income earners: 35.3 percent of the income earners had incomes of $25,000 or above and 6.8 percent earned less than $1,000. The median income for all males was $17,752, with mean income at $22,684. Of the 98.2 million females aged 15 or older, 89.3 million reported income: 24.1 percent of the income earners had incomes of $25,000 or above and 16.5 percent earned less than $1,000. The median income for all females was $11,435, 64.4 percent of that of men. The ratio of median incomes (female/male) stayed almost constant in 1985-1987 (65.0 percent, 65.1 percent and 65.6 percent), but it increased by almost 6 percentage points during 1970, and has again increased by almost 5 percentage points since 1980.

The percentage of the population below poverty level stayed fairly constant between 1975 and 1979, with narrow fluctuations around the mean of 11.7 percent. Between 1979 and 1983 it increased to 15.2 percent, and fell back to the 1980/81 average of 13.5 percent in 1987.[5]

Education

Educational attainment in the United States is generally high. In 1986, 19 percent of adults 25 years of age or older had completed at least four years of college. Between 1980 and 1986, the percentage of the adult population with 4 years of high school or more rose from 55 percent to 75 percent.[6]

In 1985/86, expenditures on public and private education from preprimary through graduate school amounted to about $269 billion. This total was expected to increase to about $309 billion in 1987/88. In recent years, total expenditures for education have represented about 7 percent of GNP.

Funding for education comes from federal, state, local and private sources. The federal contribution for education at all levels as a percentage of total expenditure for education fell between 1980 and 1987 from 10.7 percent to an estimated 8.7 percent in 1987, according to Department of Education data. The proportion of State and local contributions also decreased slightly, from 64.8 percent in 1980 to an estimated 64.2 percent in 1987.

Over the past 10 years, the proportion of private students in elementary and secondary schools rose from 11 percent in 1977 to 12 percent in 1987. During the same period, the proportion of college students in private institutions increased from 22 percent to 23 percent. The share of private expenditures in total expenditures on education (channeled mainly to private educational institutions) has increased at all levels of education while changes in government contributions have affected mainly public institutions.

The average years of formal education differ only slightly among metropolitan, non-metropolitan and farm residents. For residents over 18 years of age the average years of schooling completed (1986 data) were: metropolitan, 12.7 years; non-metropolitan, 12.4 years; and farm residents, 12.5 years (Cramer and Jensen, 1988).

The Agricultural Sector

The land area of the United States is 9,363,000 square kilometers. Total arable land in 1985 was 1,878,810 square kilometers (*Food and Agricultural Organization Yearbook*, 1986). Although only a small percentage of the U.S. population is engaged in farming, many more are engaged in food-related activities. The food and farm system as a whole makes a large contribution to the U.S. economy.

The share of production agriculture in GNP has dropped over the last 60 years, to an average of less than 3 percent in 1985-1987. However, when extended to cover the farm and food system, agriculture contributes approximately 18 percent of the nation's GNP (Cramer and Jensen, 1988). Production agriculture contributed 8 percent to GNP in the late 1930s. By 1950, its contribution had fallen to 4.8 percent as mechanical power replaced man and animal power. In 1960 it was 3.7 percent, by 1970 it was 2.5 percent, and the decline continued until 1980 (Figure 23.3). The percentage share has since risen slightly, from 2.4 percent in 1980 (1982 prices) to a peak of 2.8 percent in 1982, before dropping again to 2.6 percent in 1985-1987.

While farmers' contribution to GNP has declined, personal consumption expenditures on food (in constant 1982 dollars) have increased 76 percent between 1960 and 1987, from $255.5 billion to $450.4 billion. The share of personal expenditure on food to disposable personal income fell steadily during the same period, from 20 percent in 1970 to 16.8 percent in 1987 (Figure 23.4). The relatively high levels of food consumption per capita, along with growing real incomes, affect consumption patterns by increasing consumption of specialized high-value food items, which probably account for much of the increase in spending on food consumption.[7]

The percentage of the civilian labor force employed in the agricultural sector has declined over time, partly reflecting the increasing capital intensity of agricultural production. The agricultural labor force as a percentage of the total civilian labor force fell from 4.2 percent in 1970 to only 2.7 percent in 1987.

A major increase in farm productivity has accompanied the technological changes that reduced labor inputs and shifted much of the "value added"activities to non-farm firms. Increased inputs such as fertilizers and pesticides increased yields. The index of aggregate farm output (1977=100) increased from 76 in 1960 to a high of 118 in 1981, a 55.3 percent increase. The index fluctuated from 96 to 118 between 1981 and 1987. For feed grains, the 1960 to 1985 increase was 94 percent and for food grains the 1960-1981 (peak year) change was 118 percent. The increase was due to a combination of factors. Government policy and technical innovation played a significant role.

FIGURE 23.3 Farm Share of GNP, 1970-87

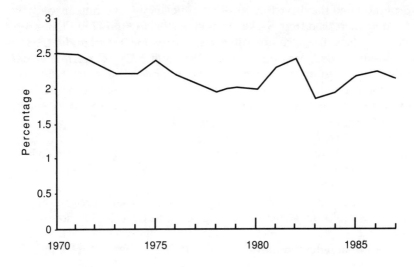

Source: USDC, Bureau of Economic Analysis.

FIGURE 23.4 Personal Expenditure on Food as a Percentage of
Disposable Personal Income, 1970-87, (Constant 1982$)

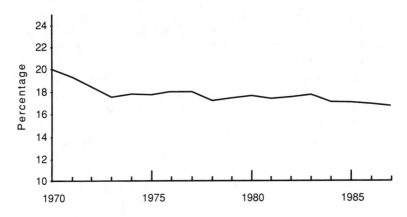

Source: The National Income and Product Accounts of the United States,
Statistical Abstract of the United States.

Despite agriculture's small share in the aggregate GNP, the sector has attracted considerable attention and policy action due to the sensitive nature of food and the drive for food self-sufficiency. Government payments to farmers increased from $1,286 million in 1980 to $16,747 million in 1987 (including the value of commodities transferred to farmers under the payment-in-kind (PIK) program in 1983, 1984 and 1985). Government payments are a significant share of farm income. While in 1980, total government payments to farmers were only 1.8 percent of total cash receipts from all crops (including non-supported ones such as fruits, vegetables, tree nuts, etc.), by 1986 that share had increased to 18.6 percent. In 1987 government payments as a percentage of farm income declined slightly to 12.1 percent. The cost of the government programs is even higher if one considers the price distortions and subsequent consumer costs of some of the commodity programs (e.g., the tobacco program).

Agriculture has received an increasing share of an increasing federal budget. Federal budget outlays for agriculture were $8.8 billion in 1980 (1.5 percent of total federal outlays). In 1983 they rose to $22.9 billion (2.8 percent of total federal outlays). In 1986, 1987 and 1988 the relevant numbers were $31.4 billion (3.2 percent), $27.4 billion (2.7 percent) and $22.4 billion (2.1 percent) respectively.[8]

Agricultural Trade

Domestic markets experienced relatively constant growth from 1960 to 1987. Export markets were much more volatile. As a result, agriculture's share in net exports has declined, from 13 percent in 1970 to 10 percent in 1987 (Figure 23.5). The combination of a dramatic increase in farm production and a low elasticity for the demand of farm output has made development of export markets essential.

A large share of the value of U.S. farm production comes from exports. For corn, the share of production that is exported was 12.5 percent in 1970, 29.3 percent in 1975, and 35.5 percent in 1980. It reached a peak of 45.5 percent in 1984, then declined to 24.1 percent in 1987. Soybean exports were 38.5 percent of production in 1970, 35.9 percent in 1975, reached a high of 46.7 percent in 1981, and dropped to 42.0 percent in 1987. For wheat, 54.8 percent of production was exported in 1970, 64 percent in 1981. The export share of wheat declined to 37.7 percent in 1985, then rose to 54.2 percent of total production in 1987. Cotton exports were 38.2 percent of production in 1970, increased to 53.2 percent in 1980, reached a peak of 87.2 percent in 1983 and declined to 44.6 percent in 1987.[9]

Developments in the U.S. commodity markets have a strong influence in world markets. For instance, U.S. wheat production was 13.9 percent of world production in 1984, 13.4 percent in 1985 and 10.8 percent in 1986.

FIGURE 23.5 U.S. Net Exports by Product Category, 1970 vs. 1987

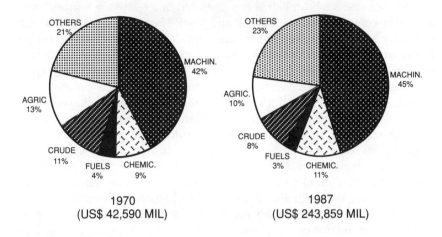

1970
(US$ 42,590 MIL)

1987
(US$ 243,859 MIL)

Note: Agric.-included food, live animals, tobacco, animal and vegetable and fat.

Source: USDC, Bureau of the Census, 1989.

Wheat exports, on the other hand, were 35.6 percent of world exports in 1984, 29.4 percent in 1985, and 31.1 percent in 1986. For corn, U.S. production as a share of world production was 42.5 percent, 46.9 percent and 44.0 percent in 1984, 1985 and 1986, respectively, while export shares for the same years were 70.1 percent, 57.8 percent and 53.9 percent of world exports. U.S. soybean producers accounted for 54.4 percent of world production in 1984, 58.9 percent in 1985 and 53.9 percent in 1986; export shares were 70.1 percent, 57.8 percent and 70.1 percent, respectively.

Soils and Climate

Most of the U.S. agricultural area lies between 30 degrees and 49 degrees north latitude, in zones with a temperate climate, with well delineated seasons and with productive soils. Climatic diversity in the United States is great, with wide variations in rainfall and temperature. While large areas are well suited to intensive farming, equally large areas are devoted to extensive farming. This combination provides an environment that supports many different crops and farming systems. With the exception of some tropical products, such as cassava and palm oil, U.S. farmers have the potential to produce nearly every grain and oilseed crop, as well as livestock and poultry, a large variety of fruits and vegetables, and tree crops. Economics of production and marketing, together with climatic limitations, determine the crops grown in each of the regions.

Grain-Growing Regions

The corn and soybean belt is concentrated in a relatively narrow band across the center of Indiana, Illinois, and Iowa. These deep fertile soils with generally adequate rainfall levels represent the largest contiguous region in the world ideally adapted to corn and soybean production. Although there are periods of weather adversity, the soil structure and topography provide a resilience to adversity, and serious crop failures are rare.

States to the west and south of this corn and soybean belt have less reliable rainfall patterns and their crops often require supplemental water through irrigation. The soils and climatic conditions of the west and southwest regions are better suited to the production of grain sorghum and wheat. The dryland wheat farming areas often use a fallow system in which the land is left idle in alternate years to accumulate water reserves.

The states to the north and east of the corn belt are still productive corn producing areas. Minnesota, Michigan, Ohio, and Wisconsin are all important in the production of corn; however, the shorter growing season reduces their importance in terms of soybean production. Oats and wheat are relatively more important in the cropping patterns of these Lake states than in the states of Iowa and Illinois. Barley is grown primarily in the northern plains states.

Although some of the soils in the southeastern United States are less fertile than those of the midwest, the region is still highly productive, with generally adequate rainfall. The soils and climate are well adapted for soybean production, with some land effectively devoted to corn. The low lying lands of the Mississippi Delta are highly fertile, with high rainfall. These states bordering the lower Mississippi River are adaptable to production of a wide range of crops but the primary grains are rice and soybeans. California also has an area of soil and climate well adapted to certain varieties of rice production.

Land Tenure

Farmland comprised 43.6 percent of the total U.S. land area in 1987. The number of farms has been declining steadily for many years from 2,521,420 in 1975 to a projected 2,158,800 in 1988 (Figure 23.6). Although the average decline over this period has been approximately 1.2 percent per year, in reality the trend follows a cyclical pattern. For example, during 1977 the number of farms declined by 1.7 percent; however, by 1981, the number of farms actually increased by 0.1 percent over the previous year. After 1981, the decline in farm numbers began again — reaching an annual 2.8 percent decline in 1986. Since 1986 the rate of decline in farm numbers

FIGURE 23.6 Number of U.S. Farms, 1975–88

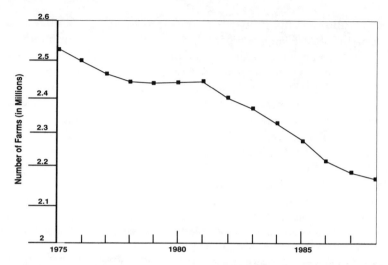

Source: National Agricultural Statistics Service.

has been less sharp. The number of farms declined by only 0.8 percent between 1987 and 1988.

Since total farm area has been declining at a slower rate than the decline
in farm numbers, farm size increased between 1982 and 1987. The average
size of farms in 1982 was 173 hectares. This increased to 187 hectares per
farm in 1988 (*Agricultural Statistics*, 1988). The average value per hectare in
1987 was $1,549, down from $1,936 in 1982. The average size of farms is
somewhat misleading, because 40 percent of U.S. farms have less than 20
hectares and 7 percent are over 405 hectares (Wallace, 1987). Large farms
account for approximately 4 percent of the total farm numbers but nearly
half of total farm cash receipts. Smaller farms with sales less than $10,000
account for nearly half of all farms but less than 4 percent of the total cash
farm receipts.

In 1984, over 60 percent of all farm units had less than $20,000 in farm
product receipts. Most of these low-income farmers (the $20,000 repre-
sents receipts, not net income) supplement their incomes with off-farm
jobs by family members. Among the 2,087,759 farm operators in 1987, 54.5
percent considered farming as their principal occupation. Although the
others qualified as farmers under the census definition, they had other
principal occupations. In 1987, 1,115,560 farm operators reported working

off the farm. This included 737,206 who reported working 200 days or more.

U.S. agriculture is still primarily a single-proprietor and family-oriented system. In 1982, 86.9 percent of all farms were individual or family operated; 10 percent were operated as partnerships; 2.7 percent were incorporated; and .5 percent were other types of organizations.

Ownership is highly concentrated in some areas of agriculture. Rental of additional land has become a primary source of expansion, especially on farms where mechanized grain production requires a large area to achieve economies of scale. Although there are about 6.2 million owners of farmland in the United States (including individuals, partnerships, and corporations), only 5 percent of these owners control 53 percent of the farmland, and 1 percent of the largest owners own 32 percent. This national proportion varies by region, from the northwest where 5 percent of the owners own 70 percent of the land to the corn belt where 5 percent of the owners own 27 percent of the land. Farmland ownership is most concentrated in the Pacific and mountain regions and least concentrated in the lake states and corn belt (Reimund, et.al., 1986). Approximately half of the farmland in the United States in 1982 was owned by individuals, partnerships, or corporations whose primary occupation or business was farming. An additional 14 percent was owned by farm operators whose principal occupation was something other than farming. The remaining 36 percent was owned by non-farmers.

There has been a marked increase in part-ownership of land since 1950. At that time, 15 percent of farm operators were part-owners, operating about 40 percent of the land.[10] By 1978, 29 percent of farm operators were part-owners and they operated 54 percent of the land. By contrast, the percentage of full tenants and the percentage of land they farm have declined, from 28 percent farming 18 percent of the land in 1950 to 11.6 percent farming 11.5 percent of the land in 1982 (*Agricultural Statistics*, 1988).

Farm Debt

Guither and Halcrow observed in 1988 that farm debt reached crisis levels in 1983-1985, but declined somewhat in 1986:

> In 1985, more than 200,000 farmers out of 1.7 million in a survey by the U.S. Department of Agriculture were in financial stress or vulnerable due to a high debt load and an inability to generate enough cash to pay their bills. The federally-sponsored farmer-owned Farm Credit System reported a record $368 million in loan losses in 1984, with heavier losses expected in the years to follow. Nearly one-third of all farms with an-

nual sales over $40,000 were facing some financial difficulties. This would force many farmers to discontinue their own operations, with losses to farm lenders and readjustments in the structure of farm debt.

Total farm debt (including operator households and loans from the Commodity Credit Corporation), which peaked at over $220 billion in the summer of 1983 after having risen each year since 1945, dropped to about $188 billion at the end of 1986. Debt excluding CCC loans, the relevant debt for studies of financial stress, which peaked at $205 billion (on a year-end basis) in 1983, declined by $36 billion to about $169 billion at the end of 1986. Debt secured by farm real estate had declined more slowly than the non-real estate debt.

Labor Force

In 1987 about 5.7 million people lived in households tied to the farm business, compared with 5.0 million in the farm population as conventionally defined by residence. The new economically defined group, called the farm entrepreneurial population, included people who depended on farming but did not necessarily live on the farm.

About 3.3 million of the farm entrepreneurial people were in the labor force, with an unemployment rate of 2.8 percent. Over 50 percent of people in the households classified as farm entrepreneurial worked in nonagricultural industries.

Family members, farm operators and unpaid workers continued to provide the major portion of agricultural labor. In 1987, the average farm employment was 2,897,000 of which 1,846,000 were family members and 1,051,000 were hired. However, the share of hired labor increased gradually from 1950 to 1987. In 1950, hired labor accounted for 23 percent of total farm employment. This share rose to 27 percent in 1960, fell slightly to 26 percent in 1970, rose again to 35 percent by 1980, and reached 36 percent in 1987.

Many hired farm workers are part-time or seasonal workers. The 1987 Agricultural Work Force Survey indicates that a total of 2,463,000 persons worked on farms in 1987 for an average of 112 days per worker.

Migrant farm workers make up a small but important sub-group of the domestic hired farm work force. Migrants supplement local labor during peak use seasons when the demand for farm workers frequently exceeds the supply of farm workers living in the local area. However, the total number of migrants and their share of the domestic hired labor force shrank significantly between 1959 and 1985. The number of migrants fell 67 percent from 477,000 to 159,000, and the percentage of migrant farm workers in the hired farm work force declined from 13.3 percent to 6.3 percent.

The presence of undocumented, often illegal, aliens working in seasonal agricultural employment has become a national policy issue. Legislation passed in 1986 attempts to reduce the flow of illegal aliens into the United States by imposing strict hiring requirements on U.S. employers. The law is also designed to help agricultural employers who have relied on illegal aliens in the past to adjust to a legal work force. By December 1, 1988, alien workers who had worked in seasonal agricultural work for at least 90 days in the year ending May 1, 1986 could apply for legal resident status. In the event of a shortage of agricultural workers in seasonal agricultural services, a special replenishment program allows additional entry for foreign workers.

The effect of the new immigration law on the hired farm work force will depend on a number of factors including the strictness of law enforcement, degree of compliance among agricultural employers, the number of aliens admitted under the special agricultural worker and replenishment programs, and the degree to which these workers, once legalized, continue to work in agriculture. It is worth noting that hired labor and migrant labor are relatively minor issues for highly mechanized grain production.

Productivity and Efficiency

U.S. agriculture is highly productive. However, this productivity is difficult to evaluate accurately because productivity measures either refer to a selected group of inputs and outputs or they aggregate across inputs and outputs in a non-comparable fashion. To obtain meaningful estimates of agricultural productivity, values must be placed on each of the inputs and outputs. These values vary with time and space. Measures of productivity per unit of labor or land are commonly used but have potential hazards for they may have been offset by changes in other inputs and do not capture effects of changes in quality or quantity of other inputs. For example, effects of environment, social structure, or changing technologies are seldom captured in partial productivity measures for land or labor.

It is even more difficult to make productivity comparisons across countries and cultures. For example, input-output ratios for corn production in Argentina seldom take account of the fertilizer generated by extended crop rotations where a hectare may be in corn only once in 5 or 6 years compared to U.S. corn farms where the nitrogen component is obtained from chemical sources and the hectare of land is in corn or soybean production continuously. Productivity indices combining inputs and outputs are value-weighted to account for price effects. The inputs and outputs selected for inclusion in the index and the weights assigned vary over time.

If these caveats are kept in mind, some useful comparisons can be made over time and across countries using the ratio of the index of total agricultural output to the index of total inputs used in farm production.

Influence of Technology. The most important influence on productivity in U.S. agriculture is technology. During the 20th century, at the same time that the amounts of land and human labor input have decreased, agricultural productivity has increased dramatically through the use of power machinery, mechanical equipment, commercial fertilizers, chemicals for disease and pest control, and prepared livestock feeds (Loomis and Barton, 1961).

In 1986, the output production index for total farm production was 111 (1977=100). The index was 110 for all livestock and livestock products, and 109 for all crops. During this 1977-1986 period the index of cropland used for crops dropped to 94 and the index of crop production per hectare increased to 116.

The index of labor hours required for livestock production was 60 and for all crops was 82. These figures suggest that labor efficiency of livestock production improved more than crop production between 1977 and 1986. The labor index per hour in livestock production was 183 in 1986 compared to 133 for labor per hour of crop production. The index for labor per hour in feed grains was 150; for food grains it was 135.

The index was 87 for all inputs used in farm production; 77 for farm labor, 90 for farm real estate, 74 for power and machinery, 118 for agricultural chemicals, 111 for feed, seed, and livestock purchases, 102 for taxes and interest and 141 for miscellaneous (*Cost of Production Economic Indicators*, 1987).

Table 23.1 shows productivity increases between 1940 and 1984. T.L. Wallace attributes these improvements to technology, but suggests that the upward trend may not continue because of drops in public research funding:

> Total productivity growth in farm production was slow and erratic between 1880 and 1940, but between 1940 and 1981 total productivity more than doubled. A major change in mix of inputs has occurred. For example, farm labor declined by 78 percent between 1940 and 1980, (but) the index of labor productivity rose almost ten-fold. Agricultural chemical inputs rose more than 13 times, mechanical power and machinery inputs tripled, and farm real estate declined slightly. Clearly, there was substitution of new technology (in the form of purchased inputs) for labor and for land. There is concern that these productivity rates may not continue because of a pronounced real drop in public research funding which has generated so much past farm and food sector productivity.

TABLE 23.1 Indices of U.S. Farm Output, Input and Productivity, 1940-1984
(1967=100)

Year	Total output	Total input	Productivity
1940	60	100	60
1945	70	103	68
1950	74	104	71
1955	82	105	78
1960	91	101	90
1965	98	98	100
1970	101	100	102
1975	114	100	115
1980	122	106	115
1981	138	105	128
1982	136	102	132
1983	111	98	112
1984	130	99	131

Source: Wallace, Tim L., Agriculture's Futures: America's Food System, Springer-Verlag, New York, New York, 1987.

Some projected productivity growth rates to the year 2000 are more nearly in line with the 1925-50 period (1.3 percent per year) than of the much higher rates realized since 1950.

Cost of Production Studies

Cost of production measures for grain crops are based both on estimates of cash expenses for producing the crop and on estimates of cost per metric ton in a given year. Though both methods of measurement have limitations, the following discussion by crop will provide information on both.

Corn. Total cash expenses for corn produced in Mountain States averaged $376.01 per hectare in 1987. When divided by average yield, 1987 costs were $56.69 per metric ton. When all costs including land were considered, the cost per metric ton increased to $84.25. Total costs including land varied among regions from a low of $520.35 per hectare in the Southeast to a high of $627.65 in the Lakes States and Corn Belt.

Soybeans. Total cash expenses for U.S. soybean production were $108.06 per planted hectare in 1987. Converted to cost per metric ton, cash expenses were $102.15. When all costs including land were included, the cost per metric ton rose to $180.78. Total costs including land varied among regions from a low of $318.51 per planted hectare in the Delta region to a high of $445.98 per planted hectare in the North Central region.

Wheat. Total cash expenses for wheat in 1987 were $169.05 per planted hectare. Divided by average yield this resulted in a cost per metric ton of $84.88 for cash expenses. When all costs including land were included, the cost per metric ton increased to $141.46. Total costs including land varied among regions from a low of $232.72 per planted hectare in the Southern Plains to a high of $748.34 in the Southwest. These wide differences in costs per planted hectare are somewhat mitigated by differences in yields.

Barley. Total cash expenses for barley in 1987 were $190.63 per hectare averaged across the United States. Total cash expenses per metric ton were $79.00. When all costs including land were included, the cost per metric ton was $122.17. Total costs including land ranged from a low of $264.81 per planted hectare in the Northern Plains to a high of $464.38 in the Northeast.

Rice. Total cash expenses for rice were $768.84 per planted hectare in 1987. Residual returns to management and risk (excluding direct government payments) were negative, -$368.05. Total costs including land ranged from a low of $889.30 in the Delta region to a high of $1,267.04 per hectare in California. There is a wide range of yield per planted hectare between the high and the low cost areas. Average yield in California for 1987 was 7.9 metric tons per planted hectare in contrast to average yield in the Delta region of 5.4 metric tons per planted hectare. Total cash expenses per metric ton produced in 1987 were $125.44. Total costs including land were $166.45 per metric ton (*Cost of Production Economic Indicators*, 1987).

Financing Agricultural Production

Financing for agricultural production in the United States is typically broken down into two major categories: real estate debt and nonreal estate debt. Real estate debt is secured by farm real estate and is commonly used for the purchase of farmland and buildings. It typically has a maturity of 5 to 30 years and is quite often written with variable interest rates. Nonreal estate debt is used to finance operating inputs including seed, fuel, and fertilizer, and to finance the purchase of machinery and equipment used to operate farming enterprises.

Real Estate Debt. Real estate debt accounts for well over one-half of all farm debt in the United States (Table 23.2). The Federal Land Bank, a borrower-owned cooperative, holds the largest volume of farm real estate debt, but its loan volume has declined faster than that of most other lenders. The second most important source of farm real estate debt are individuals and others. Individuals who sell their farmland often provide financing in the form of land contract sales.

TABLE 23.2 Debt Outstanding, Excluding Operator Households, by Lender, December 31, 1983-1989, United States

Lender	1983	1985	1986	1987	1988 [a]	1989 [b]
	Million U.S. Dollars				Billion US Dollars	
Real Estate						
Federal Land Banks	45,026	41,204	34,773	29,867	28	27 to 30
Farmers Home Administration	8,718	9,540	9,482	9,249	9	6 to 9
Life Insurance Companies	11,834	11,035	10,199	9,231	9	8 to 10
Commercial Banks	8,494	10,443	11,677	13,307	14 [c]	13 to 16 [c]
CCC Storage Facility	888	307	123	46		
Individuals & Others	29,847	25,160	22,218	19,086	17	15 to 17
Total	104,807	97,689	88,472	80,786	76	74 to 78
Nonreal Estate						
Commercial Banks	37,075	33,738	29,678	27,589	28	28 to 30
PCAs & FICBs	19,392	14,002	10,581	9,271	9	8 to 10
Farmers Home Administration	12,855	14,714	14,425	14,123	13	8 to 12
Individuals & Others	18,566	15,070	12,143	10,916	12	12 to 13
Total	87,888	77,524	66,827	61,899	62	60 to 64
Total Debt	192,695	175,213	155,299	142,685	138	134 to 142

[a] Preliminary.
[b] Forecast.
[c] Less than $500 million.
Source: Agricultural Income and Finance: Situation and Outlook Report, Economic Research Service, United States Department of Agriculture, AFO-33, May, 1989.

Commercial banks are another important source of farm real estate loans. Their loan volume has been growing since 1983. Development of a new secondary market for farm mortgage loans—Farmer Mac—is expected to further enhance the commercial banks' market share of farm real estate debt.

Two federal government lending agencies, the Commodity Credit Corporation (CCC) and the Farmers Home Administration (FmHA) provide farm real estate loans. Loans from the CCC have been for grain storage facilities. This source of funding is rather minimal at present. The FmHA is considered a lender of last resort, as it can lend only to farmers who do not qualify for commercial credit. Some FmHA loans are loaned at subsidized rates; others are based upon the government's cost of borrowing. Government lending agencies provide only about 10 percent of farm real estate loans and most of these are extended to farmers with a weak financial position.

Non-Real Estate Debt. Commercial banks are the dominant supplier of nonreal estate farm debt. While total volume has declined in recent years, market share has increased modestly. Production Credit Associations (PCAs) and Federal Intermediate Credit Banks (FICBs) have experienced a sharp drop in total volume and market share.

The FmHA has held a relatively stable dollar volume of nonreal estate debt during a declining market; its market share rose from 14.6 percent in 1983 to almost 23 percent in 1987. Recent efforts to curb FmHA lending are creating a decline in both dollar volume and market share.

The category "individuals and others"consists primarily of credit provided by merchants and dealers who supply operating and capital inputs to farm operators. The significant decline in this source of nonreal estate credit appears to be demand driven — producers are less willing than in the past to finance the purchase of operating inputs.

CCC loans, a special type of operating credit not included in Table 23.2, operate as follows: a loan rate is established annually for selected agricultural commodities — wheat, rice, corn, sorghum, barley, oats, soybeans, and upland cotton. Farmers who participate in the farm programs for these commodities can acquire a CCC loan at the established loan rate for each commodity. If the market price for the commodity is above the loan rate, the farmer can repay the loan with interest and sell the commodity on the open market; if it is below, the farmer can forfeit the commodity to the CCC in full payment of the loan. Seen in this light, CCC loans act simply as a floor on the price farmers will receive for participation in the various government programs.

Table 23.3 identifies the net outlays for CCC price-support loans from 1980 through 1990. The net amount of CCC loans varies widely from year to year and depends primarily upon the relationship between market prices and loan rates. The time framework in which CCC loans are extended

TABLE 23.3 Net CCC Price-Support Loans,United States,1980-1990

Year	$ Million
1980	66
1981	1174
1982	7,015
1983	8,438
1984	-27
1985	6,272
1986	13,628
1987	12,199
1988	3,579
1989 [a]	-153
1990 [a]	1,011

[a] Estimate.
Source: Agricultural Outlook, ERS-USDA, July, 1989.

and repayment or forfeiture is required can influence the demand for credit from commercial lenders. For example, if CCC loans are provided at the time of planting, this may reduce farmers' demands for operating credit from commercial sources.

For additional information on government credit institutions, see Cramer and Jensen, Chapter 10.

Production of Major Agricultural Products

Food and feed grains are a major source of income for U.S. farmers. Livestock and livestock sales comprise the majority of agricultural sales. In 1986, for example, cattle, hogs, and dairy products accounted for 42 percent of total sales of agricultural products; corn, soybeans, and wheat accounted for 20 percent; and the remaining 38 percent of sales were of fruits, vegetables, and other grains (Figure 23.7).

The distribution among products varies dramatically by agricultural region. In Illinois, animal product sales equalled the national figure of 42 percent, but the distribution among animal products was skewed toward hog sales. Sales of corn, soybeans and wheat greatly exceeded the national figure of 20 percent, accounting for 66 percent of total sales.

Marketing of Agricultural Products

The United States does not have government-operated marketing boards. Demand and supply are the major forces directing prices in most

FIGURE 23.7 Total Value of Sales in the United States by Major
Agricultural Product Category, 1986

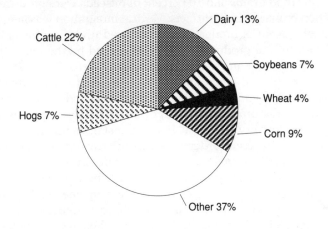

Source: Agricultural Statistics, 1988.

commodities. The influence of government on prices is generally indirect either through price supports, commodity loan programs, supply control, or authorization of marketing boards operated by producers and/or processors in individual industries.

Although the federal government has an important influence on sales of agricultural products and in some cases actually owns the product being sold, the general pattern is that privately owned products are sold, in a competitive environment, to processors, wholesalers, retailers, or consumers within the limits of government regulations and enforcement of contracts. The extent of competition differs among regions and crops.

The market channel is often complex, and movement of the physical goods through the channel is paralleled by a flow of pricing and transactions. Marketing firms as well as farmers provide both physical and transaction functions. In the pricing area, there is wide diversity among crops. At one extreme, grain prices are set minute by minute by public outcry on the floor of the Chicago Board of Trade. Although many factors enter into that price, it approaches the theoretically competitive price determination more closely than any other marketing technique. The Chicago Board of Trade provides a base that is utilized worldwide on the assumption that it represents the best estimate of the value of the product at any point in time.

Use of contracts and other forms of vertical integration has been increasing over time. Though contract-setting is common in the broiler industry and for horticultural crops, the practice is not strongly established for feed grains, food grains, or oilseeds. At the time of a study by Cramer and Jensen providing data for 1980, only 7 percent of the U.S. feed grains, 8 percent of the food grains and 10 percent of oilseeds were produced under production or marketing contracts. Vertical integration was negligible. In contrast, 89 percent of broilers were produced under contract and for another 10 percent the production unit was owned by the marketing or processing firm (Cramer and Jensen, 1988 page 27, Table 2.5).

Prices in horticultural crops are often set by contract in advance of planting. Many fruits and vegetables are produced under contracts with growers in which the processing plant, or a marketing cooperative, contracts for quantity, quality and price. In many instances this contract is organized under a marketing order voted in by growers and authorized through legislation. Marketing orders often include supply control, as well as price quotas, and allocate product among distribution channels.

In grain production, the use of contractual arrangements, though rare, is increasing. Many firms that require specialized varieties or seek close control of quality characteristics are extending their use of contracts to corn, wheat, and soybeans. Dry millers requiring certain corn varieties have found it profitable to contract with farmers for regular dent-type corn.

Notes

1. Statistics on GNP are taken from the following sources: *Statistical Abstract of the United States*, 1989; *Economic Report of the President*, 1989; *National Income and Product Accounts of the United States*, 1986; and *Survey of Current Business*, 1989.

2. Office of Management and Budget, *Historical Tables, Budgets of the United States Government*, published annually. See also the 1989 issue of the *Statistical Abstract of the United States*.

3. See *Statistical Abstract of the United States*, 1989; *Economic Report of the President*, 1989; and *Historical Tables, Budget of the United States*, 1989.

4. 1982 dollars. For 1988, quarterly data are expressed at seasonally adjusted annual rates.

5. For additional data and for the computation of the poverty level, see *Statistical Abstracts of the United States*, 1989.

6. For these and other statistics on U.S. Education, see *Digest of Education Statistics*, U.S. Department of Education, National Center for Education Statistics, September 1988.

7. A comprehensive set of projections for per capita consumption of food as well as of other indices related to agricultural performance can be found in the two FAO publications by N. Alexandratos listed in the references.

8. The 1988 number is an estimate. See *Statistical Abstract of the United States, 1989* and *Historical Tables, Budget of the United States Government, Executive Office of the President,* 1989.

9. Data represent marketing years. See *World Development Report,* 1986.

10. Part-owners are defined as farm operators who own part of their land and rent additional land.

24

Grain Production

Grain Sub-Sector Performance

Total harvested area of principal crops in the United States has varied around an average of 124.9 million hectares, ranging from a low of 113.8 million hectares in 1969 to a high of 143.3 million hectares in 1981 (Figure 24.1). Increases in total area have come from land clearing and drainage, reclamation of eroded land and irrigation in regions of limited rainfall. Decreases have come from land retirement programs, urban expansion and increased non-agricultural uses for land. The relatively small net change between 1960 and 1987 and the lack of any upward trend indicates that there have been no major acreage expansion programs such as those seen in Argentina and Brazil.

Figure 24.2 shows the 1987 allocation of harvested area of principal crops among the major grains. Corn, soybeans and wheat were nearly equal with 19 percent to 20 percent each; together, they accounted for 38 percent of the total harvested area. The most significant change over the past 20 years has been the increase in soybean area from only 8 percent of the area of principal crops in 1960 to its current share of 19 percent.

Corn

The United States is a world leader in corn production, with an annual production usually approaching 50 percent of the world total. The annual supply available for domestic use and exports consists primarily of current production and carryover stocks. Imports have been small and do not significantly affect annual supply. In recent years, the U.S. area planted to corn has accounted for about 24 percent of the area planted to principal crops declining to less than 20 percent in 1987 (Figure 24.2). On the average, about 3.6 million hectares are harvested for silage and forage and the balance is harvested for grain.

Area, Yield and Production. Corn area harvested for grain has varied considerably over time. Prospects for huge carryover stocks of corn in 1961 led the government to resume area restrictions, and harvested area dropped by 5.6 million hectares between 1960 and 1961 (Table 24.1). Restrictions on planted area remained in effect until 1973. From 1978 to 1982, the area harvested for grain averaged almost 30 million hectares. In 1983, it dropped to 20.9 million hectares, the lowest level in more than a hundred years, when over 12.6 million hectares of corn were taken out of production under the Payment-In-Kind (PIK) program. With expectations of higher prices following the short crop of 1983, farmers increased total area planted and harvested area rebounded to 29.1 million hectares in 1984. Total harvested area increased again in 1985, to 30.4 million hectares, then dropped to 24.0 million in 1987.

FIGURE 24.1 Total Harvested Area of Principal Crops

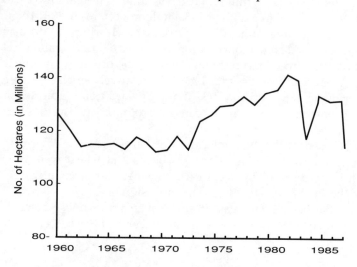

Source: Agricultural Statistics, 1988.

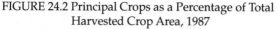

FIGURE 24.2 Principal Crops as a Percentage of Total
            Harvested Crop Area, 1987

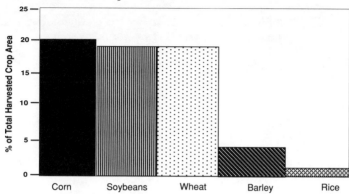

Source: Agricultural Statistics, 1988.

Grain yields have more than doubled since 1960. The increases were dramatic prior to 1973 when area restrictions were in effect. Average yield reached 6.1 metric tons per hectare in 1972. From 1973, a period of area expansion curbed the trend toward higher yields because marginal, less productive land was placed in corn production. The yield record of 1972 was not broken until 1978 when average yields surpassed 6.3 metric tons per hectare for the first time in history. Average yields continued to increase, though bad weather causes an occasional reduction (including a notable reduction to 5.1 metric tons in 1983). Yields reached an all-time high of 7.5 metric tons per hectare in 1986 followed by another record in 1987.

The general increase in yields over time is due to changes in technology and improved production practices, including development of improved high-yielding hybrids, increased rates of fertilization, higher seeding rates, and improved control methods for weeds, insects, and diseases.

Several studies have focused on the factors that have contributed to increasing corn yields in the United States (Butell and Naive, 1978) (Sundquist, et al., 1982) (Thompson, 1978). Sundquist, et al. evaluated three factors that contributed to higher yields between 1954 and 1980: average nitrogen application rate, July precipitation in the Corn Belt, and technology. Their study revealed that the contribution of non-nitrogen technologies has been constant over time, approximately .06 metric tons per hectare per year. In contrast, the contribution of nitrogen has fallen over time. Increases in the nitrogen application rate contributed approximately .13 metric tons per hectare per year during the 1950s and 1960s. Nitrogen application rates approached optimum levels during the 1970s, and the marginal impact of additional nitrogen was very small.

TABLE 24.1 Corn Area, Yield and Production, United States 1960-1984

Crop Year	Area Harvested for Grain	Average Yield per Hectare	Grain Production
	Million Hectares	Metric tons	Million Metric tons
1960	28.9	3.4	99.2
1961	23.3	3.9	91.4
1962	22.6	4.1	91.6
1963	24.0	4.3	102.1
1964	22.4	3.9	88.5
1965	22.4	4.6	104.2
1966	23.1	4.6	105.8
1967	24.6	5.0	123.5
1968	22.7	5.0	113.0
1969	22.1	5.4	119.1
1970	23.2	4.5	105.5
1971	26.4	5.5	143.4
1972	23.3	6.1	141.7
1973	25.1	5.1	144.1
1974	26.5	4.5	119.4
1975	27.4	5.4	148.4
1976	29.0	5.5	159.7
1977	29.0	5.7	165.2
1978	29.1	6.3	184.6
1979	29.3	6.9	201.4
1980	29.6	5.1	168.6
1981	30.2	6.8	206.2
1982	29.4	7.1	209.2
1983	20.9	5.1	106.1
1984	29.1	6.7	195.0
1985	30.4	7.4	225.5
1986	28.0	7.5	209.6
1987	24.0	7.5	179.6

Source: Agricultural Statistics, United States Department of Agriculture, various issues.

The total supply of corn increased dramatically during the late 1970s and early 1980s, reaching a record 264 million metric tons in 1982/83. The record production levels in 1981 and 1982 led to a rapid accumulation of carryover stocks — a record 78.7 million metric tons at the end of the 1982/83 marketing year. Cuts in area under the PIK program in 1983, combined with the effects of a severe drought in the summer of 1983, reduced corn production by about one-half. Carryover stocks fell from 88.9 million metric tons in September of 1983 to 25.4 million metric tons in September of 1984.

Illinois and Iowa compete for first place in corn production. In 1987, Iowa produced 18.5 percent of U.S. production; Illinois ran a close second with 17.0 percent. Nebraska, Minnesota and Indiana followed with 11.5 percent, 9.0 percent and 8.9 percent respectively. The top five states accounted for nearly 65 percent of the nation's corn production in 1987.

Much of the world's inventory of corn is carried across crop years by U.S. farms and elevators, in response to government programs and anticipated price increases. October 1 carryover has ranged from a low of 9.2 million metric tons on October 1, 1975, to a high of 121.9 million metric tons in October, 1987. The ratio of stocks to production often approaches 1:2. In 1987/88, stocks increased to create a ratio of 3:4. Farm stocks of corn, at their peak in January for most years, account for about 70 percent of the total inventory. Farmers clearly provide a major portion of the corn held in storage.

Since most farms allocate area to more than one crop, the average size of corn farms does not provide a useful comparison among crops. However, the U.S. Census of Agriculture reports the number of farms growing corn and total area of corn harvested. This ratio indicates relative size of the corn enterprise in all regions of the United States. The U.S. average may be misleading in terms of commercial operations in the corn belt, where average corn area is larger. Data from the 1982 Census of Agriculture shows the nationwide average area of corn per farm was 41 hectares. By way of comparison, in 1982 the average area of corn harvested in Illinois was 63 hectares per farm.

Corn is the feedstock for several processing industries. The largest is the livestock industry where on-farm feeders and feed manufacturers utilize about 60 percent of total disappearance and account for about 80 percent of domestic use. Industrial use for sugar, starch and alcohol (wet milling), is growing rapidly, at an annual rate of about 10 percent. The dry milling industry has also shown growth during the 1970s and 1980s, but still accounts for less than 5 percent of total domestic use.

Exports provide a major market for U.S. corn, varying with U.S. production and world demand and supply. Exports averaged 13.7 million metric tons for the decade of 1960/61 to 1970/71, about 13 percent of total disappearance. For the decade of the 1970s, exports averaged 38.1 million metric tons, about 25 percent of total disappearance. Exports declined during the 1980s, with a small recovery in 1987/88 to 43.2 million metric tons (22.5 percent of total disappearance).

The primary destination of U.S. corn exports has shifted from Western Europe to the Pacific rim countries, primarily Japan (Table 24.2). In 1987, 31.5 percent of U.S. exports went to Japan, 13.1 percent to the U.S.S.R., and

TABLE 24.2 U.S. Corn Exports by Destination, 1987

Destination Country	Quantity (mmt)	U.S. Exports (%)
Japan	12.9	31.5
U.S.S.R.	5.3	13.1
S. Korea	4.5	11.1
Mexico	3.3	8.2
China (T)	3.0	7.4
EC-12	2.3	5.7
Rest of World	11.7	28.7
Total	40.8	100.0

Source:"Foreign Agricultural Trade of the United States,"Calendar Year 1988 Supplement, Economic Research Service, USDA.

11.1 percent to South Korea. Increased production of feed grains (including wheat) in Europe, increased livestock production in Japan, and increased competition from Argentina have all caused a shift in destinations from Western Europe to Japan.

Soybeans

The United States produces approximately 51 percent of the world's soybeans. Harvested area has increased significantly since 1960, reaching a high of 28.5 million hectares in 1979 (Table 24.3).

Area planted to soybeans has been about 19 percent of the area planted to principal crops in recent years. In 1987 soybean and corn areas were about equal as production restraints were placed on corn, and price advantages shifted more land into soybeans (Figure 24.2).

Soybean yields have increased slowly but steadily. A record national average of 2.3 metric tons per hectare was achieved in 1985. Yields declined slightly in 1986, returning to 2.3 metric tons per hectare in 1987. The yield per hectare varies among states due to differences in soil and climate. For example, in 1987 the average yield for Illinois was 2.6 metric tons per hectare in contrast to the national average of 2.3 metric tons per hectare. Soybeans in the Midwest are traditionally grown in rotation with corn.

Illinois and Iowa alternate as the lead state in soybean production. In 1985 Illinois produced 10.4 million metric tons; Iowa was a close second with 8.4 million metric tons. In 1987, the relative positions were reversed: Iowa produced 9.3 million metric tons, Illinois 8.8 million metric tons. The other major soybean producing states, with more than 2.7 million metric tons each, include: Minnesota, Indiana, Missouri and Ohio. These six states together accounted for 69.2 percent of the total 1987 production.

TABLE 24.3 Soybean Area, Yield, and Production, United States 1960-1984

Crop Year	Area Harvested for Grain	Average Yield per Hectare	Grain Production
	Million Hectares	Metric Tons	Million Metric Tons
1960	9.6	1.6	15.1
1961	10.9	1.7	18.5
1962	11.2	1.6	18.2
1963	11.6	1.6	18.2
1964	12.5	1.5	19.1
1965	13.9	1.6	23.0
1966	14.8	1.7	25.3
1967	16.1	1.6	26.6
1968	16.8	1.8	30.1
1969	16.7	1.8	30.8
1970	17.1	1.8	30.7
1971	17.3	1.8	32.0
1972	18.5	1.9	34.6
1973	22.6	1.9	42.1
1974	20.8	1.6	33.1
1975	21.7	1.9	42.1
1976	20.0	1.8	35.1
1977	23.4	2.1	48.1
1978	25.8	2.0	50.9
1979	28.5	2.2	61.5
1980	27.4	1.8	48.9
1981	26.8	2.0	54.1
1982	28.1	2.1	59.6
1983	25.3	1.8	44.5
1984	26.8	1.9	50.6
1985	24.9	2.3	57.1
1986	23.6	2.2	51.8
1987	22.8	2.3	51.8

Source: Agricultural Statistics, United States Department of Agriculture, various issues.

Most of the world inventories of soybeans are carried across crop years by U.S. farms and elevators in response to anticipated price increases. September 1 carryover has ranged from a low of 0.8 million metric tons in October, 1965 to a high of 14.6 million metric tons in October, 1986. The ratio of stocks to production has been as high as 1:4. About 50 percent of January 1 inventory has been held on farms.

The size of enterprise for soybean production differs among states as a result of differences in cropping patterns and the rotational schemes. There is a major contrast between the Northern Cornbelt and the Southern

Delta regions, although both are important in total production. The national average calculated from census data shows 51.4 hectares of soybeans per farm reporting soybeans harvested in 1982; the average for Illinois was 54.7 hectares.

In the United States, soybeans are used almost entirely for processing into oil and meal. Only insignificant quantities are used for food products. Exports still provide a major market for soybeans. About 30 percent of the total crop is exported as raw beans. When soybean products are included in the export figures, the export proportion of production increases to nearly 40 percent. Soybean exports increased dramatically from 1960 through 1987. The most rapid growth occurred in the 1970s. A peak of 25.3 million metric tons was exported in 1981. The decline in total exports and world market share since 1981 are the result of many economic factors in world trade, including competition from other nations.

The primary destination of soybeans exported from the United States differs by port region. In 1987, exports from the lake ports went primarily to Western Europe. Japan was the primary importer from the Gulf ports, receiving 19.1 percent of the exports from that region. The Netherlands was another important destination of exports from the Gulf, with receipts nearly equivalent to Japan's. Taiwan was the third largest market for Gulf soybeans, with a 9.7 percent share. West Germany and Spain each imported 7.3 percent of the Gulf's soybeans, making the European Community the primary market for Gulf soybeans in 1987. Total U.S. exports from all ports follow a similar pattern, with the Netherlands importing slightly more than Japan (Table 24.4).

TABLE 24.4 U.S. Soybean Exports by Destination, 1987

Destination Country	Quantity (mmt)	U.S. Exports (%)
Netherlands	4.0	18.4
Japan	3.9	18.2
China (T)	1.9	8.7
Spain	1.7	8.0
Fed. Rep. Germany	1.4	6.7
South Korea	1.1	5.3
Mexico	1.0	4.8
Rest of World	6.5	30.0
Total	21.6	100.0

Source:"Foreign Agricultural Trade of the United States,"Calendar Year 1988 Supplement, Economic Research Service, USDA.

Wheat

There are many classes of wheat, with different uses and production regions. However, most statistics treat wheat as a single crop for reporting purposes, and that same procedure will be followed here.[1]

The United States produces approximately 11 percent of the world's wheat crop. Wheat accounts for about 19 percent of the harvested area of principal crops (Figure 24.2). It is the second largest crop in the U.S. in terms of volume produced and the third largest in terms of farm cash receipts. The area harvested has varied from a low of 17.7 million hectares in 1962 to a high of 32.6 million hectares in 1981 (Table 24.5). Area harvested increased during the early 1970s, dropped dramatically in 1977 and 1978, then fluctuated around an average of about 27 million hectares over the next decade.

Although yield differs dramatically from year to year depending upon weather conditions, the broad geographic region over which wheat is produced averages many of the climatic problems. U.S. average wheat yield has increased over time, reaching 2.3 metric tons per hectare in 1979, 28 percent above the 1960 yields. During the 1980s, yield has fluctuated around an average of 2.4 metric tons per hectare (Table 24.5).

Production since 1960 has fluctuated from a low of 29.7 million metric tons in 1962 to a high of almost 75.8 million metric tons in 1981. Production is concentrated in the Northern Plains, Southern Plains and Mountain regions. Kansas and North Dakota reported the largest production of wheat in 1987, jointly accounting for 30.2 percent of all the wheat produced in the United States. Other states reporting more than 2.7 million metric tons produced in 1987 include: Montana, Minnesota, Oklahoma, South Dakota, Texas and Washington.

Inventories of wheat are influenced significantly by government programs. The June 1 carryover has ranged from a low of 9.3 million metric tons in 1974 to a high of 51.7 million metric tons in 1986. The ratio of stocks to production has fluctuated from a low of 1:5 in 1974 to a high of 1.3:1 in 1982. About 40 percent of October 1 stocks are typically held on farms.

Size of enterprise can be estimated by dividing the total area of wheat by the total number of farms reporting wheat. The 1982 census of agriculture shows a total of 445,743 farms, with an average area per farm of 64.4 hectares. Size of enterprise in wheat is generally larger than for corn or soybeans because wheat is grown more extensively in dryland areas. For example, Kansas averaged 260 hectares per farm in 1982.

In the domestic market, wheat is used primarily for milling purposes. A wide variety of products are made with different combinations of the various classes of wheat. Millers and bakers often need blends of qualities and classes. Wheat is also used as a feed grain when low wheat prices make it competitive with other feed grains such as corn and sorghum.

TABLE 24.5 Wheat Area, Yields, and Production, United States 1960-1984

Crop Year	Area Harvested for Grain	Average Yield per Hectare	Grain Production
	Million Hectares	Metric tons	Million Metric tons
1960	21.0	1.8	36.9
1961	20.9	1.6	33.5
1962	17.7	1.7	29.7
1963	18.4	1.7	31.2
1964	20.2	1.7	34.9
1965	20.1	1.8	35.8
1966	20.1	1.8	35.5
1967	23.6	1.7	41.0
1968	22.2	1.9	42.4
1969	19.1	2.1	39.3
1970	17.7	2.1	36.8
1971	19.3	2.3	44.0
1972	19.1	2.2	42.0
1973	21.9	2.1	46.6
1974	26.5	1.8	48.5
1975	28.1	2.1	57.9
1976	28.7	2.0	58.5
1977	27.0	2.1	55.7
1978	22.9	2.1	48.3
1979	25.3	2.3	58.1
1980	28.8	2.3	64.8
1981	32.6	2.3	75.8
1982	31.5	2.4	75.3
1983	24.9	2.6	65.9
1984	27.1	2.6	70.6
1985	26.2	2.5	66.0
1986	24.6	2.3	56.9
1987	22.6	2.5	57.3

Source: Agricultural Statistics, United States Department of Agriculture, various issues.

Feed wheats are generally the softer wheats grown in the Midwest and South. The higher-protein, stronger-gluten wheats needed for most baking industries are grown primarily in states west of the Missouri River.

Exports of wheat have generally accounted for 50 percent to 60 percent of total production, though in 1985 and 1986 they fell below 50 percent. Between 1977 and 1985, export volume changed little but the distribution of exports by port region changed substantially. The Great Lakes ports lost exports to the other three ports. In 1987, the Gulf region had the largest

share of total exports (52 percent). The Pacific coast was second with nearly 40 percent. The other port regions had small volumes destined for few
countries.

The U.S.S.R. was the largest importer of wheat from all U.S. ports in
1987, importing 4.8 million metric tons, nearly 16 percent of total U.S. exports; 98 percent of this moved through the Gulf ports. The next largest importer was Japan, which imported 3 million metric tons or 9.8 percent of
total U.S. wheat exports (Table 24.6).

Barley

Barley is a minor crop in the United States when considered on the basis
of total area, accounting for approximately 4 percent of the total area of
principal crops (Figure 24.2). The U.S. accounted for only 6.4 percent of
world production in 1987.

The 1988 barley crop is estimated at 6.3 million metric tons, 45 percent
below 1987 production and the smallest since 1953, reflecting fewer hectares and a dramatic drop in yield. This is a reversal of a generally upward
trend over the past ten years. The total supply reflects imports of about 0.3
million metric tons, a contrast with the other major grains.

Barley area changed little between 1960 and 1987, though harvested
area dropped dramatically from the mid-1970s to the early 1980s, from 4.2
million hectares in 1973 to only 3.0 million in 1980 (Table 24.7).

After 1982, area returned to the 4 to 5 million hectare range of the early
1960s. Per-hectare yields of barley increased steadily between 1960 and
1982, from a low of 1.6 metric tons in 1961 to a high of 3.1 metric tons in
1982. The ten-year average from 1960 through 1969 was 2.1 metric tons per
hectare. The average yield from 1980 through 1987 was 2.8 metric tons per
hectare.

TABLE 24.6 U.S. Wheat Exports by Destination, 1987

Destination Country	Quantity (mmt)	U.S. Exports (%)
U.S.S.R.	4.8	15.8
Japan	3.0	9.8
Egypt	2.4	7.9
Morocco	2.0	6.6
South Korea	1.9	6.2
Algeria	1.9	6.3
China	1.9	6.3
Rest of World	12.6	41.1
Total	30.6	100.0

Source: "Foreign Agricultural Trade of the United States", Calendar Year 1988 Supplement, Economic Research Service, USDA.

TABLE 24.7 Barley Area, Yield, and Production United States 1960-1984

Crop Year	Area Harvested for Grain	Average Yield per Hectare	Grain Production
	Million Hectares	Metric tons	Million Metric tons
1960	5.6	1.7	9.3
1961	5.2	1.6	8.5
1962	4.9	1.9	9.3
1963	4.5	1.9	8.6
1964	4.2	2.0	8.4
1965	3.7	2.3	8.5
1966	4.2	2.1	8.5
1967	3.7	2.2	8.1
1968	3.9	2.4	9.3
1969	3.9	2.4	9.3
1970	3.9	2.3	9.1
1971	4.1	2.5	10.1
1972	3.9	2.3	9.2
1973	4.2	2.2	9.1
1974	3.2	2.0	6.5
1975	3.5	2.4	8.3
1976	3.4	2.4	8.3
1977	3.9	2.4	9.3
1978	3.7	2.6	9.9
1979	3.0	2.7	8.3
1980	3.0	2.7	7.9
1981	3.6	2.8	10.3
1982	3.6	3.1	11.2
1983	3.9	2.8	11.1
1984	4.5	2.9	13.0
1985	4.7	2.7	12.9
1986	4.9	2.7	13.3
1987	4.0	2.8	11.5

Source: Agricultural Statistics, United States Department of Agriculture, various issues.

Barley is grown in fairly small units, and production is concentrated in a relatively small number of states, primarily the Northern Great Plains states. In 1987, North Dakota and Montana produced 44.4 percent of the total U.S. barley production. Minnesota and Idaho produced 21 percent. The average area per farm reporting barley was 44 hectares in 1982.

Barley is divided into two classes — feed barley and malting barley. These are generally not substitutes due to intrinsic properties and price differences of $18.37 to $68.89 per metric ton. Feed use has increased in the past decade, reaching a high of 7.2 million metric tons in 1985/86. Feed use declined in 1986/87. Feed use usually accounts for a little over 60 percent of domestic disappearance. Food, alcohol and industrial use of barley increased consistently through 1980/81, and has fluctuated since then around an average value of 3.3 million metric tons, about 34 percent of total domestic use.

Exports provide an important, but uncertain, market for U.S. barley producers. Barley is exported both as a feed grain and as malting barley for the malting industries in other countries. During the past 13 years, exports have ranged from a low of 0.5 million metric tons in 1985/86 to a high of 3.0 million metric tons the following year. Exports as a percentage of total supply have ranged from 2.1 percent to 14.5 percent. In years of low production domestic users (primarily food and industrial) import supplies from other countries. For example, when production fell from 13.3 million metric tons in 1986/87 to 11.5 million metric tons in 1987/88, imports jumped from 0.2 million metric tons to 0.3 million. Imports have represented a significant proportion of total supply in most years, larger than that for any of the other grains.

The primary destinations of barley exports in 1987 were as follows: Saudi Arabia, 78.9 percent; Israel, 6.7 percent; and Algeria, Iraq and Poland, between 2 percent and 4 percent each (Table 24.8).

Rice

The United States produces less than 2 percent of the world's annual rice crop, approximately 7 million metric tons per year, compared to world production of 340 million metric tons. The area devoted to rice production

TABLE 24.8 U.S. Barley Exports by Destination, 1987

Destination Country	Quantity (mt)	U.S. Exports (%)
Saudi Arabia	2,351,960	78.9
Israel	198,901	6.7
Algeria	123,780	4.2
Iraq	100,294	3.4
Poland	71,846	2.4
Rest of World	135,510	4.5
Total	2,982,291	100.0

Source: "Foreign Agricultural Trade of the United States," Calendar Year 1988 Supplement, Economic Research Service, United States Department of Agriculture.

in the United States averages less than 1 percent of the total harvested area of principal crops (Figure 24.2). The total value of rice is small compared to other grain crops. Rice usually ranks about sixth in farm cash receipts from grains, behind corn, wheat, soybeans, sorghum, and barley.

Planted area of rice has ranged from 0.6 million hectares in 1960 to a high of 1.5 million hectares in 1982. Harvested area has fluctuated in response to the economic environment and profitability, although government policy has had a dominant influence as well. When area allotments were suspended in the mid-1970s, rice area increased dramatically, reaching 1.5 million hectares in 1981. Reintroduction of government supply control resulted in a 42 percent reduction in area harvested (0.9 million hectares) by 1983 (Smith et al., 1989).

Rice yields have remained relatively stable, fluctuating around a base period average of 5.2 metric tons per hectare until 1981. Following 1981, yields increased strongly, peaking in 1986 at 6.3 metric tons per hectare. Changes in genetic potential of the different varieties led to regional shifts in production. The introduction of a semi-dwarf variety less susceptible to drought encouraged expansion of planted acreage in California, which reached a record of 243,000 hectares in 1981.

Rice production is highly concentrated. In 1988, five states (Arkansas, California, Louisiana, Texas and Mississippi) contributed over 97 percent of the total U.S. production. Arkansas is now the dominant state, producing over 2.7 million metric tons annually. California ranks second. Since different types of rice are produced in different states, production by type is even more highly concentrated in geographical regions.

U.S. producers and marketing firms have also provided a disproportionate share of rice storage across crop years as indicated by their share of ending stocks (Table 24.9). Only 24 percent of U.S. stocks on January 1 (a 15-year average) is stored on farms.

As with wheat and barley, there are different classes of rice (long, medium and short grain) with different uses and different growing regions. The production of long-grain rice has increased relative to the other two types, accounting for nearly 75 percent of total U.S. rice production in 1988.

The number of farms producing rice declined from 1960 to 1969. Since then the number has increased steadily, even though the total number of farms of all types has continued to decline. The increase in numbers of rice producers, to a high of more than 12,000 in 1988, is a response to market conditions and policy changes that made rice production more profitable than alternative crops.

In 1982, farms growing rice had an average area of 150 hectares per farm according to the 1982 census of agriculture. With the exception of Mississippi, all the major producing states showed an increase in the area of rice harvested between the 1974 and the 1982 census. In 1987, however, harvested area per farm dropped dramatically due to limits placed on govern-

TABLE 24.9. U.S. Share of Total World Rice Production, Exports, and Ending Stocks, 1970-1988 (percent)

Year [a]	Production	Exports [b]	Ending Stocks
1970	1.3	16.5	3.4
1971	1.3	22.4	2.4
1972	1.4	18.9	1.6
1973	1.3	22.2	2.0
1974	1.6	28.1	2.2
1975	1.7	24.2	6.3
1976	1.6	21.4	7.2
1977	1.2	23.6	3.9
1978	1.6	19.0	3.6
1979	1.7	23.4	3.6
1980	1.8	23.2	2.5
1981	2.1	21.1	7.6
1982	1.7	19.3	13.8
1983	1.0	18.3	3.2
1984	1.3	17.0	3.8
1985	1.3	16.7	4.6
1986	1.3	19.0	3.4
1987	1.3	19.1	2.3
1988	1.5 [c]	18.7	2.2 [c]

[a] Based on aggregate of differing global marketing years.
[b] Calendar year.
[c] As of October 1988 for U.S. production and ending stocks.
Source: Randell K. Smith, Eric J. Wailes and Gail L. Cramer, *The Market Structure of the U.S. Rice Industry,* Arkansas Agricultural Experiment Station Bulletin 921, 1990, page 61.

ment payments and a strong compliance with the voluntary area reduction program.

Only mill by-products find their way into livestock feed. Much of the domestic use of rice is for government programs and distribution, including school lunch programs, programs for the elderly, relief for disaster areas, and institutional distribution programs.

Although the U.S. produces less than 2 percent of the world's rice, it is the second-largest rice exporter, providing nearly one-fifth of total world exports. The five leading rice exporters, based on a five-year average from 1985 through 1989, are Thailand, the U.S., Pakistan, EC 12, and the People's Republic of China. The relative position of the People's Republic of China as an exporter has fluctuated from year to year. The U.S. has contributed a major share of world exports since 1970, though its percentage share of world rice exports has declined from a high of 28.1 percent in 1974 to fluctuate around 18 percent or 19 percent in more recent years. Total rice exports in the 1987/88 crop year were 2.3 million metric tons.

TABLE 24.10 U.S. Rice Exports by Destination, 1987.

Destination Country	Quantity (mt)	Percentage of total
Canada	72,817	2.9
Haiti	100,177	4.0
Belgium/Lux	118,502	4.8
Spain	99,437	4.0
Switzerland	89,539	3.6
Iraq	508,569	20.4
Turkey	72,799	2.9
Saudi Arabia	201,477	8.1
Bangladesh	111,463	4.5
South Africa	83,212	3.3
Liberia	98,726	4.0
Others	937,086	37.5
Total	2,493,804	100.0

Source: *Foreign Agricultural Trade of the United* States (FATUS) Calendar Year 1988 Supplement.

Government programs are an important influence on rice exports, and the rice export market is heavily dependent upon government assistance. Before 1972, government exports accounted for more than half of the total U.S. rice exported. Government exports fell to 28.7 percent of U.S. rice exports for the balance of the 1970s, and in the 1980s have averaged 17.5 percent.

U.S. rice exports are widely distributed among many countries of destination. In 1987, 20.4 percent of the U.S. exports went to Iraq and 8.1 percent to Saudi Arabia. None of the remaining countries of destination received more than 5 percent of U.S. rice exports in that year (Table 24.10).

Notes

1. A detailed breakdown by class, use, and producing region is available in the publication *Wheat Movements in the United States*, University of Illinois, 1981, and in a 1979 USDA publication, "U.S. Wheat Industry," (Heid, 1979).

25

Grain Marketing

Marketing Channels

The marketing channels for all grains have much in common. The grains often pass through the same facilities and marketing firms generally handle more than one type of grain. The principal marketing channels for grains are (1) first receivers—in most cases country elevators, or inland and river subterminal elevators that assemble grain into economically sized lots for transport; (2) processing plants, which buy from several points in the market channel; and (3) exporters, who assemble grain from the production areas and from river and inland subterminals into lots that meet the volume and quality specifications of foreign buyers.

The physical market channels for corn, soybeans and wheat are similar. The grain produced on the farms is delivered to the first handlers, either at harvest or out of farm storage (wheat, usually at harvest; corn, usually out of farm storage). The first handler is generally a country elevator, although processors and some river and inland terminal elevators also receive grain directly from producers. The country elevator absorbs the major surge in the market channel at harvest time and distributes the grain throughout the marketing year into the remainder of the market channel. For domestic use, corn and soybeans frequently move from country elevators to processing plants. In the case of wheat, major terminal elevators are a more important part of the channel, serving to assemble larger volumes for shipment to milling locations. For the export channel, many country elevators do not have enough volume to benefit from the large-volume transportation rates available to larger elevators, so they frequently ship grain to subterminals that have access to large-volume rail rates or to river terminals with access to barge loading facilities. Grain from the subterminals is then moved to export ports according to principles of minimum cost and necessary logistics (Figures 25.1 to 25.4). Port elevators load the ocean vessels and move the grain into the world markets. Most of the rice goes directly from producers to rice mills.

The interstate pattern of grain shipments and receipts between domestic points varies by grain and with different crop years as demand and supply conditions change (Figures 25.5 to 25.8). The surplus production regions for corn and soybeans are primarily in the Midwest. The deficit areas are the feeding regions of the Southeast, processing plants for soybeans in the eastern United States and export points.[1]

The country elevators handle receiving, grading, pricing, drying, storage and shipping of grains. As trucks arrive from the producers, elevator employees sample the grain and evaluate its quality, primarily on the basis of grades established by the United States Department of Agriculture (USDA). No official grades are required for grain purchased by the country elevator, so elevator employees, rather than licensed inspectors, determine the quality of the delivered grain. Once the quality is determined, it is used as a basis for price, using a system of discounts that vary from elevator to elevator and in some cases over time. In many instances the discount rate for a particular defect in the grain responds to the relative supplies of different qualities. For example, if producers deliver grain containing a lot of broken corn and foreign material, the discount may be increased, since country elevators may have difficulty finding an outlet for the broken materials.

The elevator manager determines where the incoming truckload of grain shall be dumped and stored, and whether it will need drying or can be stored and blended. Prices are generally published or announced by the elevator before the farmers deliver the grain. The published price is based on a particular USDA grade, such as No. 2 corn or No. 1 soybeans. If the elevator takes title to the grain on a direct sale from the farmer, the elevator manager will generally use the futures market to cover the risk of price changes, and will make the decision as to when and where the grain will be sold. Delivery from the country elevator is either by truck or by rail. The longer hauls are generally made by rail. However, some long hauls are made by trucks on a back-haul from some distant point.

Price formation at the farmer/country elevator level is generally under the control of the elevator manager.[2] Independent manager decisions on price are usually established daily and are based on one or more of the following factors: (1) bid price from a processor, terminal elevator, exporter or other buyer; (2) cash and futures prices reported by the Chicago Board of Trade; (3) prices offered by competitors; (4) previous sales or delivery commitments or contracts; and (5) expectations of future price changes. Once the selling price is established through evaluation of factors 1 through 5, the elevator subtracts a target margin and offers that as a flat price to producers. Actual margins, ex post, often differ from target mar-

FIGURE 25.1 Pattern of Corn Flows to Port Areas in 1977

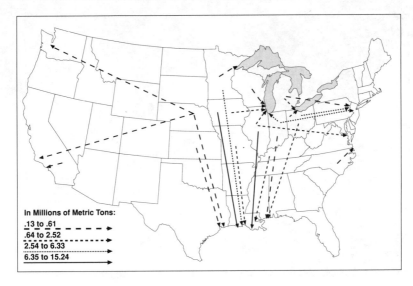

FIGURE 25.2 Pattern of Soybean Flows to Port Areas in 1977

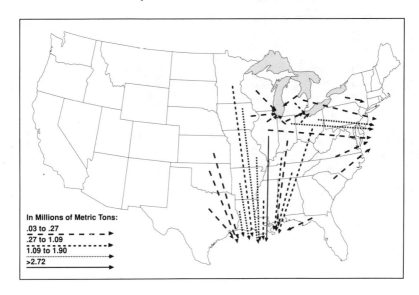

FIGURE 25.3 Pattern of Wheat Flows to Port Areas in 1977

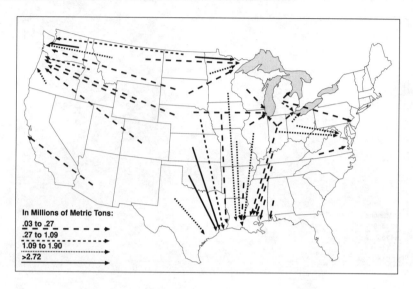

FIGURE 25.4 Pattern of Barley Flows to Port Areas in 1977

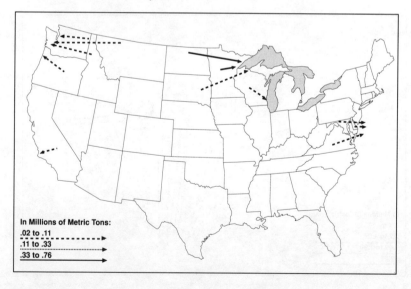

FIGURE 25.5 Pattern of Corn Flows to Domestic Destinations in 1977

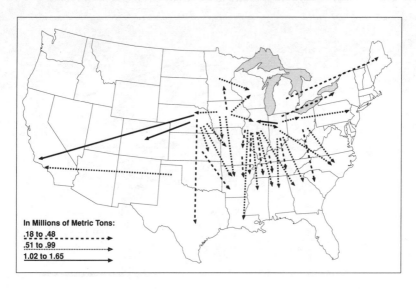

FIGURE 25.6 Pattern of Soybean Flows to Domestic Destinations in 1977

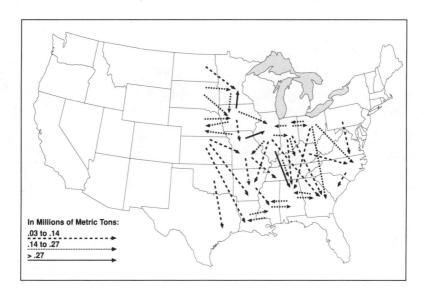

FIGURE 25.7 Pattern of Wheat Flows to Domestic Destinations in 1977

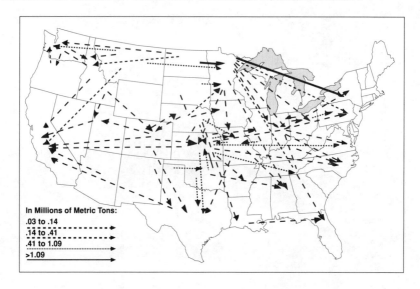

FIGURE 25.8 Pattern of Barley Flows to Domestic Destinations in 1977

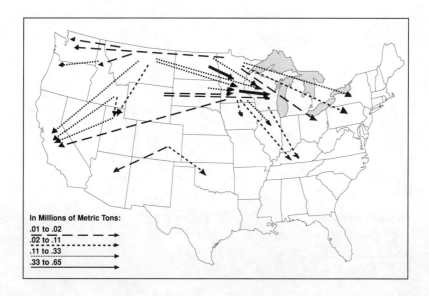

gins as a result of price and basis changes between the date of purchase and date of sale.

Other participants in later stages of the marketing channel follow a similar line of reasoning in establishing their prices. Soybean processors use cash and futures prices of final products (meal and oil) and subtract competitively-determined crushing margins. Export elevators subtract transportation costs and handling margins from international prices to offer bids to country and terminal elevators.

The interaction between demand and supply is best illustrated with a description of an export sale. Most purchases of grain by importing firms start with a private or public tender, giving details of the desired quantity, quality, and approximate delivery time. Potential sellers review these tenders, evaluate their supplies and potential purchases, and respond with a bid offer stating their contract terms. The sellers' terms do not necessarily coincide with those of the tender. For example, a lower selling price may be offered for grain delivered from a certain port or at a slightly different time, providing the importer with a choice of terms, offset by differences in price. The buyer selects the combination of price and delivery terms that are most economically favorable. Once the exporter knows that the importer has selected his bid (sometimes before a firm contract has been made), he covers the sale by purchases in the futures market, purchases in the cash market, grain already in storage, or a combination of the three. Once future delivery at a fixed price is established, the seller tries to accumulate grain at the lowest possible cost delivered to the port while controlling the risk of adverse price changes that could turn a positive merchandising margin into a negative one.

Each step in accumulating grain and arranging for transportation influences the cash and futures prices throughout the market. Bids and counter bids for cash grain are mixed with speculative actions in the market. Within limits that might be created by government-established price supports, prices move in response to the aggregate demand and supply. Prices are mobile at each point in the market channel, from the farmer's selling price to the final price at a processing plant in the importing nation. Prices for storage, domestic transport, ocean transport, handling, and merchandising, as well as prices for final products, are in constant motion and are an influence in the determination of the prices for grain in a rapidly changing market.

If demand for grain momentarily increases at one geographical location, the firm that requires the grain raises its buying price slightly. Many grain handlers respond by shifting supplies from a lower- to a higher-priced market. Even a difference as small as $0.20 per metric ton can have significant impact on profits because of the large volumes involved. Withdrawal of grain volume from the lower-priced market puts an upward pressure on prices. As supplies arrive at the higher-priced market, prices adjust downward until the system is back in balance. Arbitrage, the process of buying grain in one market and selling it in another, maintains an equilibrium, with each price related to all other prices in the local, national and international markets. Arbitrage can occur over time, form and space dimensions. Uniform grades, a sophisticated market structure, a large volume in the market and trading institutions such as the Chicago Board of Trade make arbitrage effective in equating forces of supply and demand.

Transportation

The United States has a highly developed transportation system integrating rail, truck, and barge through a sophisticated infrastructure that is probably unequalled in any other grain-producing nation. The organization and operation of this transportation system influences the direction of grain flows, the ultimate destination of the grain, and the prices at which it can be delivered to the domestic and export markets. Competition between regions, between markets, between and among modes of transportation, and between the U.S. and other countries also influences grain prices and flows.

Public funds provide the highway system and the waterway system, with user charges assessed to cover the costs of maintaining these facilities. Nearly all freight moves in privately owned rail cars and on privately owned roadbeds; public subsidies are much smaller and less direct than for either truck or barge.

The river systems in the United States flows generally toward port areas. The Mississippi river system, which covers most of the grain producing states, flows to a large and well-developed port system in the Gulf of Mexico. A second river system flows into ports at the Pacific Northwest. An extensive railroad system developed in the mid to late 19th century with government encouragement is currently over-built in terms of the number of railroads and the number of miles of lines, but is being rationalized through private and public actions. Actions taken in recent years have led to a decrease in the number of competing railroads, the number of lines, and total miles. Grain can be moved rapidly to domestic and export destinations over a highly developed interstate road system that is readily accessible to grain trucks.

Grain destinations, transportation modes, and rate levels vary regionally due to differences in geography and environmental endowments. The main production areas for corn, wheat and soybeans are the Eastern Cornbelt, Western Cornbelt, Upper Great Plains, and Great Plains (Figure 25.9).

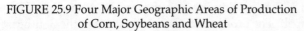

FIGURE 25.9 Four Major Geographic Areas of Production
of Corn, Soybeans and Wheat

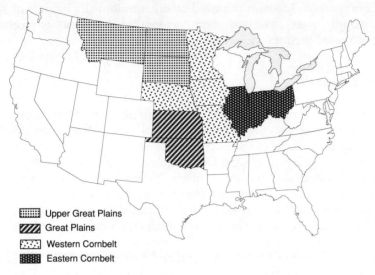

- ▦ Upper Great Plains
- ▨ Great Plains
- ░ Western Cornbelt
- ▦ Eastern Cornbelt

Eastern Cornbelt

The primary export markets for Indiana, Ohio and Illinois corn and soybeans are the Gulf ports, the East Coast ports, and the port of Toledo, Ohio.

The Gulf ports are the largest export market for the Eastern Cornbelt. Most of the export-bound grain is transported to the Gulf by barge. The grain is usually trucked to barge loading points on the Illinois River, the Mississippi River, or the Ohio River, and then carried on barges to Louisiana or Mississippi ports. Most of the grain is trucked to the river from the country elevator, although some grain moves directly from the farm. Some of the grain for this area is transported to the Gulf by 100-car unit trains. Many of the unit train shipments originate in eastern Illinois and western Indiana.

Much of the grain, particularly soybeans, is shipped to in-state processors. The corn and soybean processors are served by the grain elevators with truck shipments or small multi-car rail shipments.

Shipments of grain through the East Coast ports have declined considerably during the past few years due to changes in the transport rate structure and to falling European demand. When shipped to the East Coast, the grain is usually transported via unit trains of about 100 rail cars per shipment to either Baltimore, Maryland or Norfolk, Virginia.

Toledo, Ohio, has a small export market, served by nearby elevators. Assembly is by truck or multi-car rail shipments of 3-10 cars.

The Southeastern U.S. is an important and growing domestic market for the Eastern Cornbelt. The broiler and cattle industries in the Southeast use corn for feed. The corn is moved by rail, traditionally in units of 3-10 cars; recently, there have been more larger-sized shipments as receivers have increased their receiving capacity.

Western Cornbelt

The primary export market for grains from Iowa and Minnesota is the Gulf, via barge shipments on the Mississippi River. Most of the grain is trucked to the river, although a significant amount is transported by rail to barge loading points in 25- and 50-car units. Corn and soybeans are also transported by rail directly to the Gulf via 75-car trains from western Iowa, western Minnesota and Nebraska. During the past decade, unit train shipments of corn from these areas to Pacific Northwest (PNW) ports have increased considerably due to increased demand for corn in the Far East, and to favorable unit train and ocean vessel rates.

As in the Eastern Cornbelt, the corn and soybean processors in Iowa, Southern Minnesota and Nebraska are served primarily by truck and by multi-car rail shipments. Also, southern feed markets (mostly in Arkansas and Texas) are a common destination for rail shipments of corn.

Great Plains

The two primary markets for wheat grown in the Great Plains are the Gulf export market and the processing market. Wheat for the Gulf market is shipped mainly by rail. Until the early 1980s, nearly all the wheat was shipped under single-car tariff. Since then, multi-car and unit-train service have become prominent. The Arkansas and Missouri rivers also carry wheat to the Gulf. However, barge rates on the Missouri are high due to the high costs of negotiating shallow waters and narrow, winding segments. Barge rates for the Arkansas River are lower than those for the Missouri, but the barge loading site for Great Plains wheat (Catoosa, Oklahoma) is a long distance from where most of the wheat is grown.

Processors (wheat millers) in the Kansas City area receive much of the Great Plains wheat via truck and multi-car rail shipments. Mills along the Eastern Seaboard also use Great Plains wheat, and are often served by unit trains.

Upper Great Plains

Wheat grown in Montana and North Dakota is often moved by rail to the Pacific Northwest (PNW) ports. Until 1982, all rail shipments from the

Upper Great Plains were under single-car tariffs. Tariffs for multi-car and unit train shipments have been offered and used extensively during the past seven years. In the western Great Plains, wheat for the PNW ports may be trucked to Lewiston, Idaho, then transferred to barges on the Columbia/Snake river system.

An alternative rail destination is Duluth, Minnesota, where the wheat is either milled, exported through the Great Lakes, or transhipped to processors in Buffalo, New York by ship or train.

In the eastern Upper Great Plains, Minneapolis serves as a transhipment point for truck- or rail-barge combination shipments on the Mississippi River. However, the demand for this alternative is not large unless barge rates are unusually low.

Competition in Grain Transportation

A short review of the types of competition within the grain transportation industry will be useful in assessing the competitive environments of the grain-producing regions. Two common types of competition are intermodal (choice between different transportation modes) and intramodal. Intermodal competition exists when the grain can be either trucked or transported by train to a destination or when a shipper has the option to use either rail or barge transportation; and intramodal when more than one railroad or more than one trucking company operates within a region.

Indirect competition between regions and between markets is prevalent but less apparent than inter- and intramodal competition. In interregional competition, two or more regions are competing for the same market. Illinois corn, for instance, competes directly with Iowa corn for the Gulf export market. In turn, railroads in Iowa compete with railroads in Illinois even though they may not be in close proximity. Competition between markets involves the extent to which the grain can be used in different types of markets. For example, two major markets for Illinois corn are the processing and export markets. The competition between these markets for corn results in competition between the transportation firms serving the markets.

Regional Competitive Factors

Eastern Cornbelt. Four major railroad companies as well as regional railroads provide an extensive rail system. In addition to grain shipped for export, large feed and processing markets use the region's grain. Because of these competitive factors, transportation rates are very responsive to supply/demand factors, and are often close to the cost of service.

Western Cornbelt. The competitive structure of grain transportation in the Western Cornbelt is similar to the Eastern Cornbelt in that rail competes heavily with barge, and many destinations are feasible markets for truck and rail shipments. The major difference between the Eastern and Western Cornbelt is that a larger share of the Western Cornbelt's grain is shipped in unit trains.

Great Plains. The structure of competition is largely in the form of inter-railroad competition between three major railroads, although the Arkansas River offers effective competition for the southern Great Plains area. Rail rate levels are generally higher than those in the Cornbelt, but a "market dominance"environment does not seem to exist.

Upper Great Plains. The Upper Great Plains area has the least amount of transportation competition of the major grain-producing regions. Barge loading points are a relatively long distance from production, the wheat is primarily grown for only one market (export), and only one major railroad operates through much of the area. Consequently rail rates within the region are relatively high.

Drying and Storage

Grains are harvested during a relatively short time period, but consumed at a fairly constant rate throughout the year. Thus, storage is needed at some point in the marketing channel. Drying and storage are about equally divided between farms and commercial facilities. About 60 percent of the U.S. grain storage capacity is on the farms where grain is produced. In the major producing states, usually about 65 percent of the corn crop is stored on the farm at harvest. The balance is either sold directly from the field or placed in off-farm storage. About 80 percent of the volume stored on the farm is artificially dried on the farm before storing. This leaves the farmer in control of quality and distribution decisions for much of the grain that will eventually enter the market channel.

Farmers have expanded their drying and storage capacity for several reasons: to maintain physical control of the grain on the farm; to make better use of excess labor, previous investments and equipment in the off-season; to avoid high charges for drying at the country elevator; to obtain more flexibility in the choice of final market; and to gain more control over the price at which the grains are sold.

While storage capacity has expanded at all stages of the market, the greatest growth has been near the point of production. The distribution of storage between farm and off-farm locations can be seen if we look at stock positions for each grain following harvest. A higher proportion of corn and soybeans than of wheat, barley, and rice is stored on the farm (Table 25.1).

Although all grains receive some artificial drying, only corn and rice need drying of a large proportion of the crop. Because of high-speed harvests, and the producers' desire to avoid potential field losses, the two crops are generally harvested at moisture levels above optimum for stor-

TABLE 25.1 Production and Ending Stocks of Principal Grains, United States, 1987

Year Ending	Corn	Soybeans	Wheat	Barley	Rice
Production[a] (million mt.)	179.4	51.8	57.3	11.5	5.8
Total stocks[b] (million mt.)	261.7	47.8	34.5	14.3	6.6
% of production in storage	146%	92%	60%	136%	90%
% on farm	96%	45%	25%	37%	23%
% off farm	50%	47%	35%	99%	67%

[a] Ending Month: Corn, Sept. 30; Soybeans, Aug. 31; Wheat, May 31; Barley, May, 31; Rice, July, 31.
[b] Ending Month: Corn, Nov. 30; Soybeans, Nov. 30; Wheat, May 31; Barley, June 1; Rice, Dec. 31.
Source: Agricultural Statistics, 1988. U.S. Department of Agriculture.

age or processing. On-farm and off-farm drying takes place immediately following harvest. About half of the corn harvested is dried on the farm. The other half is either dried at the country elevator or blended with drier grain somewhere in the market channel to reduce the average moisture to contract limits. Since U.S. grades permit the delivery of any moisture content, high-moisture grain is often found throughout the market channel during cold weather when storage is not a problem. As the weather gets warmer, all corn must be dried to 15 percent moisture or less if it is to be kept in satisfactory condition.

Some soybeans and wheat are also dried in farm and commercial dryers though the total volume is small compared with the other grains. Drying is done primarily as an emergency measure when weather conditions make it difficult to wait for field drying or under double cropping, when the grain is harvested early to open the field for planting the second crop.

Rice is stored at farms, marketing facilities, and in mills. There are no regularly published data for storage capacity, but estimates show that in 1988 on-farm rice storage capacity was approximately 1.8 million metric tons, about 30 percent of the total stored in that year (Smith et al., 1989). On-farm storage of rice seems to have declined recently as a result of lower production levels and the marketing loan program. The number of commercial warehouses has been increasing since the mid-1960s, and the number of commercial rice dryers associated with these facilities has also increased. In the five-state area where most of the rice is grown, the number of facilities has increased continuously since 1965; the largest increase has occurred in the larger capacities (above 24,500 metric tons).

Price Determination in U.S. Markets

Price plays a major role in the grain markets in allocating grain supplies over the dimensions of time, form, and space. The interaction of the forces of supply and demand determines the overall level of prices at the central markets. These prices are reflected back through the market channel to individual firms to direct the flow of grain. Price differentials are the variations above or below the central price that regulate the flow of grain over time, space, and form.

Price Differences Over Time

Though grain crops are harvested in a relatively short period of time, they are consumed throughout the year at a fairly constant rate. Seasonal prices control the flow of grain to market and provide the incentive for storage so that processors and consumers can maintain a uniform rate of use. Storage rates are reflected in price differentials over time; the price of bin space is related to the volume of all farm stocks of grain and the volume of current sales. Within daily market movements, price differentials shift grain from destinations of momentary surplus to destinations of momentary deficit. Supplies of grain in storage in production areas can be readily redirected from domestic processors to distant domestic markets, to the Gulf Coast or to the East Coast, or any of the other market points.

The Futures Markets. The futures markets are used to shift risk from buyers and sellers of cash grain to speculators who are willing to absorb these risks. Futures markets also ration supply and provide the incentive for maintaining inventory reserves between harvests. Farmers' use of the futures markets is limited but increasing. When properly used the futures markets become a marketing tool, enabling the producer to establish a price for the grain before committing his resources to production.

Price Differences Over Space

Seasonal price differences present special problems to farmers in selecting their marketing strategies. Grain moves from the point of production to consumption centers in response to price differentials greater than the cost of transportation. Because of the large volumes of grain that are moved, these price differentials are often as little as a fraction of a dollar per metric ton. The resulting price pattern in the U.S. generally resembles the theoretical iso-price maps. Although market imperfections distort the pattern, the transportation rates are clearly reflected in long-run averages.

Much of the variation among individual elevators in prices paid to farmers may be due to market imperfections, including a lack of information, inadequate competition, and lack of availability of transportation at competitive rates. Country elevators establish their bid prices to producers primarily by subtracting transportation and handling costs from bid prices at demand centers such as ports or processors, or by basing their price relative to the Chicago Board of Trade cash or futures prices.

Price Differences Over Form

Only a small proportion of raw grain is consumed directly as human food. It must undergo several changes in form in response to opportunities for increasing its value by more than the cost of transformation. Price serves as the allocating mechanism among the various users who require different qualities or forms. Price also provides the necessary incentive for changing the form of the raw grain into processed or semi-processed products, and generates the incentives for quality differences. Cleaning, blending, and drying (all undertaken to alter the quality or condition of the grain) are done primarily in response to economic incentives generated by price discounts or premiums in the market channel. Price discounts or premiums also influence the farmers' selection of grain varieties and their choice of harvesting, handling and marketing alternatives. Grain grades, in combination with market-determined price differentials, communicate buyer preferences to sellers, producers, and plant breeders. In cases where grain grades do not reflect certain economically-important quality characteristics, it becomes more difficult and costly to communicate buyer preferences; however, the desired quality characteristics may still be considered in the market transaction.

Producers' Pricing Strategies

In making their pricing decisions, U.S. farmers have many choices available. Some of these options allow farmers to separate their pricing decisions from the physical delivery of the grain, giving them greater control over the price at which they sell.

The options most commonly used are cash-on-delivery or the forward contract. About half of the total harvest of corn, soybeans, and wheat is sold to the country elevators using the cash-on-delivery option. The farmer delivers the grain to the elevator either immediately after harvest or following a period of on-farm storage, and is paid in cash.

About 10 percent to 15 percent of the harvest is sold under forward contract. The forward contract is a written or oral agreement between the elevator and farmer for delivery of the grain at a future date with price specified in the agreement. This enables farmers to lock in a price prior to harvest. The elevator hedges its purchase on the futures market, locking in a margin for itself at the same time that it guarantees the farmer a price when the grain is delivered at harvest. The farmer could obtain a similar price guarantee by entering the futures market directly, but the elevator has an advantage by having specialized management with greater experience in futures trading.

The farmers may choose to use a "delayed price" or "price later" option. This option gives the farmer complete flexibility as to the time of sale, but the title passes to the elevator when the grain crosses the elevator's weighing scale. The farmer can select any price any time within the period specified in the contract (usually up to one year), but the elevator retains title to the grain and physical control of it. This gives the elevator greater flexibility in its marketing strategies, enabling it to maximize income. The delayed price technique has increased in popularity in recent years as farmers have become more sophisticated in their marketing strategies. However, to date probably less than 10 percent of farm marketings are made under the delayed price option.[3]

Farmers also have the option of storing grain at the country elevator, where they retain title and ownership rights. The elevator in turn must issue a warehouse receipt, which is supervised by federal or state inspectors under the General Warehouse Law. This assures the farmer that his grain is physically stored in the elevator, available at any time for his marketing decision. As a result of several elevator bankruptcies and defaults on warehouse receipts, a number of state-level insurance schemes have been developed to reimburse producers for losses on grain stored at country elevators. State regulations protect producers by requiring elevators to "cover" their positions on the delayed-price grain.

Pricing Alternatives for Country Elevators

Country elevators have many of the same options as the producers. They may use forward contracts with processors to lock in a flat price or they may use the futures markets in order to offset potential risks of increases or decreases in flat prices. In these cases most elevators price on the "basis" i.e., the difference between the cash and futures prices. They may also use "back-to-back" sales, in which they obtain from the morning markets a bid for a fixed quantity to a specific buyer, subtract their merchandising margin, and then announce a bid price that will hold for the day to any producer delivering grain. As mentioned earlier, prices are almost always set at some base grade established by the U.S. Department of Agriculture (No. 2 for corn, No. 1 for soybeans). Farmers receive the price for that grade minus any discounts for poor quality, and charges for drying or storage. The elevator's costs of transportation, storage, merchandising, and general operation are usually incorporated in the merchandising margin—the difference between their selling price and their bid price to the local farmer.

Pricing in the Rice Industry

The pricing pattern for rice is different from that for other grains. The industry has organized pricing methods as well as direct sales agreements between the producer and the mill. The method of marketing differs among states. Producers in Arkansas and California market their rice primarily through cooperatives. Louisiana and Mississippi rely more on direct sales or a bidding process. Marketing agencies (independent firms or cooperatives) are active in the southern rice-producing states. They do not handle the physical commodity, but deliver samples of rice to buyers representing various mills. The buyers submit a sealed bid for each lot of rice. After receiving their bid, producers are usually given 24 hours to respond to the offer. Once an offer has been accepted, the selling agency transfers ownership to the buyer. The buyer pays transportation costs for the rice. Most rice marketing agencies charge a flat rate fee per unit for their services.

In 1987, the Louisiana Farm Bureau Marketing Association, which has a marketing sales desk for handling members' rice, handled an estimated 20 percent of Louisiana's rice marketing. In the same year, Arkansas' three independent rice marketing companies marketed an estimated 6 percent of production. Between 40 and 50 percent of Mississippi rice production was marketed by the bid and acceptance method. Texas marketing associations, with 17 sales desks, marketed more than one-third of Texas' total rice output. California was the only state not using marketing associations, primarily because of the dominance of marketing cooperatives.

Rice marketing cooperatives in California and Arkansas use a seasonal pool for storage and payment to their producer members. Approximately 70 percent of the rice production of these two states is marketed in this manner. Rice is delivered to the cooperatives and commingled according to quality and grade. Producers receive a partial payment at the time of delivery. Additional payments are made after final sale of the cooperative pooled lots. Several cooperatives are integrated from producers through the marketing of milled rice to consumers. Profits from drying, storing, milling and marketing are returned to the producers.

Rice is also sold through private contract between the producer and the mill or at a public sale. About 25 percent of the rice marketed in 1984 was sold this way.

Strategies for Reducing Price Risk

In a market operating in a competitive environment with freely floating prices on a daily and even minute-to-minute basis, every buyer or seller runs the risk of adverse price movements. These short-run price changes can generate losses or windfall profits for the producer greater than the annual net income per metric ton. When large volumes of grain are involved, these risks can also be disastrous for marketing firms if they have no protection against an open position in the cash or the futures markets. The risks and their associated probabilities must somehow be incorporated into the cost of marketing.

In general, U.S. grain marketing firms do not speculate on grain prices. Instead, they try to secure a favorable margin at the outset of their cash transactions to buffer themselves against adverse price changes. To secure a favorable margin and reduce risk, they use a variety of strategies, including back-to-back sales, forward cash contracting, brokerage, and hedging and pricing through futures and options markets. When choosing among these strategies, the firm makes a decision between risk and expected return.

Back-to-Back Sales. In back-to-back sales, the grain marketing firm usually takes physical possession of the grain, but resells it immediately after buying it, retaining ownership of the grain only momentarily. Usually the firm has a good idea of the price it will receive for the grain and sets its bid to the farmer (or whatever firm it buys grain from) accordingly. Thus, the price risk to the grain firm is minimal. However, the firm may entail handling costs for the grain (e.g., loading it into a rail car or putting it through an elevator facility). The grain handling firm typically earns a greater margin for back-to-back sales than for direct delivery because its risks are slightly greater.

Forward Cash Contracting. This method is similar to back-to-back sales in that the firm sells grain for a fixed, flat price that is usually known when the grain is purchased. The difference is that under forward contracting the physical commodity is delivered at a later date. Grain marketing firms who sell under forward contract may use the contracted price as a guide for bids to farmers. Very often grain elevators buy grain from farmers before harvest under forward cash contracts. Firms that offer forward contracts usually reduce the risk of an adverse price change during the period between entering the contract and eventually disposing of the grain by either converting their forward purchases to forward cash sales, or by hedging their forward purchases on futures markets.

Both farmers and elevator managers use flat price forward contracts to eliminate the possibility of losses due to price declines prior to delivery. One disadvantage of this method is that it also eliminates the opportunity for windfall profits from price increases. The farmer also faces the risk that a crop failure would make it impossible for him to deliver the amount of grain stipulated in the contract. To reduce this risk to the farmer, some elevators make allowances in their contracts for extenuating circumstances. Other refinements to the system provide additional protection against inability to deliver contract quantity and quality.

Grain Brokerage. Some grain marketing firms brokerage grain. That is, they arrange sales between farmers and grain processors or other large grain handling facilities. The intermediate grain marketing firm formally buys from the farmer and sells to the grain processor, without taking physical possession of the grain, which is "direct delivered"to the processor by the farmer. The processor pays the intermediate grain marketing firm, and the firm in turn pays the farmer, retaining a small fixed brokerage fee of about a dollar per metric ton. In this type of transaction, there is no price risk to the broker.

Futures and Options. In the U.S., the futures market is the most widely recognized institution for spreading the risks involved in the buying and selling of grain. In the grain futures market, large numbers of small buyers and sellers are willing to accept the risk of losses for the opportunity to achieve windfall gains.

A simple illustration of the way the futures market works is the example of the country elevator manager who has purchased 127 metric tons of corn from a producer and wishes to hold it for delivery several months in the future. The manager decides to sell a futures contract for the corn on the Chicago Board of Trade. This contract may be purchased by any private individual or group (including grain firms) wishing to speculate that the price of the corn contract he has purchased will increase before its delivery date. If the price should fall before the final cash sale and delivery, the elevator is protected: the futures contract he sold earlier will also fall in price. He can repurchase that contract at the reduced price, and thus receive a margin of profit on the futures transaction that (ideally) will offset his losses on the cash sale. The country elevator has thereby substituted a small risk of a small change in the basis by transferring a larger risk of changes in flat price to an unknown individual in the market, who purchased the contract on the Board of Trade.

Grain-marketing firms use futures and options markets in a number of ways to protect themselves against adverse price changes. For example, they may use futures markets to voluntarily "lock-in"a price for their anticipated sales well in advance of any actual sales, in a manner similar to forward cash contracting. One major difference between this use of futures markets and forward cash contracting is that it is costly to enter into a futures position as the firm must place margin money (usually a small percentage of the contract's value) with a futures brokerage firm, and the firm must "mark-to-the-market"gains or losses to their futures position. Another difference is that in futures contracting the grain marketing firm usually encounters some form of "basis risk". Basis risk pertains to the predictability of the difference between the futures price and the price at the grain marketing firm's relevant cash market. The more variable and the less predictable this price difference or basis is, the less attractive forward pricing through futures becomes.

Basis-Priced Contracts. Grain-marketing firms may trade futures contracts to protect themselves against adverse price movements and to preserve flexibility in the forward contracting and pricing of cash market transactions. Basis-priced contracts (contracts made with reference to futures prices) are the standard form of contracting for grain exports. They are also used to a lesser extent at other stages of the U.S. grain marketing channel.

The parties to a basis-priced contract agree to two separate components in pricing a sale. One component is the choice of the futures contract that will later determine the absolute price of the sale. The other component is the basis, or the amount over or under the relevant futures price that the seller will earn for making the sale. Included in the basis are transportation costs between the futures market and the delivery location and any premiums or discounts for differences between the quality of the commodity sold and the quality of the commodity specified in the futures contract. Basis-priced contracts allow the parties to the sale to fix one dimension of a sale in advance, but maintain flexibility, and a certain amount of risk, in the other dimension.

Typically, both parties to a basis contract take a futures position equal to the amount of the sale when they enter into the contract. The seller takes a short futures position and the buyer takes a long position. However, the price at which the buyer and seller enter into futures positions need not be the same. Herein lies the pricing flexibility and risk. The final clearing of futures positions may occur in a number of ways. The way in which it does occur is specified in the contract. The buyer's purchase of futures may be done for the seller's account, thereby fixing both the seller's and buyer's net returns/costs. Or, close to the time of delivery, the seller and buyer may exchange futures positions for cash. The price of the futures contract at that time then determines the buyer's and seller's net returns/costs.

Options on Futures Contracts. Some firms take options on futures contracts. This strategy is closely related to forward contracting. For a market-determined premium, an option contract gives its owner the right to buy or sell a futures contract for a certain price (the exercise price) for a limited period of time. Options provide the opportunity to set price floors and ceilings, while maintaining the possibility of capturing favorable price movements. They are thus a form of forward-pricing insurance. When deciding whether to use options or futures for forward pricing, the grain-marketing firm must weigh the benefits of pricing flexibility against the costs of the options.

Hedging. Grain-marketing firms use futures markets to determine when it is prudent to store or sell grain. If the market is providing a return to storage, as evidenced by a large enough difference between futures prices for nearby and later maturities, the firm may hold grain in storage by hedging stocks on futures markets without fear of adverse price movements (except for basis risk). Such a hedge would require selling futures contracts for the distant maturity equal to the amount of grain to be held in storage. If the futures market does not provide a return to storage that compensates for the cost of storing grain between maturities, there is no rationale for storing the grain. However, a firm may still decide to store the grain if it wishes to implement a speculative strategy or to meet anticipated commercial needs at a lower cost than would be incurred by selling grain stocks in the present, and buying later on the spot market.

State and Federal Warehouse Laws[4]

In the United States, state and federal warehouse laws provide legal protection for farmers storing grain in commercial facilities. Since farmers often wish to retain ownership of grain in anticipation of price rises after harvest, storage is a necessary function. The least cost pattern of storage would allocate most of the grain marketed to commercial warehouses, because economies of scale favor large commercial bins over small farm bins. Use of commercial warehouses by farmers is encouraged by assurance that the quantity and quality of grain deposited for storage will be maintained. This assurance is provided by federal and state warehouse laws that regulate public grain warehousepersons.Despite these regulations and the accompanying inspection of facilities, farmers have frequently suffered losses associated with financial failures of country elevators.During recent bankruptcy proceedings, farmers received payment for less than seventy percent of the full value of grain warehouse claims at elevators filing for bankruptcy. Additional changes in Illinois law provide an insurance fund for protection against losses, especially where illegal actions by a manager might leave insufficient grain in the warehouse to cover the amount represented by outstanding warehouse receipts.

Early warehouse laws required the warehouseperson to maintain the integrity of each lot of grain absent the consent of the owner. Since grain identified by federal grades is, in general, a homogeneous, interchangeable commodity, the need for segregating each lot unnecessarily increased the cost of storing grain. This increased cost of storage was in turn passed on to the farmer and inefficiencies resulted.

These inefficiencies have been largely eliminated. Changes advocated by the grain industry and later by state regulatory agencies have been adopted and warehouse laws now commonly permit the practice of comingling grain of the same kind and grade. In fact, Illinois law allows a warehouseperson to refuse to accept grain when the identity of the lot is to be preserved.

Under current warehouse laws, each farmer-seller receives a warehouse receipt for the quantity and quality of grain delivered. The elevator manager is held responsible for maintaining on the premises the quantity and quality represented by the sum of all individual warehouse recepits, but need not maintain the identity of each individual lot. Thus, costs of storage are reduced and the increased efficiency is reflected in lower storage charges to the producer.

These lower charges are an incentive to farmers to store in commercial facilities rather than to build storage bins on the farm. Thus, the changes just described in warehouse laws altered the balance between on-farm and off-farm storage capacity, reduced investment in farm storage bins, increased the value of stored grain warehouse recepits and improved the profitability of storing grain in commercial facilities.

The warehouse laws become especially important when elevators encounter financial problems and are unable to pay producers and other creditors. Although state warehouse receipts were intended to protect farmers with grain stored for them by commercial grain warehousepersons, several bankruptcy cases in Illinois and other states left insufficient funds to repay holders of warehouse recepits as well as unsecured creditors. From January 1974 through May 1982 Illinois farmers lost $1.5 million through elevator insolvencies and bankruptcy proceedings. Neither bankrutpcy regulations or warehouse receipts were sufficient protection for farmers. Changes to the Illinois Public Grain Warehouse Act and the Grain Dealers Act provided increased protection for producers. The Illinois warehouse law as amended gave the Department of Agriculture authority to raise bonding requirements for dealers of questionable financial position to restrict speculation, and to require annual financial statements prepared according to generally accepted accounting principles. These requirements increased costs to elevator managers but reduced the risk of financial loss to farmers storing grain in commercial warehouses. This increased protection was still considered inadequate by many and on August 16, 1983, Illinois passed the Grain Insurance Act. Collection of mandatory premiums from grain warehousepersons created an insurance fund that has provided more complete protection. According to statements by Illinois Department of Agriculture officials, "every claim by farmers that has been adjudicated valid has been paid in full, either by elevator assets or from the insurance fund," since the implementation of the Act.

Most other grain exporting nations also have warehouse laws to protect storers of grain. For example, Argentina has seperate warehouse regulations for grain for internal consumption and for export. Warehouse receipts are registered by the Junta Nacional de Granos and delivery is guaranteed to purchasers in case of loss of grain, theft or failure to comply with warehouse regulations. The strength of the warehouse law protects farmers with grain in commercial storage, but other losses are handled under general bankruptcy regulations.

Notes

1. See *Corn, Soybean, Wheat* and *Barley Movements in the United States,* 1977 and 1985 for further detail on patterns of grain movement.

2. In the case of some larger firms, prices are determined at regional or national headquarters.

3. Except in Ohio, where survey results showed that, following a strong program promoting delayed pricing, over a quarter of the corn, soybeans, and wheat were marketed under delayed price as early as 1974.

4. Excerpted from "Effects of Regulation on Efficiency of Grain Marketing" by Lowell D. Hill. Originally published in the Journal of International Law, Vol. 17, no. 3, Summer 1985. Reprinted with permission.

26

Policies and Group Actions That Affect the Grain Markets

Policies That Affect the Grain Markets

National policies provide the rules and regulations within which individual managers must make their decisions. These policies exist on several levels and may restrict or encourage improvements in market performance.

The types of policies reviewed here are: general economic policies; domestic farm price and income policies; policies to limit the effects of adverse weather; transportation policies; market regulations; credit and capital policies; trade policies; export enhancement programs; and demand expansion policies.

General Economic Policies

The United States' interrelationship with a changing world economy influences greatly the interrelationship of the federal government's macroeconomic policies with the behavior of the farm sector. A combination of factors has created a more unstable U.S. agricultural sector within a more unstable macroeconomic and international environment over the last 20 years. Some of these factors are:

- The increased integration of the U.S. economy within the world economy, with the reduction of trade barriers, the (partial) abolition of capital controls and the move to a floating exchange rate system.
- The increased integration of agriculture within the general economy in the U.S., with the abolition of interest rate subsidies and the

two-tier price system, and reduction of export subsidies.
* The changes in U.S. monetary policy (in October of 1979), from controlling the nominal interest rate to controlling the money supply within target growth rates.

In addition, the agricultural sector's dependence on the cost of capital and on exports makes it vulnerable to fluctuations in the domestic and international capital and currency markets.

During the early 1970s, the move to flexible exchange rates, the ensuing drop in the value of the dollar, and the monetary expansion initiated by the Federal Reserve in its effort to inflate away the effects of the 1973 energy price shock led to high increases in real prices of farm commodities. At the same time, an expansionary monetary policy by other governments caused an explosion in international liquidity and a general increase in real world agricultural and commodity prices.

As the 1970s advanced, farmers with expectations of continued inflation contracted debts and expanded area and farm size, and many farmers switched to more extensive cropping practices, such as grain production.

In the early 1980s, U.S. agriculture faced macroeconomic conditions almost exactly opposite to those of the 1970s. The large federal deficits of the early 1980s, in combination with the tight monetary policy followed by the Federal Reserve, caused real interest rates to increase and reversed the decline of the U.S. dollar that had occurred during the 1970s. When the dollar increased in value, the competitiveness of U.S. agricultural exports was reduced. At the same time real commodity prices in the U.S. dropped significantly.

The option followed by the federal government in the 1980s was to sustain production at levels not justified by prevailing relative prices, and let the government budget or consumers pick up the burden of adjustment to the new conditions.

As the inflation rate declined, expectations of future inflation of land values receded. The price of land dropped dramatically, reducing farmers' capital base. Farmers were then unable to pay off loans and mortgages; consequently, many mortgages were foreclosed. Grain farmers, many of whom had large farms purchased with low equity and large mortgages, were more seriously affected than small farmers with livestock operations. Land prices gradually recovered after the low of the mid-1980s.

A further effect on farmers during the 1980s was the federal government's attempt to limit debt. The shift of programs between federal and state governments has affected grain production costs. In particular, federal funds for agricultural research and extension have been reduced.

Domestic Farm Price and Income Policies

Since the agricultural depression of the 1920s, the major federal government instruments for increasing farm income have been commodity loans and area and production controls.

The Federal Farm Board, established by the Agricultural Marketing Act of 1929, was created with the belief that with federal aid, cooperative marketing organizations could provide a solution to the problem of low farm prices. The board, with a revolving fund of $500 million, had authority to make loans to cooperative associations, to make advances to members, and to make loans to stabilization corporations to control any surplus through purchase operations.

By 1932, the board's efforts to stem the disastrous decline in farm prices that took place during the Depression had failed. As a result, the board members recommended legislation to provide an effective system for regulating area or quantities sold.

The Agricultural Adjustment Act of 1933 was intended to restore farm purchasing power of agricultural commodities to the prosperous 1909-1914 level. The major instruments to carry out the program were production controls and commodity loans. This approach to increasing farm income has dominated legislation through the 1980s.

Program goals have centered around the concept of parity, a term first used in the 1933 act. Parity seeks an equality of income or living standard between agriculture and industry or between persons living on farms and persons not on farms.

Since the 1933 act, agriculture has changed dramatically. The U.S. national agricultural policy has also evolved. U.S. agriculture is currently more heterogeneous, with greater diversity in farm size, more specialization in production, and differences in production of the same commodity across the country.

The formation of agricultural and food policy in the 1980s was strongly influenced by developments of the previous decade. During the 1970s, producers expanded area and production to meet an expected long-term upswing in export sales. When prices fell in the 1980s, and many farmers lost income, new farm policies were developed in an effort to pull agriculture out of its recession.

The Food Security Act of 1985. The Food Security Act of 1985 has provided the foundation for U.S. agricultural and food policy since 1986. New legislation will be considered again in 1990. The principal features of the 1985 legislation dealing with price and income support are contained in the commodity programs, described in the following paragraphs. The act also included provisions for conservation, agricultural trade, research, extension and teaching and some miscellaneous programs.

Commodity Programs. The commodity programs are designed to deal with excess production and unstable prices through production controls and price supports. The major policy instruments to carry out these programs are: area bases or allotments, commodity loans, target prices and deficiency payments, the farmer-owned reserve, market orders and agreements, and direct government purchases of farm commodities.

The major crops covered by some form of price and income support are: wheat, feed grains (corn, grain sorghum, barley and oats), cotton, rice, peanuts, soybeans, sugar, and tobacco.[1] In addition, the 1985 act continued price supports for dairy products, wool and mohair.

The Department of Agriculture's Agricultural Stabilization and Conservation Service (ASCS) administers the programs with producers. It also administers conservation and forestry cost-sharing programs. Payments are made through the Commodity Credit Corporation (CCC), a wholly owned Federal corporation. CCC borrows money from the Treasury to make payments to farmers and repays the Treasury with receipts from loan payments or sales and with appropriations voted by the U.S. Congress.

Budgeting and controlling costs of the price and income support programs has been difficult since they have come to be regarded as entitlement programs. Any producer of the crops covered by the programs may choose to participate (with the exception of tobacco producers, for whom participation is mandatory). Outlays for deficiency payments are directly related to the market price of the crops, the target prices, and the extent of participation.

A gap between loan rates and target prices has inspired high participation among producers and has incurred high costs for the federal government. For example, the percentage of base area included in the corn program in 1986, 1987 and 1988 ranged from 86 percent to 90 percent. In 1989, 80 percent of the base area was enrolled. The base area of wheat in the program ranged from 85 percent to 87 percent. The financial incentives to participate were so high that most farmers could not afford to remain outside the program.

Under the 1985 act, efforts were made to contain costs by gradually reducing target prices, setting limits on the size of the farmer-owned reserve, setting loan rates on the basis of recent average market prices, and restricting increases in program yields and area bases.

In spite of these efforts, the cost of price and income supports rose to $25.8 billion in fiscal year (FY) 1986. Outlays were $22.4 billion in FY87, $12.5 billion in FY88 and are estimated at $13.8 billion in FY89 and $11.6 billion in FY90.

The high loan rates set in the 1981 Agricultural and Food Act led to increased prices for U.S. grain. The result was to encourage increased production in other countries, thus reducing the U.S. share of world trade in grains. Efforts were made in the 1985 Act to correct this situation by setting loan rates on the basis of five-year average market prices and giving the Secretary of Agriculture authority to lower the loan rates still more to keep U.S. commodity prices competitive in the world market. The lowering of loan rates on the major commodities in 1986 led to lower market prices and an increase in the volume of exports.

Commodity price and income support programs are often criticized for misallocating agricultural and other resources. For example, the value of the area base or allotment which provides the right to produce a particular commodity and receive program benefits increases land values and rental payments. It also encourages production of the crop for which a base is available even if market signals suggest substitution of other crops. Thus, an excessive amount of resources may be committed to producing price-supported commodities, resulting in surplus production.

Commodity price and income support policies have also affected the grain marketing and merchandising industry. Under the farmer-owned reserve program, farmers who hold their program commodities under loan may store them on their farm and receive storage payments. Over time, the government storage payments pay the cost of constructing and maintaining these farm storage facilities. In the 1970s, most government-owned storage bins at sites located in the producing areas were eliminated. In order to store grain surpluses, the government had to lease storage from private grain marketing firms. Government storage payments provided substantial income to farmers and grain marketing firms storing under government contract.

In 1987, the CCC paid farmers an annual rate of $.265 per bushel for storing corn, sorghum, barley, oats and wheat. Rates were negotiated with grain companies. The storage and handling rates negotiated for bulk grains resulted in annual storage rates ranging from $.12 to $.445 per bushel, with an average rate of about $.34 per bushel for country elevators and $.37 per bushel at terminal elevators. The average handling rate was about $.19 per bushel at country elevators and $.15 per bushel at terminal elevators (*Farm Programs, An Overview of Price and Income Support and Storage Programs*, 1988).

Conservation Programs. Conservation programs clearly affect crop production. The most recent of these programs, the Conservation Reserve Program authorized in the 1985 Act, authorized placement under long-term contract of up to 18.2 million hectares of highly erodible cropland. Participating farmers receive annual rental payments for keeping the land placed in the program out of production, on condition that they meet program requirements for providing permanent cover or planting trees. By the end of 1989, about 12.1 million hectares had been accepted for the 10-year program.

Policies to Limit the Effects of Adverse Weather

In the past two decades, farmers have had access to three forms of federal assistance to counter the potential financial problems created by adverse weather. These are:

(1) Federal Crop Insurance,
(2) Disaster relief payments, and
(3) Farmers Home Administration Emergency Disaster Loans.

Each program has different criteria for eligibility, different levels of subsidy, and affects different groups of farmers or farm commodities. A review of each program follows.

Crop Insurance. The 1980 Federal Crop Insurance Act expanded the use of government crop insurance into virtually all crops and geographic areas. The program is available each year to all producers. A major objective of the program is to replace low-yield disaster assistance programs. However, this objective has not been met. Participation has been lower than anticipated, with an annual average of about 14 percent of eligible area. For the program to be cost-effective, almost 70 percent of eligible area needs to be enrolled.[2] To entice producers to participate in the program, 30 percent of insurance premiums paid are subsidized up to 65 percent of the guaranteed yield.

Analysis of a 1983 nationwide survey of corn, soybean and hog farmers in Illinois and Indiana indicated that they ranked all-risk crop insurance as the least important of financial alternatives in response to risk (Patrick et al., 1985), and that less than 6 percent of the farmers surveyed used this type of insurance. Furthermore, all-risk crop insurance was ranked below government emergency credit. These findings indicate farmers' lack of interest in Federal Crop Insurance (FCI). The findings may also indicate a possible substitutability between FCI and government emergency loans, such as Farmers Home Administration disaster loans. This substitution could be one reason for the low participation in FCI.

In 1988, in response to the low rates of farmer participation in FCI, the U.S. Congress established the 25-member Federal Crop Insurance Commission, to ensure a thorough review of FCI and provide recommendations for improving the program. The Commission is to consider: (1) reasons for the low participation rates; (2) adequacy of the insurance coverage FCI provides; (3) identification of the states and commodities for which lack of participation is the most serious; and (4) ways FCI can be modified so as to eliminate the need for costly disaster payment programs such as the 1988 Disaster Assistance Act.

Many studies indicate the usefulness of crop insurance to one degree or another. The extent to which crop insurance is a cost effective tool in risk management depends on the location of the farm operation, yield variability, crops grown, and diversity of farm enterprises. However, the studies also indicate that, for the scenarios evaluated, crop insurance is underutilized. One explanation for this is that in each of the studies, government emergency loans were not included in the accounting. As the 1983 survey by Patrick et al. suggested, farmers may be relying on FmHA disaster loans as a means of income support during periods of poor yields instead of purchasing crop insurance. If farmers believe that FmHA disaster loans perform the same function as crop insurance, but at essentially no cost, why should they purchase crop insurance? Another argument against crop insurance is that crop insurance indemnity payments may make farmers ineligible for FmHA disaster loans.

Disaster Relief Payments. Until Federal Crop Insurance was expanded in 1980, the majority of financial risk was alleviated by a disaster assistance program administered by the Agricultural Stabilization and Conservation Service (ASCS) of the USDA. This program provided direct payments to farmers during periods of low yields due to poor growing conditions. The program differed from FCI in that farmers did not incur any sign-up or coverage costs, and only needed to apply for payments after losses were suffered (Miller and Trock, 1979). To qualify for direct payments under the program, farmers' yields had to be below a given threshold, which varied from year to year between 60 percent to 75 percent of normal yields.

Despite evidence from several studies that many producers were better off with the ASCS Disaster Payment Program than with expanded Federal Crop Insurance, the high cost of direct disaster payments led the federal government to shift to crop insurance (King and Demek, 1985). From 1974 to 1977, producers received approximately $450 million per year in compensation for losses. This amount is considerably less than the estimated cost of crop insurance, given expected participation rates.

Since the expansion of Federal Crop Insurance in 1980, there has been no continuously available disaster payment program. However, the ASCS is still empowered to make adjustments in regular commodity programs to compensate producers for extraordinary losses (*Natural Disaster Assistance Available from the U.S. Department of Agriculture*, 1984). The nature of these adjustments varies greatly depending on growing conditions, area, and commodity, but from time to time payments may be made available to producers when planting is prevented or yields are abnormally low due to adverse weather conditions.

Indirect disaster assistance is available to farmers who participate in the land set-aside program authorized in the 1985 Food Security Act and who own livestock. The U.S. Secretary of Agriculture is given the latitude to provide non-cash assistance to farmers in areas that are experiencing poor growing conditions. During years of persistent drought or other unusual weather conditions, the Secretary can grant farmers in individual counties

the right to use set-aside area to feed livestock, either by grazing or by baling and feeding. The ASCS in each county must determine the impact that
releasing set-aside areas would have on local hay and forage markets and
make the decision whether to open these areas to grazing, baling or feeding. Although no direct payments are made, the farmer with livestock can
make substantial feed savings, and thus increase his income over what it
would have been without release of the diverted areas. Under this program provision, farmers can receive benefits even if the disaster is local-
ized, since diversion areas can be released on an individual county basis.

A severe and widespread drought in the U.S. in 1988 led Congress to enact the 1988 Disaster Relief Act, to ensure that farmers would receive an
adequate level of gross income in spite of the adverse growing conditions.
The relief program that resulted from this act has cost an estimated $4 bil-
lion. Despite arguments that disaster relief programs of this kind undermine the drive to have farmers participate in crop insurance, policymakers
felt compelled to give farmers additional income relief from the effects of
the drought.

In addition to this comprehensive disaster assistance program, many
counties throughout the U.S. cornbelt allowed farmers who participated in
the set-aside program and fed livestock to bale and graze set-aside areas.
Though farmers could not sell the forage they made, they were permitted
to feed it to their own livestock, thus reducing the amount of forage they
would otherwise have to purchase.

Emergency Disaster Loans. The Farmers Home Administration
(FmHA) emergency disaster loans are designed to help farmers recover
from production and physical losses incurred as a result of natural disasters such as drought, floods, and hailstorms. Interest rates for the loans are
subsidized. The loans themselves vary in size and duration depending on
the type and severity of the disaster.

The FmHA loan portfolio for emergency disaster loans increased dramatically between 1976 and 1981, from $900 million to $10.4 billion, partly
as a result of widespread natural disasters in 1978 and 1980. It reached a
peak of $10.8 billion in 1982, then began a gradual decline, dropping to
$8.8 billion by 1987. As of June 30, 1987, the emergency disaster loan pro-
gram had become FmHA's largest loan program, in terms of both dollar
amounts and numbers of loans outstanding.

As with all FmHA loan programs, delinquencies of FmHA emergency
disaster loans have been an ongoing problem. As of December 31, 1987,
delinquencies amounted to $5.3 billion, over 59 percent of all outstanding
emergency disaster loans, and 78.7 percent of all outstanding principal
loans under the program. This is by far the largest dollar amount of delinquencies of any FmHA loan program. Furthermore, over 89 percent of
these delinquent loans have been delinquent over three years. As many of

these delinquencies are not collectible, they will be reflected in future loan write-offs by the agency.

Escalating loan delinquencies have resulted in increasing loan losses for the emergency disaster loan program. In 1976, total loan losses from emergency disaster loans were $4.9 million. In 1987, losses amounted to $485 million, a hundredfold increase in loan losses in only 11 years. These losses account for 43 percent of all FmHA loan losses.

To combat increasing loan losses associated with the disaster loan program, the FmHA has recently tightened eligibility requirements. The recent sharp decline in loans under the program can be attributed to the changes in eligibility as well as to generally favorable growing conditions in 1987. In 1988, the FmHA loaned only $30 million for disaster assistance, the least amount it has loaned for this purpose since the 1950s.

Under present provisions a farmer becomes eligible for an emergency loan if revenues have been reduced by more than 30 percent of normal revenues, as a result of natural weather conditions (for the purpose of this program, revenues include marketing receipts, crop insurance, and disaster assistance, if any). If a farmer can prove that revenues have been reduced below the threshold, a loan up to 80 percent of losses, or $500,000, whichever is less may be obtained. Interest rates for these loans are 4.5 percent, substantially below the market rate of interest. Repayment terms vary according to the type of loss, use of funds, available collateral, and the borrower's repayment ability. The average repayment period is seven years, and the maximum amortization period is 40 years.

The 1985 Food Security Act set forth additional eligibility requirements. Beginning in 1987, no emergency loans were to be made for production losses that could have been covered by FCI. However, producers prevented from planting crops due to natural farm disasters would be covered regardless of FCI participation. Under the act, emergency loans are no longer available to farmers who can get credit elsewhere (Glaser, 1986). Both of these stipulations were waived in the Disaster Assistance Act of 1988, but remain in effect for all years after 1988 (Disaster Assistance Act of 1988, 100th Congress, 1988).

Transportation Policies

The government has reduced its intervention in the transportation industry in the past decade. Until 1980, all rail rate changes had to be approved by the Interstate Commerce Commission (ICC), and these changes came slowly. Most rate changes today are made quickly, confidentially, and without intervention. This deregulatory attitude has also applied to right of ways. Commercial users of inland waterways and port waters now help pay for development and maintenance costs through user fees.

New regional and short-line railroads have received little subsidization from federal or state agencies.

Many new regulatory policies and procedures are still in their infancy. However, at this point, there does not appear to be any strong movement toward further regulatory changes.

Rail Transportation. The railroad industry's grain rate structure was heavily regulated during most of the past century by the ICC. The basic grain rate structure was a "rate-break"system; rates were based on a combination of gathering rates to a terminal market and proportional rates from the market. The ICC established the terminal markets, which were usually located at river crossings. While the gathering rates to the markets were based on mileage, the additional rates from the terminal market to consuming areas were proportional rates *exclusive* of transit charges. The ICC approved all rate changes. General or ex parte rate changes approved by the ICC were used to slowly adjust the entire structure according to costs.

Supporters of the rate-break system argued that it gave grain shippers certainty about future rate levels, enabled control of monopolistic behavior by railroads, and allowed control of rate discrimination. Proponents of regulatory reform argued that the system could not respond to changing grain-flow patterns and increased competition from trucks and barges.

During the late 1960s and 1970s, primarily because of barge and truck competition, railroads were allowed to form more point to point rates and to introduce multi-car and unit train rates.

The Staggers Rail Act of 1980. The railroad companies' low earnings and operating losses over many years led to bankruptcies and liquidations during the 1970s. The Staggers Rail Act (SRA) was designed to enable railroads to increase earnings through rate flexibility, cost reductions, and new contract arrangements.

Under the act, a rail carrier may offer any rate that falls within a wide range determined by the variable cost of the shipment. In most contested-rate cases, the burden of proof of the reasonableness of the rate is placed on the protestant (usually the shipper). As the ICC has reduced its surveillance since the 1980 Act, and has made only a small number of rulings in favor of the shipper (particularly for grain), rail companies have been effectively free of rate regulation. Since 1980, grain rates, particularly throughout the U.S. cornbelt, have been well below the ICC threshold level, and have been fairly responsive to supply/demand conditions.

The SRA allows rate levels and other provisions of a shipment to be contracted on a confidential basis between the shipper and carrier or between the receiver and carrier, with the stipulation that the number of railroad owned or leased cars placed into this type of contract service may not exceed 40 percent of the railroad's fleet of a particular type of car. The contracts must be filed with and approved by the ICC. Shippers may file complaints to the ICC on the grounds that the contract impairs the ability of the railroad to meet its common carrier obligation, that the railroad has practiced "discrimination" among shippers, or that the contract constitutes a destructive competitive force. However, few grain contracts have been repealed.

Since 1983, contracts for grain shipments have been used extensively. A 1986 survey of Illinois and Indiana grain shippers, for example, indicated that almost half of them had formed rate contracts directly with railroads.

The supply of railroad cars has improved in the last two decades. In fact, the large accumulation of cars during the 1970s, coupled with a decrease in demand for cars during the 1980s, has led to a nationwide car surplus. However, it is possible, if not probable, that a new period of car shortage will begin again during the next few years. The critical question for this new era of car availability involves car ownership and/or control. There is an increasing concern about large shippers and receivers controlling too much of the car fleet through direct ownership and through contracting. A current case before the ICC will determine how easy it will be for shippers and other non-railroad companies to shift cars between different railroads' lines and thus avoid the 40 percent statutory maximum of cars that can be controlled through contracting.

Other rail regulation issues relate to mergers, abandonments, and short-line formations. End-to-end mergers, particularly in the West, have been readily allowed during the 1980s to provide single-line service. On the other hand, mergers between railroads serving the same general routes have often faced varying degrees of ICC opposition, depending on the case at hand.

The abandonment rate of low-density branch lines during the 1980s has equaled the rate of the 1970s. There is little empirical evidence that the rate of abandonment has increased due to deregulation. Rather, each abandonment case has been, and will continue to be, mostly a function of profitability to the railroad.

As abandoned track miles increased, several grain companies and grain-related groups began purchasing the abandoned segments to assure continued service. The rate of new short-line company formation has increased dramatically during the past few years. Between 1970 and 1984, 118 new companies were formed (excluding companies that took over trackage from previously formed short-line railroads), at an average rate of about eight new companies per year. Between 1984 and 1986, 66 new short-line companies were formed at an average rate of 26 per year. A majority of the new companies that have formed over the past two decades are succeeding. Very few are receiving subsidies.

Numerous factors determine whether a short-line operation is feasible. Usually, the biggest advantage that a short-line (or new carrier) has over existing carriers is that it faces lower labor costs. However, lines on which only grain is hauled are usually not good short-line prospects. Of the 66 new companies formed during the 1984-1986 period, twelve companies hauled grain as one of their principal payloads, but only five companies hauled grain exclusively.

A major factor determining the way in which railroads are regulated is how the ICC chooses to interpret the law. For example, an attempt at deregulation was made in 1976 through the Railroad Revitalization and Regulatory Reform Act. However, details on which ICC rulings could be made were not clear. In effect, the ICC has chosen not to decide issues in favor of regulation. Since 1980, the ICC has taken a pro-deregulation stance on most issues, and its present attitude is to allow free market forces to prevail. This situation may change, following the appointment of new ICC commissioners.

The regulation most likely to be seriously considered in the future concerns switching agreements and joint rates between railroads. To discourage shipments on a competitor's line, some railroads have set high switching charges for grain moving from a competitor's line onto their own. While there have been many complaints about this type of rate-setting activity, there are no pending cases or legislation that would have an immediate impact.

Barge Transportation. The grain barge industry is fairly free of government regulation. Over 80 percent of barge rates are determined through private contract. Agreements are made between shippers and carriers or receivers and carriers, and typically cover long-range (annual) needs. Barge rates may also be set through public auction at the St. Louis Merchants Exchange. These rates are usually for near-term shipments, and are settled openly. In effect, the Merchants Exchange represents a spot market for small-quantity shipments. A third type of rate-setting does not require an "explicit"rate settlement because in this situation the shipper or receiver owns barges and thus is also the carrier. Large grain companies such as Cargill and Archer Daniels Midland (ADM) often use their own barges.

Barge rates have been highly responsive to supply/demand conditions. For example, as grain exports grew during the 1970s, new barges were built at a fast rate to meet the increasing demand. When exports fell during the 1980s, barge rates fell drastically due to the large supply of barges and decreased demand. The drop in barge rates also caused rail rates to fall.

The only significant government policy directly affecting the barge industry during the past 10 years has been the imposition of an inland waterways user fee (in the form of a tax on diesel fuel), which is charged to commercial users of the river system in order to recoup some of the government costs of operation and maintenance of the public waterways. About 25 percent of these costs are currently being recovered. The user fees have led to higher barge rates, with the result that some grain traffic has shifted to other modes.

Truck Transportation. Few government regulations apply to the pricing, entrance, and exit aspects of the grain trucking industry. Because the barriers to entry into this industry are low, truck rates approach costs. The costs are such that grain is often trucked, rather than shipped by rail, over distances of less than 150-200 miles. This threshold distance tends to decrease in areas of the Midwest where rail competition is high, and to increase in areas of low rail competition such as the Upper Great Plains.

Market Regulations

The grain market is affected by state and federal regulations and by regulations made by trade associations for their members. It is also affected by export contracts and international trade agreements.

State Regulations. State government regulations provide much of the institutional and regulatory environment of the domestic grain market. Many state regulations relate to the operation of grain warehouses. Typical statutory schemes require warehouses to be licensed, bonded and insured and often include detailed provisions dictating the form of warehouse receipts.

The states also regulate the forms and terms of contracts under which grain is traded. For example, delayed pricing (see Chapter 25), a marketing tool farmers frequently use to retain the opportunity for price speculation without the responsibility of storing the grain and maintaining its quality, is regulated by statute in the major grain-producing states. Misuse of delayed pricing has sometimes been associated with financial failure of country elevators and subsequent financial loss to farmers. Some states have established contingency funds to reimburse farmers who have delayed-priced their grain with elevators that subsequently filed for bankruptcy.

Federal Regulations. Grain grading standards, antitrust legislation, health and safety regulation, domestic price and income supports, and support of export prices all affect the grain marketing industry.

Grading standards have been in force since the early part of this century. The U.S. Grain Standards Act of 1916 authorizes the Secretary of Agriculture to develop standards for grain and specifies that all *export* grain must be graded by official inspection personnel under these standards. The development of uniform national standards has decreased the cost of marketing; the ensuing welfare gains have been distributed between farmers and consumers.

The efficiency of grain marketing has been affected by the existence and enforcement of antitrust legislation. Large traders within the grain industry are sensitive to their potential liability and attempt to avoid actions that might be construed as violations of the antitrust laws, even at the expense of opportunities to increase profit or efficiency. On the other hand, some segments of the industry enjoy limited immunity from antitrust legislation. The Capper-Volstead Act of 1922 limits the antitrust liability of farmer-owned cooperatives, enabling them to be more competitive with other marketing firms. The result is a shift in the balance of power among firms and an alteration in the structure and performance of the industry. Grain exporters have gained some antitrust immunity through the Export Trading Company Act of 1982. The goal of the act, to tap the export potential of U.S. businesses and expand export markets, is facilitated in part by encouraging the formation and operation of export trading companies and associations. A certification process is available through which these companies may obtain exemption from potential antitrust liability relating to the pricing and marketing activities described in the certificate. After a certificate is obtained, few legal barriers to export should remain.

Federal health and safety regulations are enforced by the Occupational Safety and Health Administration (OSHA). OSHA enforces regulations to protect the health and safety of workers in grain elevators. In general, these regulations have increased costs and reduced efficiency in elevator operations. A study conducted by the Midwest Research Institute showed that enforcement of the OSHA standard on dust control in grain elevators would result in annual recurring costs of over $269 million during the first ten years of the program. However, the establishment of OSHA regulations reflects a determination by society that worker safety justifies the additional costs.

Domestic farm income and price support programs and export price supports obviously have a profound influence on the grain marketing industry. The domestic programs have been described in detail elsewhere (see above) and will not be discussed here. Export price supports for wheat were enacted through the Agricultural Act of 1970. The Act provided U.S. export subsidies on wheat with the objective of supporting prices paid to farmers while keeping export prices low enough to be competitive on the world market. These subsidies, however, effectively insulated foreign buyers from rising prices when export sales increased.

Rules of Private Trade Associations. U.S. trade associations generate regulations that govern members of the grain industry in domestic markets. The National Grain and Feed Association has a system for trade arbitration that dates back to 1901. The association's trading rules, adopted in 1902, govern all disputes of a financial, mercantile or commercial character. Under these rules, arbitration is compulsory among members. Efficiency in marketing is encouraged by rules that permit the purchase and sale of "enormous volumes of grain and feedstuffs... largely on the basis of mutual trust."[3]

Export Contracts and International Trade Agreements. Grain traders in the United States, as well as traders from other countries, are bound by various international contracts when they sell grain in the international markets. Organizations such as the Grain and Feed Trade Association of London (GAFTA) have developed standardized contracts that simplify transactions and reduce the cost of exchanges of title. Such contracts are very detailed and attempt to cover all possible contingencies. They are used by all major grain importing and exporting countries. Despite these attempts to avoid misunderstandings, contract disputes do occur. When this happens, an agreement to submit to arbitration can speed resolution. The 1958 United Nations Convention on Recognition and Enforcement of Foreign Arbitral Awards has aided in the enforcement of arbitration awards in the international grain trade. Courts of most major grain trading nations respect contractual obligations for arbitration and arbitration conditions are generally included in these contracts. U.S. courts are favorably disposed to recognize such agreements and once it is determined that a valid contract containing a valid arbitration clause exists, the agreement cannot be easily revoked.

The Reciprocal Trade Agreement Act of 1934 emphasized the importance of world markets for U.S. products and fostered several new marketing strategies. One of the most important elements of the act was the authorization of bilateral trade agreements. Trade agreements with over thirty countries opened new markets and provided the conditions necessary for expanded trade.

Because of continued concern over trade barriers, an international agreement was developed to move the participating nations toward freer trade. The resulting document, the Charter for the International Trade Organization (ITO), was never ratified. The General Agreement on Tariffs and Trade (GATT) replaced the ITO as an international agreement for the conduct of trade: "The provisions of GATT include: unconditional most-favored-nation treatment of member nations, no quantitative restrictions, free transit of goods, equality of internal taxation of imported and domestic goods, simplification of custom procedures, and periodic examination of members' subsidies."[4]

To moderate the erratic effects of foreign grain production and demand on domestic and world prices, the United States has periodically entered into bilateral or long-term agreements (LTA) with several countries, including Poland, Mexico, and most recently the People's Republic of China (PRC) and the U.S.S.R. The only LTA still in effect is the one with the U.S.S.R.

The preceding examples illustrate the workings of an incentive-response model in which informed decisionmakers respond to economic incentives within the limits created by the regulatory environment. The orderly, regulated system for marketing grain and the availability of sophisticated pricing and marketing institutions lowers the cost of marketing and accounts for much of the economic progress in developed economies.

Credit and Capital Policies

Monetary Policies. Monetary policies of the Federal Reserve System influence U.S. grain markets in a variety of ways. Monetary policies that limit availability of credit in financial markets increase the cost of borrowed funds, thereby driving up production costs for farmers who borrow money to finance their operations. Restrictive monetary policies may also increase the costs of borrowed funds for input suppliers, who in turn pass these costs on to farmers. Monetary policies also have a direct bearing on exchange rates and consequently on the demand for agricultural products.

Banking and Credit. In the United States, public policies and programs have helped create a banking system with numerous small commercial banks geographically dispersed throughout the country. This structure has led to financing at the "grass roots"level. Farmers often borrow money from local institutions which, though often intimately familiar with the local area, tend to be somewhat inefficient. Recent merger and acquisition activities, resulting from an easing of federal rules and regulations, have allowed for considerable consolidation within the banking industry.

Farmers have access to two specialized agricultural lenders, the Farm Credit System and the Farmers Home Administration (FmHA). Because these institutions receive some government subsidization, they can provide more and/or cheaper credit to producers. While availability of this subsidized credit has probably contributed to excess agricultural production, including grain production, farm commodity programs and price supports probably play a much larger role in creating the excess capacity.

Restrictions on Foreign Ownership of Farm Assets. Several states either prohibit or limit foreign ownership of agricultural assets. Though these restrictions probably affect the demand for agricultural assets, and the cost of acquiring them, they appear to have a limited impact on the total volume of grain produced.

Tax Policies. Use of the cash method of reporting income favors expanding grain production, but to a much lesser extent than other public policies that favor increased production. Most producers have the option of reporting income for federal tax purposes under either the cash or the accrual method. The cash method recognizes income only when it is converted to cash. Increases in inventory are not considered a part of taxable income until the inventories are sold. The net effect is that farmers who expand over time, and build inventories in the process, pay less taxes than they would under the accrual method.

U.S. Trade Policies That Affect the Grain Markets

During much of the 1970s, the overall export outlook for grains and other agricultural commodities was strong, and U.S. agricultural trade policy received little attention. Competition among exporting countries was relatively mild and emphasis was placed on domestic supply programs. During the 1980s, the United States has faced strong international competition in agricultural trade, and the federal government has taken action to strengthen the competitive position of the agricultural sector.

One of the few changes the government made in the 1970s and still applies was to set up an export sales reporting system, following unexpectedly large purchases of U.S. grain by the U.S.S.R. U.S. exporters must report large daily sales of corn, soybeans, wheat, barley, oats and grain sorghum to the U.S. Department of Agriculture the day after the sale is made, and must submit weekly reports for these commodities and for rice, regardless of the size of the sales.

U.S. agricultural trade policy changed in the early 1980s when U.S. exports began a several-year decline. In 1985, the U.S. government took action to reverse this decline, by urging the elimination of international trade-distorting and unfair trade practices during the Uruguay round of the General Agreement of Tariffs and Trade (GATT) and by enacting the Food Security Act of 1985.

The 1985 GATT round was of major concern to the agricultural trade community as agriculture would be included for the first time since the Kennedy round two decades before. In those twenty years, a number of producing and trading countries had developed and implemented trade-distorting policies. At GATT, the United States sought to eliminate all trade-distorting and unfair trade practices. It was originally hoped that these practices could be eliminated by converting the various quotas, subsidies, and levies into producer and consumer subsidy equivalents as reference points across countries from which negotiations to reduce the practices could begin. The Economic Research Service of the U.S. Department of Agriculture has estimated these equivalents for a number of countries (see Chapter 27).

The U.S. has recently proposed a "tariffication"system, under which all nontariff import barriers would be converted into tariffs. Negotiations would center on the reduction of these tariffs across countries. The U.S. has not advocated elimination of all domestic subsidies, such as income support policies, but only elimination or reduction of those policies that have a distorting effect on world agricultural trade.

A trade policy provision legally required by the United States in trade with all free world countries and embraced in GATT is the most-favored-nation provision. This provision basically prevents differentiated treatment among various exporting countries and a given importing country or among various importing countries and a given exporting country. The goal of the most-favored-nation provision is to foster the efficient use of resources in international trade resulting from open markets for products across countries.

Export Enhancement Programs. Two types of program mandated under the Food Security Act of 1985 are used to promote the export of agricultural products: programs to counter unfair trade practices, and commercial export credit programs. Examples of the first type are the Export Enhancement Program (EEP) and the Targeted Export Assistance Program (TEA). Examples of export credit programs include the short-term and intermediate Credit Guarantee Programs (GSM-102 and GSM-103).

The EEP is an export subsidy program administered by the U.S. Department of Agriculture, that provides a means to expand U.S. agricultural exports and to compete against unfair trade practices by other exporting countries.[5] A country is eligible to import a commodity under the EEP if U.S. exporters are facing subsidized competition for that commodity from other suppliers. Commodities are eligible if exporting them under the EEP would increase exports beyond what would have occurred without the program, if a net increase to the overall economy will result, and if the program remains budget neutral. A proposal must be written for each commodity and each country specifying which competitor country is being targeted because of unfair competition. Commodities currently eligible include the following grains and grain products: wheat, wheat flour, barley, barley malt, rice, sorghum and vegetable oil. There is no EEP for corn.

The EEP is designed to challenge the system of intervention prices and export restitutions that the European Community (EC) uses to subsidize agricultural exports. The EEP allows a subsidy or "bonus"to be paid to U.S. exporters so that they may offer an export price to importers that is competitive with prices of other subsidizing exporters. The bonus is calculated for a given value for each sale and is given to the U.S. exporters in the form of generic commodity certificates that can be redeemed through the Commodity Credit Corporation (CCC).

In order to keep U.S. rice competitive on the world market by allowing U.S. prices to better approximate world prices, a marketing loan for rice was mandated, the first of its kind applied to grains or oilseeds. Producers may repay their non-recourse loans at less than the loan rate if world prices fall below the loan rate, thereby eliminating the price floor induced by the loan rate itself.

The TEA program is a market development program designed to promote the exports of those U.S. agricultural products whose producer groups are disadvantaged by the unfair trade policies of other nations. Under the program, the CCC provides generic certificates to non-profit producer groups and other eligible participants as partial reimbursement for export promotion expenses. Examples of export promotion activities include advertising, consumer promotion and nutrition education.

The CCC also administers the GSM-102 Export Credit Guarantee Program and the GSM-103 Intermediate Export Credit Guarantee Program. These programs "provide protection to U.S. exporters or their assignees — the latter must be banks or other financing institutions in the United States — against nonpayment by foreign banks when export sales of U.S. agricultural commodities are made on a deferred payment basis".[6] The short-term GSM-102 program is geared toward repayment periods of one to three years. The repayment period for GSM-103 may extend from three to ten years. Only a few countries are eligible for GSM-103 credit as it is designed to ease the initial transition of developing nations from concessional to commercial sales.

Demand Expansion Policies

Export Demand Expansion. The U.S. government is directly involved in expanding the demand for U.S. grain overseas through the Foreign Agricultural Service (FAS) of the USDA, as well as indirectly through the United States Agency for International Development (USAID). To promote the sale of U.S. grains, FAS uses employees based in its Washington, D.C. office and employees serving overseas as agricultural counselors, attaches and assistant attaches in most of the U.S. embassies abroad. FAS is responsible for the activities of the many overseas Agricultural Trade Offices, and also oversees world crop conditions, policies and markets.

Various commodity organizations promote agricultural exports, typically in concert with FAS, as well as often with state commodity commissions and producer groups. These organizations frequently have strong political power gained through intense lobbying of Congress. The major commodity organizations for grains are the American Soybean Association (ASA), the U.S. Feed Grains Council (FGC), the U.S. Wheat Associates (USWA), and the U.S. Rice Council.

These organizations represent the objectives and goals of various American producers, processors, industry and exporters in international agricultural trade through overseas offices in addition to their home offices in the United States. In working with FAS to increase U.S. agricultural exports, they may use TEA funding provided by FAS to conduct trade shows or provide samples of the commodities they promote.

The states are active in export promotion. Each U.S. state has its own agricultural office. Several states also have trade offices in foreign countries as an extension of the State Department of Agriculture to promote exports of their own state's agricultural products. Recently, some states have established general trade offices in foreign countries. These trade offices differ from the overseas' state agricultural offices since agriculture is not the sole commodity they promote. State trade offices provide technical information on how to import commodities and how to contact exporters.

The primary activity of organizations involved in overseas demand expansion for grains is education of potential foreign customers on the processing and use of U.S. grain products. Many of these educational activities are supported jointly by FAS, state and national commodity organizations, and third party cooperators (typically organizations in the client countries). The FGC and ASA sponsor studies of the livestock feeding potential of U.S. grains and soybean meal, and provide demonstrations. USWA is involved with several educational and research programs that promote the export of U.S. grains, largely for human consumption. Among these are the International Grains Program (IGP) at Kansas State University and the programs of the Northern Crops Institute (NCI) at Fargo, North Dakota. The IGP offers courses in wheat milling and baking, grain grading and storage, feed milling, and soy oil processing. The NCI offers a variety of processing and marketing courses for flax, sunflowers, edible beans, malting barley, and spring and durum wheat. Both of these programs send experts overseas to train potential customers, and also host potential customers from abroad. USWA is forming another comparable institute for white wheat in Portland, Oregon. The International Soybean Program (INTSOY) at the University of Illinois, funded by USAID, is one of several programs aimed at expanding demand for soybean food products in less developed countries.

The demand expansion activities geared toward middle and upper income countries have been very effective. For instance, USWA was partly responsible for the development of the Asian demand for wheat through its activities of creating milling and baking associations in Japan and Korea in the 1950s, and training their members. The various commodity promotion organizations have also been effective because they are directly linked to government export enhancement programs through FAS. Demand expansion has met with less success in the area of promoting the adoption of new food uses for U.S. grains in less developed countries. This may be in part because of cultural resistance to unfamiliar foods.

Domestic Demand Expansion. Both state and federal government agencies promote domestic food consumption through income enhancement programs and targeted food aid programs such as the food stamp program. Rural congressmen have supported many of these programs, expecting that they would lead to an increased domestic demand for U.S. agricultural products. However, the effect of the programs in increasing demand has not been significant.

In the 1980s, the U.S. has faced increasing competition in the domestic market from imported food products, and some groups have been working to promote U.S.-grown food products at the expense of imported products in the domestic market. In particular, the American Soybean Association has recently been promoting soybean oil over the tropical oils (palm and coconut oil) on the grounds that it is healthier as it contains less saturated fat.

Groups Involved in the Grain Markets

Cooperatives

Cooperatives have traditionally been more active in marketing grain than in marketing most other agricultural commodities (Figure 26.1). Cooperatives are found at all levels of the grain marketing industry, but their presence is strongest at the origination stage (procuring grain from farmers), and weakest in grain exporting. The cooperatives attempted to gain a greater presence in grain exporting during the late 1970s and early 1980s, but since then they have largely abandoned the exporting business, though a notable few still actively "fob"grain at export ports. Most cooperatives concentrate on a narrow set of activities where they have traditionally held comparative advantage. These "bedrock"activities include marketing fertilizer, gas, oil, and grain at the producer level.

Though cooperatives receive some direct and indirect economic advantages (mainly tax advantages) through government regulations, they are basically farmer-organized, financed, and operated; some were developed through farm organizations such as the American Farm Bureau Federation.

Bunker and Jones concluded in a recent assessment that cooperatives' primary influence and power base is in the local community, and that in general cooperatives have had little influence on government policies, as few of them are organized to provide a unified opinion on state or national policies (Bunker and Jones, 1986).

Cooperatives have some clear strengths and weaknesses that affect their ability to compete in the grain marketing channel. One strength is that their members are generally farmers so they both "know"their customer, and usually command his loyalty. This strength can become a competitive disadvantage if cooperatives let loyalty and exceptional service to individual members interfere with the long-run financial viability of the organization. Often the cooperative is the last remaining commercial institution in a dying rural community, and many have felt pressured to offer the services previously offered by the businesses that have abandoned the community. Generally, this has not been a viable strategy because most local cooperatives do not have the volume or expertise to sustain ancillary services in a cost-efficient fashion.

482

FIGURE 26.1 Cooperatives' Share of Marketing and Purchasing Activities

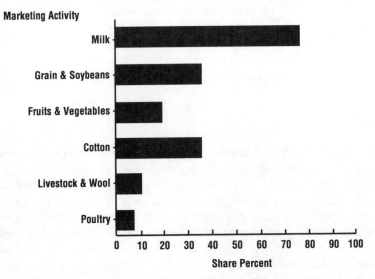

Based on net marketing business of $51.4 billion during 1982. All marketing, 30 per cent.

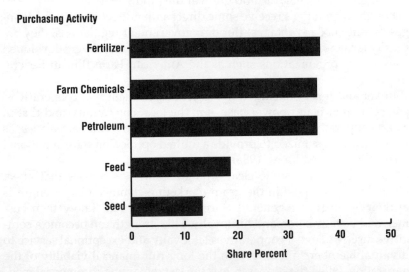

Based on net farm supply business of $16.4 billion during 1982.

Source: Wallace, Tim L., *Agriculture's Futures: America's Food System*, Spring-Verlag, New York, New York. 1987. p. 20.

In the grain exporting field, one weakness many cooperatives have is that their management is not structured to make the timely decisions and incur the financial risks inherent in grain exporting. Another weakness is that they do not have the network of information sources and business contacts needed to maintain a business with overseas clients and competitors.

The cooperatives' market share declines as the grain moves through the market channel, from 40-50 percent of the volume at the country elevator, to 20-25 percent at terminal elevators, to 20 percent at port elevators (Bunker and Jones, 1986). At the country elevator level, cooperative marketing share has grown slowly, while at the export level the number of elevators owned and their percentage of total volume have declined over time (Table 26.1).

In 1984 more than 2000 local grain cooperatives handled about 128 million metric tons of grain — 41 percent of total off-farm sales. In the same year, there were 12 major regional and 2 interregional grain cooperatives. In 1981/82, cooperatives operated 13 port elevators.

Some cooperatives have entered into joint ventures or other agreements in order to increase their marketing strength. Notably, two regional cooperatives, Harvest States (a merger of Midwest and Pacific Northwest grain cooperatives) and Growmark, a large grain marketing cooperative in Illinois and Iowa, have entered into joint ventures with private, multinational grain processing and exporting firms. Harvest States has a joint venture with Alfred C. Toepfer International, and has over a 20 percent interest in the German-based trading firm, which purchases grain from Harvest

TABLE 26.1 Exports of Grains and Oilseeds By U.S. Cooperatives. (Million U.S. Dollars)

Destination	1970		1976		1980	
	Direct	Total	Direct	Total	Direct	Total
Canada	36.8	38.4	[a]	9.1	116.0	[b]
Latin America	61.7	65.1	[a]	60.5	113.8	[b]
Europe	74.8	98.7	[a]	740.4	566.8	[b]
Asia	330.4	367.7	[a]	654.8	888.0	[b]
Africa	1.3	10.1	[a]	32.1	3.7	[b]
Oceania	1.6	1.6	[a]	2.4	1.6	[b]
Unknown	11.4	332.0	[a]	1045.3	250.6	[b]
Total	518.0	913.6	1,358.7	2,544.6	1,940.5	4,961.5
% Share of U.S. exports	13	23	9	16	7	18

[a] Direct exports not reported by destination.
[b] Total exports not reported by destination.
Source: Journal of Agricultural Cooperation, 1986, p. 42.

States for the export market. Growmark has a similar arrangement with Archer Daniels Midland (ADM). Under their joint venture agreement, most of the grain that Growmark procures is sold directly to ADM for processing or further marketing. Growmark owns ADM stock, so Growmark members profit from ADM profits.

Harvest States also has an agreement with another large regional cooperative, Union Equity, (headquartered in Enid, Oklahoma), the largest regional wheat and grain sorghum cooperative in the U.S., under which the two cooperatives have made their export facilities available to each other. Harvest States generally "fobs"at export ports. Union Equity generally fobs grain at the Gulf, but occasionally exports grain.

Grain Exchanges and Boards of Trade

Organized grain exchanges or boards of trade play an important role in the marketing system for U.S. grains. Most exchanges are voluntary, nonprofit organizations whose primary motivation is to facilitate the trading of futures and options contracts.

Grain futures contracts have been traded for over 125 years. A futures contract is an agreement between the buyer and seller to exchange a well-defined commodity or instrument at a fixed price in the future. Almost every aspect of the contract — including its size, the place of delivery, the quality of the commodity, and the month of delivery — is highly standardized and regulated, internally and externally. This is accomplished through licensing futures exchanges, registering brokers and futures commission merchants, auditing records and bank accounts, surveillance of trading, regulating trader positions, publishing market reports, and investigating complaints.

An option contract gives its owner the right to buy or sell a futures contract for a certain price, the exercise price, for a limited period of time.

Access to and use of the futures markets does not require extensive financial resources and is widespread. At the time the contracts are created no title changes hands and only minimal amounts of money are posted by both sides to guarantee performance of the contract's requirements. Futures contracts are also easily liquidated through offsetting transactions; they are different in this way from cash forward contracts. Although only exchange members can trade contracts on the trading floors, trading is open to the public through brokerage houses. Buyers and sellers of contracts represent a wide range of economic interests, including farmers, grain merchandisers, cooperative elevators and the general public as speculators. The value of futures and options contracts traded greatly exceeds the cash value of the underlying grains produced.

The exchanges and the federal government both regulate futures and options trading. The exchanges establish, supervise and enforce trading rules and standards of business conduct. They also write contract specifications, settle disputes and guarantee contract settlements. Federal regulation is carried out by the Commodity Futures Trading Commission (CFTC), a five-person administrative body appointed by the President. Its primary purposes are to ensure proper execution of customer orders, to prevent or curb unlawful manipulation, price distortion, fraud, cheating, fictitious trades or misuse of customer funds, and to assure the general solvency of the system. Recent allegations of market irregularities and resulting proposed changes in trading and reporting activities suggest the importance of the surveillance maintained by the exchanges and CFTC to insure a competitive trading environment.

The Role of Futures Markets in Grain Marketing. Futures markets are extensions of existing cash market forces and have as a purpose to make cash markets work better. They can reduce the degree of uncertainty in producer and marketing decisions by establishing a price for grain to be sold in the future. They also perform the critical economic functions of risk shifting and information dissemination, which facilitate the production and marketing of many commodities. In addition, they can facilitate the securing of capital.

One of the most important functions performed by the futures market in grains is pricing. Futures markets are a primary price discovery market. They reflect available information about actual and expected supply and demand conditions. In effect, they provide the market's expectations about subsequent cash prices.[7] By placing a premium on market information, futures markets have fostered the development of numerous useful information sources. (For a detailed discussion of futures markets and their role in risk shifting and price discovery, see Chapter 26).

In the grain industry, effective price discovery is important to simplifying the carrying of inventories. Producers, inventory holders and speculators interact to form forward prices based on expected supply and demand and current supplies to ensure that existing stocks are rationed throughout the year. In practice, large price differentials between contract months are a signal to store abundant current supplies and release the stocks at a later period.

Producers and other market participants use futures prices as a guide to production and marketing decisions. Using current futures prices as expectations of subsequent cash prices, they allocate their resources to their most productive activities. Futures prices are also used in conjunction with other marketing arrangements to establish cash prices for commodities to be delivered or received in the future. A flat-price contract based on the futures market price for later delivery is one such mechanism. More commonly, cash prices are tied to the futures price at maturity by a fixed differential.

Producers and marketing firms can use the futures market to reduce the effect of price fluctuations as the markets establish a value for their outputs and inputs and thus transfer at least part of their risk. Options permit the commodity holder (buyer) to guard against the likelihood of a price decrease (increase) while taking advantage of a price increase (decrease).

Shifting risk with futures contracts is done through hedging (see Chapter 26). The hedger sells a futures contract if he owns the commodity or buys a futures contract if he expects to purchase the commodity. For example, an elevator can hedge its purchase of corn by selling a futures contract. The hedge exists because the elevator takes opposite positions in the cash and futures markets, that is, it owns corn in the cash market and has sold corn in the futures market. Because the cash and futures prices tend to move together, their offsetting positions cause the gains in one market to be close to the losses in the other market. Though hedging does not eliminate all price risk, studies suggest that selected hedging effectively reduces the variability in returns.

Some producers or sellers prefer options to hedging. While hedging reduces the variability in returns, it also limits the ability of the user to gain from favorable price movements. Options, on the other hand, provide the opportunity to set price floors and ceilings, while maintaining the possibility of capturing favorable price movements. From the producer's or seller's perspective, options contracts offer an "insurance coverage"in which the seller is assured of a minimum price while retaining the opportunity to benefit should prices move higher. Like all insurance coverage, however, there is a cost. In the case of setting price floors, as the hedger increases the minimum price, the likelihood of receiving a price above the minimum decreases. As the price floor is lowered, the opportunity for receiving higher prices increases.

By using the futures exchanges to manage risk, the producer or market participant can lessen his financial constraints. Trading activities can effectively stabilize income or truncate the unfavorable distribution of returns. This security in the value of the commodity makes it easier to obtain loans under more attractive financial conditions.

While some irregularities have appeared recently, futures markets give the decision maker greater flexibility and scope of operation. On balance, they are effective mechanisms for facilitating and coordinating price discovery, for managing inventories, allocating resources, managing risks, and for the collection, dissemination and interpretation of marketing information.

Contracts Between Producers and Primary Traders. In cash markets, several forms of contracting exist to facilitate the marketing and pricing process. In markets where both cash and futures trading occur, a futures contract price can serve as a base for discovering the cash prices paid for specific lots of the commodity. Contracts can be established such that the price paid to producers upon delivery of the grain is "20 cents under the December"futures contract. The quote of the basis contract usually reflects discounts or premiums representing differences in location. It is generally specified in terms of specific quality characteristics of the grain (e.g., U.S. No. 2 yellow corn) and additional discounts can be applied if the delivered grain does meet these specifications.

Delayed pricing (see Chapter 26) gives the producer the opportunity to deliver grain during the harvest period and price it later. As explained earlier, with delayed pricing, the title and physical control of the grain change hands at delivery; the elevator can then market the grain to its best advantage. However, the producer retains the power to decide on what date to price the grain. This approach may be attractive to the producer because it can eliminate grain storage costs and provide him flexibility in determining price. The producer faces price level and basis risk and the added problem of possible elevator insolvency. Many grain-producing states now have regulations to protect producers by requiring elevators to "cover"-their positions on the delayed-priced grain, and have established contingency funds to reimburse farmers in the event of elevator bankruptcy.

The extent of producer use of the various types of contracting options available varies by region of the country and specific grain. Generally, all of the above contracts provide the producer with added flexibility in pricing his grain and permit a more orderly assembly and distribution of the product.

Multinational Trading Companies

Large multinational trading companies (MTCs) dominate the export stage of the U.S. grain marketing system. The dominant MTCs exporting U.S. grain are: Cargill, a private U.S. firm whose head-quarters are in Wayzata, Minnesota; Continental, a private firm owned by a French family and headquartered in New York; Archer Daniels Midland, a public U.S. company headquartered in Decatur, Illinois with a joint export venture with Toepfer, a German firm; Bunge, whose world headquarters are in Brazil; Ferruzzi, an Italian company with a joint venture with the U.S. processor, Central Soya; Louis Dreyfus, a French firm whose headquarters are in Paris; Mitsui Grain Corporation, (a subsidiary of a larger Japanese firm), headquartered in Chicago; and Marubeni America, (also a Japanese subsidiary), headquartered in New York. This list is not exhaustive. Among other firms that export U.S. grain, there are at least nine other Japanese trading companies.

All of these companies either originate grain in other countries as well as in the United States, or originate grain via their affiliated companies in other countries. The MTCs have worldwide information sources and business contacts, an ability to make large financial commitments rapidly, and the ability to shoulder, at least temporarily, large financial risks. MTCs must also have resident expertise in the operation and profitable use of commodity futures and transportation markets. It is common for MTCs to own physical grain handling and transportation facilities, though not required; all facilities may be leased or their services purchased. The advantage of ownership is that the MTC then has the services available and does not need to exercise market power to obtain them. Many MTCs own grain elevators at export ports. A number of U.S.-based MTCs own transportation equipment (rail cars, barges, and trucks) as well as domestic procurement and handling facilities in the United States.

Exporting U.S. grain is an "enterable" market as evidenced by the recent increase in the number of Japanese trading firms present in U.S. ports, especially in the Pacific Northwest. It is also a market where exit is possible as evidenced by the demise of Cook in the late 1970s. Scale economies exist in grain handling, financing, and information procurement. Because of the industry's concentration, oligopolistic performance is sometimes suspected. However, research has repeatedly shown that the industry is competitive. The prices in the U.S. grain export system direct the flow of grain from surplus to deficit areas, usually at minimum cost.

Many people question why the Japanese trading firms have been so successful in obtaining an increasing share of U.S. grain exports to Japan as well as to other Asian countries. While the Japanese firms have an advantage over firms from other countries when dealing with the monopsonistic Japanese food agency, they have no inherent advantage with respect to each other. However, the highly valued quota granted to Japanese firms for sales to the Japanese food agency is determined by the market share Japanese firms achieve in sales to third countries. It may therefore be profitable for Japanese firms to export U.S. grain below minimum cost to third countries in order to gain a greater quota of the sales to Japan. Thus, it is possible that margins in the U.S. grain marketing system are less than minimum cost because certain MTCs attempt to maximize market share in some markets rather than profits.

Notes

1. Actually, tobacco was not included in the 1985 Act but authorization for this program dates back to earlier legislation.

2. According to James Deal, former director of the Federal Crop Insurance (FCI), 68 percent of the eligible area is needed to make the program cost-effective.

3. *1979-1980 Directory-Yearbook*, National Grain & Feed Association.

4. General Agreement on Tariffs and Trade, opened for signature Oct. 30, 1947, 61 Stat. A3, T.I.A.S. No. 1700, 55 U.N.T.S. 194 (hereinafter cited as GATT).

5. Another objective of the EEP was to encourage GATT negotiations toward the elimination of unfair trading practices.

6. "Commodity Credit Corporation Regulations: Export Credit Guarantee Program, GSM-102, and Intermediate Export Credit Guarantee Program, GSM-103,"Office of the General Sales Manager, FAS, USDA, Washington D.C., August 1987.

7. For grains, current prices are closely linked to future prices through the price of storage.

27

Conclusions

From the preceding description, data and analysis, we can draw some conclusions about the performance of the U.S. grain industry that can be used as a basis for comparison with the grain industries of other countries.

Cost and Efficiency

Direct cost comparisons among countries is extremely difficult because of differences in technology, rotations, and cultural practices, and because of unstable exchange rates and varying inflation rates. Recent studies have shown that the United States is not the lowest-cost grain producing nation (Rask, 1987; Stanton, 1986; Koo, 1987). But limitations in each of these studies make a final conclusion difficult. Rask's study, for example, shows that per-hectare costs of producing corn are lower in Argentina than in the U.S. However, Rask's calculations ignore the cost of generating soil fertility in Argentina, where some farms use a crop rotation that ties up one-fourth or more of the land in legumes. While this method of generating soil fertility eliminates the cash outlay for chemical fertilizers, there is still an opportunity cost when producing nitrogen from a corn-legume rotation (see Chapter 23).

The U.S. system of production is generally more efficient on the basis of production per unit of labor. Efficiency based on other inputs gives mixed results in comparisons among countries. Different ratios of land, labor, and other inputs generate different returns to these other factors depending upon where the enterprise is located on the production function. Lower prices for labor and land, and excess labor supplies can lower total costs of production, but they shift resource use away from labor-saving technologies.

U.S. labor efficiency is the result of a high level of labor-saving technology in production, harvesting, and marketing. However, this technology is readily transferable to most other producing countries when economic incentives justify substituting capital for labor. Thus the U.S. advantage in labor efficiency or cost of production is primarily the result of relative prices of inputs and outputs.

U.S. Marketing Advantages

The major grain-producing countries use similar technologies for storing, handling, drying and transporting grain. Much of the U.S. advantage is due to its geography. Port facilities are available on all four coasts. A system of natural waterways leads directly to major port areas. A history, dating back to the 1800s, of providing public support for the development of a nationwide rail, water, and highway transport system provides an advantage not readily duplicated in other countries. For example, countries that have recently entered the export market seldom have adequate infrastructure to handle grain from the newly-developed production regions.

The U.S. grain marketing system has a second advantage, its ability to respond rapidly to changes in price relationships over time, form, and space. Small changes in prices at one or more ports or between processing locations lead to an almost immediate shift in the flows of grain. The private market system, which comprises thousands of individual firms each seeking its own economic advantage, creates strong economic pressure for responsiveness. The rapid and widespread dissemination of marketing information by public and private agencies enables firms to make rapid decisions. Mistakes in judgment by one firm are frequently offset by correct decisions by others. The economic system rewards those who make good choices.

Reliability of Supply

The United States is one of the few major exporting countries that provide storage across crop years. This creates an inventory from which to draw in periods of shortage, and thus the United States has an image as a reliable supplier even in years of low yields. Storage is provided through government agencies and by private firms. The cost of government-held reserves is borne by the public. Storage costs in the private sector are accepted in anticipation of future economic rewards. The willingness of private firms to assume the risk, the physical handling, and the cost associated with storage, increases the ability of the U.S. to respond to short supplies and higher prices in the world market. Both private firms and government agencies have flexibility to respond to changes in world demand and supply more readily than in countries where few public or private stocks exist. The U.S. should be able to maintain its reputation as a reliable supplier if it resists the temptation to impose grain embargoes on politically objectionable countries.

Marketing Costs

One method for assessing marketing costs is to examine the relationship between export prices and producer prices. Table 27.1 and Figures 27.1 and 27.2 provide information on export prices for delivered grain (corn and soybeans) at port facilities and prices paid to farmers in selected primary producing areas of the U.S. between June 1988 and July 1989. For both commodities, prices at the Gulf and the farm declined during the period. Prices for each of the commodities moved in very similar ways. Average price differentials between the Gulf and the farm were $14.33 per metric ton for corn and $15.00 per metric ton for soybeans. Marketing costs as a percent of the Gulf price were 12.3 percent for corn and 5.1 percent for soybeans.

TABLE 27.1 U.S. Farm and Export Prices for Corn and Soybeans, June 1988 - July 1989 [a]

	Corn		Soybeans	
Period	Farm	Export	Farm	Export
June, 1988	105.11	121.25	327.75	347.22
July	112.20	126.76	307.54	325.91
August	106.69	118.89	306.81	320.77
September	105.90	121.25	304.97	320.03
October	105.90	120.86	284.03	297.62
November	99.99	113.77	274.11	287.70
December	101.57	117.71	280.36	296.15
January, 1988	103.54	118.50	281.01	296.89
February	101.96	117.71	273.00	289.17
March	103.93	119.28	279.25	294.68
April	101.18	115.35	264.92	279.25
May	103.14	117.71	264.92	279.25
June	99.99	112.99	259.78	274.11
July	96.45	107.47	253.53	265.65

[a] Farm prices are a monthly average of elevator bid prices (dollars/mt), in central Illinois. This reflects No. 2 yellow for corn, No. 1 for soybeans. Export prices are a monthly average of the Louisiana Gulf prices (dollars/mt). For corn, this reflects No. 2 yellow, for soybeans, No. 2.
Source: Wall Street Journal and Grain Marketing News.

FIGURE 27.1 Corn Prices at Farm and Gulf, June 1988 - July 1989

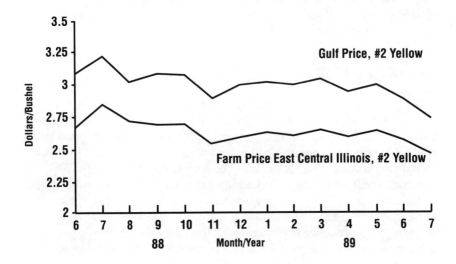

FIGURE 27.2 Soybean Prices at Farm and Gulf, June 1988 - July 1989

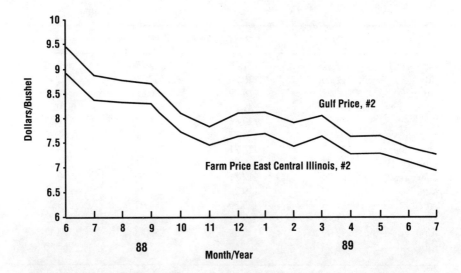

Price Relationships

The pricing efficiency of a marketing system is often measured by the ability of markets to reflect available information, the ability of the system to transmit the information from more central to local markets, and the consistency of price differentials in space, time and form with the costs of transportation, storage and transformation. Futures markets are seen as the primary institution for sellers to identify the present and expected value of grains. Most grain market efficiency studies suggest that, especially for the nearby contracts, the futures markets reflect worldwide changes in supply and demand efficiently. Price differentials between contracts play an important role in allocating available stocks to current and future consumption. Within the marketing year, these differentials are rather closely related to price of storage. Across marketing years, the price differences are influenced by current and expected supply and demand.

The ability of the system to transmit information to various points of the marketing system has not been investigated extensively for all grains. Research examining the relationships between changing prices on the futures markets and local-elevator bid prices to the farmers for selected grains suggests that price transmission is rather rapid, and lags are most likely to occur in areas with limited production where there are few market alternatives.

Often, the movement of grain from one area to another is brought about when price differences between two areas signal the need for grain in the area where the higher price is offered. Grain movements to the area where the higher price is being offered increase the supply of grain in relation to demand in that area, and the price differential narrows. Spatial price differentials change constantly in response to local supply and demand and to changes in transportation and handling costs. However, prices at various points in the marketing system are highly correlated, reflecting the effective transportation and communication systems used in the marketing of grain.

Due to shortage of information on processing costs, we cannot assess accurately their consistency with price differences in form. Synthetic cost studies of the soybean processing industry during the 1970s suggested that rates of return on investment were consistent with the market costs of capital. Since then, the processing industries have become more concentrated, and more work is needed to assess market performance.

Another problem in looking at price differences in relation to form is that grain grades and standards sometimes fail to reflect the value of the grain to the end user. This lack of correlation between grades and standards and the qualities the buyer needs makes it difficult for the pricing system to effectively transmit market signals related to form. Many of the specific characteristics of the current grading system were developed in the early 1900s; they do not necessarily reflect the present value of the grain to the end user and the final consumer. This disparity in the pricing system limits the ability of producers and other market participants to provide a specific quality of grain at an appropriate price to meet demand (Hill, 1985).

Supply and demand in the world markets generally determine the prices for export grains, though many countries use subsidies, levies, quotas, or taxes for imported or exported grain. The effect of distortions caused by domestic policies is generally felt most acutely at the borders of the country that has instituted the policies. Thus, U.S. exporters are selling grain into the same markets and at the same price level as Argentina, Brazil, Canada and Australia. The export policies of the U.S. and other countries influence these world prices indirectly. The extent and influence of the subsidies have varied over time. As stated earlier (see Chapter 26), the U.S. government is espousing a reduction in subsidies and increased reliance on free-market pricing.

Industry Concentration

Concentration in the U.S. grain industry varies at different levels in the market channel. At the level of the farm, grain production is distributed among a large number of farmers, so that no one individual is able to influence price (one of the tests of perfect competition). At the level of the first receiver, country elevators are able to influence prices within their local area. The resulting market structure is often classified as monopolistic competition. However, competition with many elevators in adjacent, easily accessible regions limits the ability of local elevator managers to independently determine prices and profits. Concentration among country elevators varies from insignificant when the total grain market is considered, to an almost pure monopoly when there is only one elevator in a 50-mile radius.

At the processing and exporting levels, the U.S. grain industry is highly concentrated. A few firms own the major processing industries for corn and soybeans, and a few large multinational corporations handle the majority of grain exports. In many cases, multinational corporations include exporting, importing, and processing firms, or they have partnerships, joint ventures, or subsidiaries that include exporting, milling and soybean crushing firms.

For the multinationals, economies of scale in production and marketing may reduce the cost of providing marketing services. But, although total profits may be high in the multinational companies, unit margins are too small to provide an umbrella that would encourage smaller firms to enter the processing and exporting business. Attempts by smaller firms and farmer cooperatives to compete in the international export markets have been largely unsuccessful.

Economies of size are less pronounced in processing industries than in multinational exporting, and some cooperatives are operating successfully in grain processing, notably in the soybean crushing industry. At the country elevator level, the ease of entry and exit and the smaller size requirements enable smaller firms and farmer cooperatives to operate successfully. If country elevator margins reflected excess profits, farmers could readily obtain a grain dealer's license and conduct their own merchandising activities. Even in the country elevator industry, the minimum economically viable size has been increasing. There have been frequent mergers among cooperatives and a number of country elevators, river elevators and cooperatives have been purchased by multinational grain firms.

Price and Income Stability

The original purpose of price supports, government storage and land retirement was to "stabilize prices."The operational objective has become one of providing floor prices to producers of the major grains. Thus, U.S. price support programs have stabilized against downward price movements. They do not stabilize price movements above the floor, but in effect truncate the lower end of the price distribution. The stabilizing effect of government programs differs by grains. Market prices of corn and wheat have frequently fallen below the support price; soybeans and barley have been less affected.

The price response to changes in demand and supply has contributed to stabilizing producer income in the aggregate. Given the elasticity of demand in domestic and export markets, a 1 percent decrease in production due to adverse weather conditions generally leads to a greater than 1 percent increase in price. As a result, aggregate gross income from crop sales may be slightly higher in years of short crops than in years of surplus production.

Most of the U.S. commodity programs link price support to production. As a result, the larger the producer, the larger the government payments. This has resulted in a bias towards larger farms and may have encouraged expansion in farm size. Maximum limits on government payments have had little effect upon limiting farm growth, because a large number of exceptions make the limits ineffective.

Effectiveness of Domestic Policies

The target price and deficiency payment price support system was designed to provide more government support in years when market prices are low and less when prices are high. However, their effect has sometimes been the opposite of that intended. In 1986, when loan rates were lowered to make grains more competitive in the world market, the CCC net outlays rose to a high of $25.8 billion. Target prices were lowered and market prices rose following the export recovery and the 1988 drought. Outlays declined to an estimated $13.8 billion in fiscal year 1989. The larger farmers have received substantial benefits from the price support systems, and with considerable variation among farms. At the peak of CCC net outlays, an average of $11,664 per farm was spent on commodity programs. Farms with $500,000 or more in sales received an average government payment of $35,963 per farm. In contrast, the smallest size category (sales of less than $40,000) received an average payment of only $1,336 per farm in 1986 (Table 27.2).

The financial benefits of past government programs have been tied directly to base acreage, thereby tending to encourage and maintain the production of grains and discourage the production of other crops, even though market price signals indicated opportunities for increasing area planted to other crops. Grain commodity programs have been successful in maintaining a stable area of grain crops. They have allowed some area adjustments in response to market signals.

TABLE 27.2 Direct U.S. Government Payments to Farmers, by Size Farm, 1986 and 1987.

Size of farms [a]:	500,000	499,000 250,000	249,000 100,000	99,999 40,000	under 40,000	All Farms
		1987 Calendar Year				
No. farms [b]:	29	71	201	286	1589	2176
Govt. Payments [c]:	1325	2705	5263	4271	3182	16746
Avg. per farm [d]:	$45,689	$38,098	$26,184	$14,933	$2,003	$7,696
		1986 Calendar Year				
No. farms:	27	72	207	291	1616	2212
Govt. Payments:	971	1889	3774	3019	2160	11813
Avg. per farm:	$35,963	$26,236	$18,231	$10,374	$1,336	$5,340

[a] Farm sizes are classed by total dollar sales of all products for each year.
[b] Numbers of farms are in 1,000 units
[c] Government total payments are in millions of dollars.
[d] Average per farm represents the average government payment per year received by farms in that particular class.
Source: USDA.

To the extent that government programs have increased the prices of raw grains, there has been an implicit transfer of income from consumers to producers. However, this income transfer has been quite small. The use of price supports for grains has had only limited effect on prices of consumer goods at retail. Most of the income transfers to producers comes through the general taxpayer. Most grain-related consumer products undergo extensive processing and marketing before reaching the retail market. The marketing margins far exceed the value of the original products, and the increased price of raw grain is almost masked by the changes in other costs between raw grain and the consumer. The net effect of higher grain prices on consumer expenditures is considered to be negligible.

Effectiveness of Trade Policies

While U.S. trade policy for grain has not been as successful as originally hoped in expanding world trade in grains and developing new product uses, it *has* been successful in recovering lost U.S. export market share. And the U.S. government has developed policies which advocate a marketing system free of trade-distorting policies, including domestic policies that contain fewer trade distortions and pressure on other countries through negotiations at the GATT to reduce their own trade distortions.

The Targeted Export Assistance (TEA) program has focused on the longer term through development of new markets for U.S. agricultural products. The Export Enhancement Program (EEP) has been successful in helping the U.S. regain lost market share in several agricultural commodities. It has also forced the European Community (EC) to reevaluate its own trade distorting system of export restitutions and import levies as increased competition from the United States has placed intense pressure on the EC budget.

Producer-Consumer Subsidy Equivalents. A major goal in the GATT negotiations is to reduce trade-distorting policies in international agricultural trade (see Chapter 26). To make cross-country comparisons of these trade-distorting policies, the Economic Research Service (U.S. Department of Agriculture) converts them into Producer Subsidy Equivalents (PSE) and Consumer Subsidy Equivalents (CSE). The equivalents were developed to facilitate equitable reductions in support across countries as an alternative to making reductions on a time consuming country-by-country and policy-by-policy basis.

The PSE gives a cross-country comparison of the dollar compensation producers would need to leave them indifferent to the removal of current government support. The CSE makes the same comparison for consumers.

ERS estimates of aggregate PSEs for the United States, EC, Japan, Canada, and Australia for 1982-1986 show a wide range of values, indicating

TABLE 27.3 Aggregate PSEs of Key Multilateral Trade Negotiations Participants, 1982-1986

Country	1982	1983	1984	1985	1986	1982-86 Average
Australia [a]	13.3	9.5	9.5	10.8	13.3	11.1
Canada [b]	20.4	25.0	30.6	34.8	43.1	31.0
EC [c]	29.0	29.5	30.4	38.3	49.8	35.4
Japan [d]	66.6	71.2	71.9	70.1	78.6	71.7
U.S. [e]	17.3	25.6	21.6	23.9	35.8	24.6

[a] 9 commodities.
[b] 13 commodities.
[c] 13 commodities.
[d] 12 commodities.
[e] 12 commodities (includes outlays by state governments).
Source:U.S. Department of Agriculture, Economic Research Service. Estimates of Producer and Consumer Subsidy Equivalents: Government Intervention in Agriculture. Staff Report AGES880127, April 1988 (3).

that the amount of compensation governments are providing to producers varies greatly between countries (Table 27.3).

The five countries or regions compared account for a large share of world agricultural trade and are key participants in the Multilateral Trade Negotiations. The PSE for the United States indicates less distortion than that for Japan but more than that for Australia.

The ERS estimates of consumer subsidy equivalents for the period 1982-1986 show a wide range in the charge of consumer costs due to government actions. The United States ranks below Japan and slightly above Canada in the percentage by which government policies have increased consumer costs. CSEs were not estimated for Australia (Table 27.4).

External Influences

No country involved in production and trade in grains can isolate itself from the influence of policies in other countries. Domestic and trade policies create distortions of the freely-determined market price. In the international arena, actions taken by one country to reduce or increase distortions create counter-actions by other countries. As the U.S. subsidizes wheat or flour exports to certain areas of the world, other exporting nations find techniques for subsidizing their sales. The importing nations are the primary beneficiaries. Income is transferred from all exporting countries engaged in this type of competition. For example, both U.S. and Eu-

TABLE 27.4 Aggregate CSEs of Key Multilateral Trade Negotiations Participants, 1982-1986

Country	1982	1983	1984	1985	1986	1982-86 Average
Canada [a]	-9.3	-10.7	-11.4	-13.0	-12.9	-11.5
EC [b]	-14.2	-14.5	-14.8	-15.4	—	-14.8
Japan [c]	-37.5	-40.1	-40.1	-37.7	-38.7	-38.8
U.S. [d]	-11.1	-10.8	-13.4	-11.8	-14.3	-12.3

[a] 11 commodities.
[b] 14 commodities, 1986 data omitted.
[c] 10 commodities.
[d] 9 commodities.
Source: U.S. Department of Agriculture, Economic Research Service. Estimates of Producer and Consumer Subsidy Equivalents: Government Intervention in Agriculture. Staff Report AGES880127, April 1988 (3).

ropean taxpayers have subsidized wheat shipments to the U.S.S.R., which has received the wheat at less than the full cost in the market.

In the short run, the grain produced in a country above immediate needs and demands will move into the export markets. Price reductions or subsidies that have the effect of lowering prices to the importer must be met by similar concessions from other exporting countries or they will be left with unsold grain. Since only the U.S. has sufficient storage capacity to carry grains across crop years, competition through prices, subsidies or qualities will only serve to lower world prices. In the short run, the lower prices will affect the volume imported.

Over the longer run, higher prices in the U.S. that provided an umbrella for the world market have clearly encouraged expansion of corn and soybean area in Brazil and Argentina. In general, supply response is not symmetrical. The price increase required to generate a 1 percent increase in output is considerably less than the price decrease required to generate a 1 percent decrease in area in Argentina and Brazil.

U.S. agricultural and trade policies, and those of other countries, have altered the origin-destination pattern of exports. Export credits and TEA programs (see Chapter 26) have influenced the amount of grain going to selected markets, including the U.S.S.R. and Egypt. It is uncertain whether total exports to all other countries have been increased. Generally, the other countries have developed marketing strategies to offset the U.S. strategy of offering export credits and assistance. Import restrictions, such as the variable levy in Europe, have also affected U.S. exports, both by limiting the volume entering the importing countries and by raising the price within the importing countries so that total consumer demand would be decreased. For example, high prices for meat products in Japan reduce the amount of meat the Japanese consume, and thus U.S. grain exports to Japan are reduced.

System-Wide Evaluation

The following conclusions can be drawn from the preceding description and analysis:

1. *Efficiency.* The technology used in the U.S. grain marketing system is similar to that used in the other major grain-exporting countries. On the whole, the U.S. system uses technology, labor, and transportation efficiently. The United States does not have a grain marketing board, and therefore has the advantage of a pricing mechanism relatively free of government administration and control. The marketing and pricing system is especially efficient in establishing value and responding to price changes and price differentials. The extensive use of the futures market throughout the grain marketing channel enhances the system's flexibility and responsiveness and reduces the cost of covering risk. The ability to arbitrage over time and space increases economic efficiency. Freely operating markets, standardization through uniform grades, and sanctity of contracts permit buying and selling between markets to keep relative prices in balance. Fluctuations in demand and supply can be instantly matched by price adjustments among numerous locations and over extended periods, thus reducing extreme price fluctuations.

2. *Storage, Handling, and Transportation.* A sophisticated system of storage, handling, and transportation eases movement of grain from farm to processor or to export elevator. Competitive intermodal rates and diversity of supplies from many independent sellers assure matching of shipments with arrival of vessels, unit trains, etc. Because large volumes of grain move through the export facilities, operating costs are reduced as substitution among lots can be made in response to price opportunities.

3. *The Futures Market.* The futures market, supported by a broad base of processing industries, assures a year-round, continuous competitive market for producers. Grain can be sold at any time of the year at a fair price based on world markets and competitive marketing margins.

4. *Safeguards for Producers and Traders.* Free market prices introduce more uncertainty into the market than administered prices. Producers or traders holding cash stocks in the field or in storage are subject to the risks of adverse price changes. The losses in value can be substantial. However, access to the futures markets, forward prices and price contracts are effective means for reducing these risks. All mar-

ket participants are free to speculate on price changes or to shift the risks to others, without the costs associated with administered prices.

Increased concentration in the grain industry — especially in the processing and exporting sectors — creates the potential for abuse of market power. However, the presence of cooperatives at all stages of the market, substitutability among grains and sources of grains for many uses, and competition for volume even among the giants, reduces much of the potential for excess profit taking.

5. *Government Policies.* The many government policies affecting the grain industry are not consistent within agriculture, or between agriculture and macroeconomic policies. No single agency coordinates agricultural policies. Monetary, fiscal, employment, and trade policies have often conflicted with agricultural policy goals. The federal government has often developed subsidies for land reclamation and irrigation simultaneously with land retirement policies. To improve efficiency, the relationships among the various programs need to be explicitly recognized when new policies are developed.

6. *Incentives for Production and Productivity.* The United States is recognized as a reliable supplier in the domestic and foreign markets despite an occasional embargo or grain shortage. Surpluses of production relative to acceptable price levels suggest that adequate incentives for increased production have been present.

Productivity has grown rapidly. However, according to some researchers, recent output levels indicate that productivity is slowing (Sundquist, et al., 1982). High price supports provide an umbrella for less efficient producers, thus lowering productivity averaged across all producers. Policies that have encouraged increased farm size have increased productivity among the middle-sized commercial farms.

Grain quality is a problem. Government incentive programs have emphasized quantity over quality ("Enhancing the Quality of U.S. Grain", 1989). The incentives for producers emphasize increasing quantity, not quality (defined as the intrinsic properties and value of the grain for final use (Hill, 1989)). The U.S. grading system for grains is not always consistent with the qualities consumers want. Several studies have demonstrated that the grading system for grains and lack of price differentials for intrinsic quality of U.S. grains have decreased our ability to compete for the high quality grain markets. (Hill, 1989).

7. *Social Goals.* U.S. farm price support legislation has included both economic and social goals. The social goals have included recognizing the importance of the family farm, achieving parity of income for farmers with nonfarm groups, and ensuring that consumers have an

adequate and stable supply of agricultural commodities at fair prices. Other government programs have mixed economic and social objectives, aiming at help for small farms, food assistance to urban low-
income groups, and rural development.

Government policies and programs often conflict with each other.
For example, social goals of equitable distribution of income among
producers and maintaining the income for small farms conflict with
goals of ensuring an adequate supply of food at reasonable cost
available to consumers of all income levels. Government assistance
to farmers through the farm credit system has led to increased farm
size, and income has been distributed away from the family farmer
toward the large commercial operations. As farm size increased, the
total number of farmers decreased, and the economic viability of
many rural communities was reduced.

Government safety regulations, enforced by the Occupational Safety
and Health Administration (OSHA), meet the social goal of protecting worker safety, but lead to higher costs for grain marketing firms.
These costs are transferred either into lower prices to producers or
higher costs to consumers.

Current legislation includes programs to encourage soil conservation, maintain the most highly erodible land in conservation uses,
and maintain wetlands for water conservation, recreation and wild-
life. Future legislation is likely to focus on maintaining water quality.
Water quality regulations may restrict the use of fertilizers and agricultural chemicals, and thus limit the area and yields of major grain
crops.

Following a policy that users of a service should pay for it, the federal government has set user fees for water-borne transportation. As
these user fees are passed on to farmers and exporters via marketing
costs to farmers and exporters, the cost of delivering U.S. grain into
world markets is increased. This works against the goal of trying to
increase U.S. exports and competitiveness in exporting grain.

In most cases, social goals related to grain policies are not explicitly
stated, and in many cases they are not recognized by those who are
involved in developing or enforcing the regulations. A more systematic approach to policy could probably enhance the overall efficiency
of the U.S. grain marketing system.

Bibliography

Agricultural Statistics, 1988, U.S. Department of Agriculture. Washington, D.C.: USGPO, 1988

Alexandratos, Nikos, editor, *World Agriculture Towards 2000*, Food and Agriculture Organization of the U.N., New York University Press, New York.

Alexandratos, Nikos. *European Agriculture: Policy Issues and Options to 2000*, FAO of the U.N. ERC/88/INF/4, 1988 Prepared for the Sixteenth Regional Conference on Europe, Cracow, Poland, 23-26, August 1988.

Barley Movements in the United States, "Interregional Flow Patterns and Transportation Requirements in 1977". Leath, Mack N., Lowell D. Hill and Stephen W. Fuller. North Central Regional Research Publication, No. 277, Agricultural Experiment Station, University of Illinois Department of Agriculture, Urbana-Champaign, Illinois, January 1981.

The Basic Mechanisms of U.S. Farm Policy. Part One: Target, Loan and Deficiency, U.S. Department of Agriculture, Economic Research Service, Miscellaneous Publication No. 1470, May 1989.

Bowers, Douglas E., Wayne D. Rasmussen and Gladys L. Baker. *History of Agricultural Price-Support and Adjustment Programs, 1933-84, Background for 1985 Farm Legislation*, Agricultural Information Bulletin No. 485, U.S. Department of Agriculture, Economic Research Service, December 1984.

Bunker, Arvin R. and James R. Jones. "U.S. Cooperatives and the Potential for Grain Exports to Eastern Europe", *Journal of Agricultural Cooperation*, Vol. 1, 1986, pp. 38-51.

Butell, R., and J.J. Naive, "Factors Affecting Corn Yields", *Feed Situation*, Washington, D.C., USDA, ESCS, May 1978.

Butler, Margaret A. *The Farm Entrepreneurial Population, 1987*, Rural Development Research Report No. 72, U.S. Department of Agriculture, Economic Research Service, June 1989.

Census of Agriculture 1982, U.S. Summary, U.S. Department of Commerce, Bureau of the Census, 1982.

Congressional Research Service Review, Library of Congress, Congressional Research Service, "Farm Problems, A Major Issue Forum."8:5, 100th Congress, May 1987.

Corn Movements in the United States, "Interregional Flow Patterns and Transportation Requirements in 1977". Leath, Mack N., Lowell D. Hill and Stephen W. Fuller. North Central Regional Research Publication, No. 275, Agricultural Experiment Station, University of Illinois Department of Agriculture, Urbana-Champaign, Illinois, January 1981.

Corn Movements in the United States, 1985. Fruin, Jerry E., Daniel W. Halbech, Lowell D. Hill and Albert J. Allen. "A Preliminary Report of Data", North Central Regional Research Publication, Staff Paper No. P89-24, Agricultural Experiment Station, University of Illinois Department of Agriculture, Urbana-Champaign, Illinois, Publication Pending.

Cost of Production Economic Indicators of the Farm Sector, Economic Research Service, United States Department of Agriculture, Reference No. ECIF57-3, 1987.

Cramer, Gail L. and Clarence W. Jensen. *Agricultural Economics and Agribusiness,* Fourth edition, John Wiley and sons, New York, New York. 1988, pp. 212-215, p. 342, pp. 439-440.

Current Population Reports, U.S. Department of Commerce, Bureau of the Census, "Population Estimates and Projections", Series P-25, No. 937, Issued August 1983.

Denbaly, Mark. *Shortrun Effects of U.S. Macroeconomic Policies on U.S. Grain Exports,* Agriculture and Rural Economic Division, U.S. Department of Agriculture, Economic Research Service, April 1986.

Digest of Education Statistics, National Center for Educational Statistics, U.S. Department of Education, Office of Educational Research and Improvement. Washington, D.C.: USGPO. September 1988.

Disaster Assistance Act of 1988, 100th Congress. Washington, D.C. August 1988.

Economic Indicators of the Farm Sector: Income and Balance Sheet Statistics, 1979. U.S. Department of Agriculture, Economic Research Service, National Economics Division, Economics and Statistics Service, Statistical Bulletin No. 650. December 1980.

Economic Indicators of the Farm Sector, Production and Efficiency Statistics, 1986, U.S. Department of Agriculture, Economic Research Service, ECIFS 6-5, June 1989.

Economic Report of the President, U.S. Council of Economic Advisors, Washington, D.C.: USGPO. 1989.

Enhancing the Quality of U.S. Grain for International Trade, Office of Technology Assessment, U.S. Congress, OTA-F-399, Washington, D.C., February 1989.

Farm Programs, An Overview of Price and Income Support and Storage Programs, U.S. General Accounting Office, GAO/RCED-88-84BR, February 1988.

Food and Agricultural Organization Yearbook, Statistics Division of the Economic and Social Policy Department. Food and Agricultural Organization of the United Nations, 1986.

Frankel, J. and G. Hardouvelis. "Commodity Prices, Monetary Surprises and Fed Credibility", *Journal of Money, Credit and Banking*, 17(4), 425-438, November 1985.

Frankel, Jeffrey. "Expectations and Commodity Price Dynamics: The Overshooting Model", *American Journal of Agricultural Economics*, 68(2), 344-350, May 1986.

Glaser, Lawrence K. *Provisions of the Food Security Act of 1985*, Agricultural Information Bulletin No. 498, U.S. Department of Agriculture, Economic Research Service, April 1986.

Hallberg, Milton C. *The U.S. Agricultural and Food System, A postwar Historical Perspective*, Publication No. 55, The Northeast Center for Rural Development, Pennsylvania State University, October 1988.

Heid, Walter G., Jr. *U.S. Wheat Industry*, Economic Statistical and Cooperative Service, U.S. Department of Agriculture, Agricultural Economic Report, No. 432, August 1979.

Hill, Lowell D. "Opportunity for Illinois to Enter the Markets for High Quality Corn", paper given at Illinois Legislators Workshop, University of Illinois, Urbana, Illinois, January 27, 1984.

_____. "Effects of Regulation on Efficiency of Grain Marketing." *Journal of International Law*, Vol. 17, no. 3, Summer 1985.

Historical Tables, Budget of the U.S. Government Executive Office of the President, Office of Management and Budget, Washington, D.C. Fiscal year 1989.

King, Robert P. and George E. Demek. "Risk Management by Dry Land Wheat Farmers and Elimination of the Disaster Assistance Program", *American Journal of Agricultural Economics*, Vol. 65, 1985, pp. 247-255.

Koo, Won W. and Todd Drennan. *Optimal Agricultural Production and Trade: Implications on International Competitiveness*, Department of Agricultural Economics, Agricultural Experiment Station, North Dakota State University, Fargo, North Dakota, Report No. 251, July 1989.

Langley, Jim and Suchada Langley. *State Level Wheat Statistics, 1949-1988*, Statistical Bulletin No. 779, U.S. Department of Agriculture, Economic Research Service, March 1989.

Loomis, Ralph A. and Glen T. Barton. *Productivity of Agriculture, United States 1870-1958*, U.S. Department of Agriculture, Agricultural Research Service, Technical Bulletin No. 1238, April 1961.

Miller, Thomas A. and Warren L. Trock. "Disaster Assistance to Farmers: Needs, Issues and Programs", Cooperative Extension Service, Colorado State University, Fort Collins, Colorado, 1979.

Monthly Labor Review, U.S. Department of Labor, Bureau of Statistics, Washington, D.C., Vol. 112, No. 7, June 1989.

National Accounts of OECD Countries, Organization for Economic Cooperation and Development, Paris, 1988.

National Income and Product Accounts of the United States, 1929-1984, Statistical Tables U.S. Department of Commerce, Bureau of Economic Analysis, Issued June 1986.

Natural Disaster Assistance Available from the United States Department of Agriculture, United States Department of Agriculture, Washington, D.C., August 1984.

Oliveira, Victor J. *Trends in the Hired Farm Work Force, 1945-87*, Agriculture Information Bulletin No. 561, U.S. Department of Agriculture, Economic Research Service, April 1989.

Patrick, G. F., P. N. Wilson, P. J. Barry, W. Boggers and D. L. Young. "Risk Perceptions and Management Responses: Producer-Generated Hypotheses for Modeling". *Southern Journal of Agricultural Economics*, Vol. 17, 1985, pp. 231-238.

Rask, Norman. "Production and Marketing Costs for Corn, Wheat and Soybeans in Major Export Countries", Department of Agricultural Economics and Rural Sociology, Ohio State University, Columbus, Ohio, January 1987.

Rausser, G. C., J. A. Chalfant, H. A. Love and K. G. Stamoulis. "Macroeconomic Linkages, Taxes and Subsidies in the U.S. Agricultural Sector, *American Journal of Agricultural Economics*, 399-417, May 1986.

Reimund, Donn A., Nora L. Brooks and Paul D. Velde. *The U.S, Farm Sector in the Mid-1980's*, Agricultural Economic Report No. 548, U.S. Department of Agriculture, Economic Research Service, May 1986.

Sands, Laura. "Drought Aid Comes Home", *Farm Journal*, September 1988.

Schuh, Edward. "The Exchange Rate and U.S. Agriculture", *American Journal of Agricultural Economics*, 5, 1-13, 1974.

Schuh, G. Edward. "The Exchange Rate and U.S. Agriculture", *American Journal of Agricultural Economics*, Vol. 56, No. 1, February 1974, pp. 1-13.

Skecs, Jerry R. and Perry J. Nutt. "The Cost of Purchasing Crop Insurance: Examining the Sensitivity of Farm Financial Risk". *Agricultural Finance Review*, Vol. 48, 1988, pp.37-48.

Smith, Randell K., Eric J. Wailes and Gail L. Cramer. "The Market Structure of the U.S. Rice Industry", unpublished bulletin, Agricultural Experiment Station, University of Arkansas, February, 1990.

Soybean Movements in the United States, "Interregional Flow Patterns and Transportation Requirements in 1977". Leath, Mack N., Lowell D. Hill and Stephen W. Fuller. North Central Regional Research Publication, No. 273, Agricultural Experiment Station, University of Illinois Department of Agriculture, Urbana-Champaign, Illinois, January 1981.

Soybean Movements in the United States, 1985. Larson, Donald W., Thomas R. Smith and E. Dean Baldwin. North Central Regional Research Publication, Agricultural Experiment Station, University of Illinois Department of Agriculture, Urbana-Champaign, Illinois, Publication Pending.

Stamoulis, K. G., J. A. Chalfant and G. C. Rausser. "Monetary Policy and Relative Farm Prices", Working Paper No. 413, University of California at Berkeley, November 1987.

Stamoulis, K. G. and G. C. Rausser. "Overshooting of Agricultural Prices", in *Macroeconomics, Agriculture and the Exchange Rate,* Westview Press, 1988.

Stanton, B. F. *Production Costs for Cereals in the European Community: Comparisons with the United States, 1977-1984,* Department of Agricultural Economics, Agricultural Experiment Station, Cornell University, Ithaca, New York, A.E. Res. 86-2, March 1986.

Statistical Abstract of the United States, U.S. Department of Commerce, Bureau of the Census, U.S. Government Printing Office, Washington, D.C., 1989.

Sundquist, W. Burt, Kenneth M. Menz and Catherine F. Neumeyer. "A Technology Assessment of Commercial Corn Production in the United States", University of Minnesota, Agricultural Experiment Station, Bulletin 546, 1982.

Survey of Current Business, U.S. Department of Commerce, Bureau of Economic Analysis, Vol. 69, No. 6, U.S. Government Printing Office, Washington, D.C., June 1989.

Thompson, L. M., "Weather and Technology in the Production of Corn in the U.S. Corn Belt"; *Agronomy Journal* 61(3), 1978.

Tweeten, Luther. "Adjustment in Agriculture and Its Infrastructure in the 1990s", paper presented to National Agricultural Symposium on "Positioning Agriculture for the 1990s: A New Decade of Change", Kansas City, Missouri, October 14, 1988.

Wallace, Tim L. *Agriculture's Futures: America's Food System,* Springer-Verlag New York, Inc., New York, New York, 1987.

Wheat, Background for 1985 Farm Legislation, U.S. Department of Agriculture, Economic Research Service, Agricultural Information Bulletin No. 467, September 1984.

Wheat Situation and Outlook Yearbook, U.S. Department of Agriculture, Economic Research Service, WS-284, February 1989.

Wheat Movements in the United States, "Interregional Flow Patterns and Transportation Requirements in 1977". Leath, Mack N., Lowell D. Hill and Stephen W. Fuller. North Central Regional Research Publication, No. 274, Agricultural Experiment Station, University of Illinois Department of Agriculture, Urbana-Champaign, Illinois, January 1981.

Wheat Movements in the United States. "Interregional Flow Patterns and Transportation Requirements in 1985". Reed, Michael and Lowell D. Hill. North Central Regional Research Publication, Agricultural Experiment Station, University of Illinois Department of Agriculture, Urbana-Champaign, Illinois, Publication Pending.

Womach, Jasper. *The 1990 Farm Bill: Issues Likely to Shape the Policy Debate,* Food and Agriculture Section, Environment and Natural Resources Policy Division, Congressional Research Service, The Library of Congress.

Womack, Letricia M. and Larry G. Traub. *U.S.-State Agricultural Data,* Agriculture Information Bulletin No. 512, U.S. Department of Agriculture, Economic Research Service, April 1987.

World Development Report, International Bank for Reconstruction and Development (World Bank), Oxford University Press, New York, New York, 1986

1979-1980 Directory-Yearbook, National Grain and Feed Association, Washington, D.C., 1979.

1987 Census of Agriculture, U.S. Department of Commerce, Bureau of the Census, Advance Report AC87-A-53-000(A), May 1989.

About the Book and Editors

Behind the statistical outcomes of international trade are the institutions and policymakers that shape performance in the international marketplace. But even though trading outcomes are a matter of record, for most nations very little is known about the institutional and policy frameworks that determine a nation's success or failure.

In this comparative study of five nations—Argentina, Brazil, Australia, Canada, and the United States—experts on international agricultural trade analyze the institutions and policies supporting the international grain trade. There is special emphasis on macroeconomic policy, marketing, the role of the public sector, and the legal and regulatory context. Until now, scarcely any published analysis has been done on Argentina, Brazil, and Australia in regard to these important topics.

This work will be essential for professionals and academics working on agriculture in these countries as well as for scholars of agricultural economics, international trade, and international political economy.

Michael J. McGarry is division chief, Agriculture Division, the World Bank. **Andrew Schmitz** is Robinson Chair Professor, Department of Agricultural and Resource Economics, University of California at Berkeley.